D1286137

Telecommunications Switching

Applications of Communications Theory
Series Editor: R. W. Lucky, *Bell Laboratories*

INTRODUCTION TO COMMUNICATION SCIENCE AND SYSTEMS
John R. Pierce and Edward C. Posner

OPTICAL FIBER TRANSMISSION SYSTEMS
Stewart D. Personick

TELECOMMUNICATIONS SWITCHING
J. Gordon Pearce

ERROR-CORRECTION CODING FOR DIGITAL COMMUNICATIONS
George C. Clark, Jr., and J. Bibb Cain

Telecommunications Switching

J. Gordon Pearce

PLENUM PRESS • NEW YORK AND LONDON

Library of Congress Cataloging in Publication Data

Pearce, J Gordon.
Telecommunications switching.

(Applications of communications theory)
Includes index.
1. Telecommunication–Switching systems. I. Title. II. Series.
TK5103.P4 621.38 80-20586
ISBN 0-306-40584-9

A Division of Plenum Publishing Corporation
227 West 17th Street, New York, N.Y. 10011

Printed in the United States of America

Preface

The motivation for this book stems from an early exposure to the book *Applied Mechanics* by John Perry. Professor Perry strove to encourage his readers to understand the applications and use of mathematics in engineering without insisting that they become immersed in pure mathematics. The following text uses this approach to the application of telecommunications switching. Readers wishing to study the derivation and proof of formulas will be able to do so using relevant references. The existence of low-cost programmable calculators frees practicing engineers from much laborious calculation, allowing more time for creative design and application of the art. The reader should not need to be able to derive formulas in order to apply them just as, to quote Professor Perry, "He should not have to be able to design a watch in order to tell time."

The material for this book has been drawn from my own experience in the field. Inevitably, however, I have used CCITT and Bell System publications for references and in some cases quotation, and I gratefully acknowledge permission for their use.

I am also grateful to Stromberg Carlson Corporation for their earlier encouragement and support without which this book would not have been possible.

Thanks are also due to Fred Hadfield for his advice and assistance in the preparation of the many figures and to my wife Ada for her support and patience as I pursued the demanding but interesting task of producing the text.

Finally, my thanks are due to Ms. Donna Wethington, who has coped with the task of deciphering and typing my manuscript so effectively.

Sanford, Florida J. Gordon Pearce

Contents

1

Introduction

1.1. Scope

Telecommunications switching has been defined by the IEEE[1] as the function of selectively establishing and releasing connections among telecommunication transmission paths. This book will describe the principles by which switching is accomplished. Telecommunications switching is one of the three major parts of the telecommunications network. The other two major parts are terminals and transmission. The telecommunication network carries telecommunications calls, while the terminals are the sources and sinks for the calls. The transmission means connect the terminals to the switching centers and the switching centers to each other to provide paths for the calls. The switching centers provide the relevant transmission paths between the transmission means in response to signals from the originating terminal. Hence they establish the connections necessary for calls. Figure 1.1 shows a basic telecommunications network.

1.2. The Parts of a Switching Network

1.2.1. Switching Systems

There are two basic functional types of switching systems: local exchanges, which interconnect local and trunk transmission means, and trunk exchanges, which interconnect only trunk transmission means. The local exchanges serve to connect the lightly used low-traffic local transmission means (which are usually dedicated to use by single terminals) to the shared interconnecting means provided by the switching center. Thus it serves as a concentrating and distributing center for telephone calls. The trunk exchanges serve to interconnect the various trunk transmission means which are dedicated to exchanges. Trunk exchanges

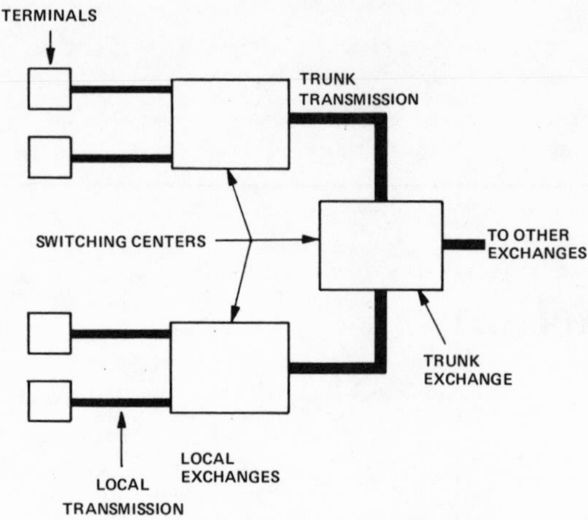

Figure 1.1. Basic telecommunications network.

switch concentrated traffic. The exchanges are arranged in an hierarchy which provides routes to permit the interconnection of subscribers on a worldwide basis.

1.2.2. Line Terminals

Line terminals are of various types, the most common of which is the residential telephone. In order for the telephone to be a part of the switching network it must comply with the parameters of the local transmission equipment and also the parameters of the switching center which serves it. Figure 1.2 shows the basic functions provided by a terminal.

Telephone Functions

A telephone converts voice signals into electrical signals and provides the means to originate and respond to telephone calls. This is effected by providing the following functions:

- A means of signaling to indicate the desired address (e.g., a calling device such as a telephone dial) for originating calls.
- A means of alerting the addressee (the called terminal) (e.g., a telephone bell) for terminating calls.
- Transducers for conveying information between the calling and called parties (e.g., a telephone transmitter and receiver).
- An interface for the local transmission equipment (e.g., a telephone transformer and terminal block) for the wired connection.

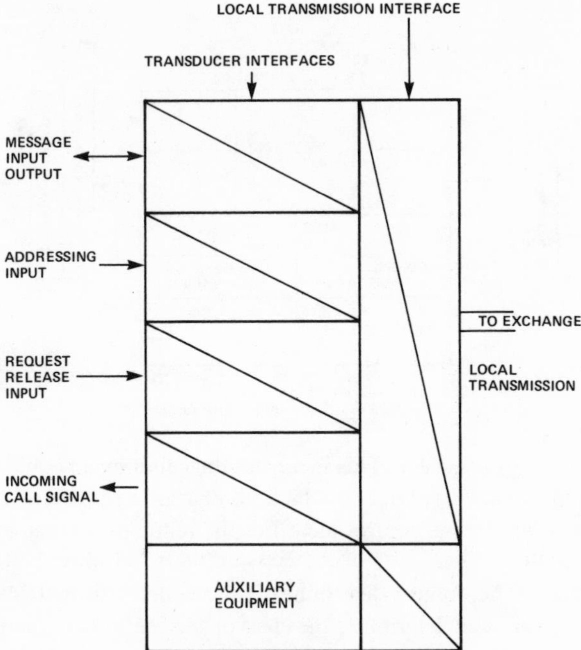

Figure 1.2. Basic terminal.

1.2.3. Switching Systems

Figure 1.3 shows the general organization of a switching system. This system includes functional units which allow it to interface with the line and trunk transmission means and to control the transmission paths involved in a telephone connection. It includes interfaces (distribution frames) between the cables which provide the local and trunk transmission and the line and trunk interfaces to the switching matrix. The distribution frames relate the equipment positions of the line and trunk interfaces to the destinations served by the street cables. The line and trunk interfaces adapt the signaling and transmission modes of the transmission equipment to that of the technology employed in the switching matrix. The system provides the relevant call routing by translating the address signals received from the terminal calling device into a code which identifies the line or trunk interfaces required to complete a call; and it also provides the features and services required by the telephone administration.

The functions of a switching system control may be summarized as follows.

1.2.3.1. Call Routing

This establishes the transmission path by the use of functional units in response to the address data.

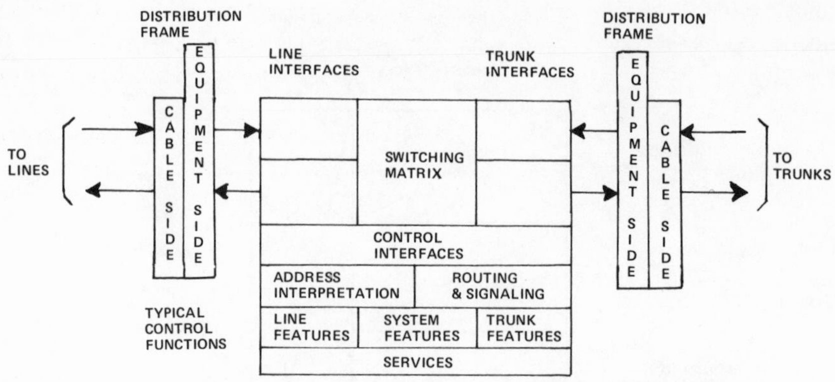

Figure 1.3. Basic switching center.

Calling class of service. This identifies the caller by an equipment number to determine the class of service to which he or she is entitled.

Called number. The control identifies the required destination by the signal address.

Local call. The control determines if the called terminal is connected to this switching center and determines the class of service and the route to reach it.

Trunk call. The control determines if the called terminal is connected to another switching center; it determines the necessary interexchange signals and routes the call to its destination.

Route selection. The control selects a path to take the call towards its destination and generates any relevant signals to send over the selected path.

Called class of service. The control provides any special actions relating to the class of service of the called party.

1.2.3.2. Call Control Functions

Call control functions supervise and otherwise control the progress of the call and provide the relevant administrative data.

Call charging. This function establishes the data for the cost of the call.

Call control. This function supervises the progress of calls.

Call progress tones. This function controls the return of relevant tone signals.

Fault recording. This function records malfunctions.

Release of connection. This function disconnects the transmission path at the end of the call.

1.2.4. Transmission

The purpose of a telecommunications network is to convey messages. This requires that all the elements involved in the transmission path have a minimal impact on the intelligibility of the message. They should not distort nor attenuate quality nor should they reduce or increase its level beyond the assigned limits. A switching network provides a means of controlled amplificaton for long-distance calls.

The need to meet rigorous transmission parameters imposes limits on the techniques and components used in the transmission path and switching matrix of a switching system.

1.2.5. Private Branch Exchanges

In addition to the provision of local telephone exchanges for the routing of traffic from residential subscribers it is necessary to supply service to businesses, motels, etc., where a group of lines will serve a single customer. There may be a number of line or trunk interfaces serving such a private branch exchange, any of which provides access to it. This service can either be provided by separate equipment located at the customers' premises (sometimes called *CU Centrex*) or it can be provided as part of the features of the local exchange to which it connects (which is called *CO Centrex*).

1.2.5.1. Equipment at Customers' Premises

Use of equipment at customers' premises is called CU Centrex in the U.S.A. In this case a private branch exchange (PBX) concentrates the traffic from its own terminals into a smaller number of lines or trunks (depending on the amount of traffic offered) to the local exchange. If the number of calls is small enough so that it does not overtax the call-handling capacity of a line interface group then it can be served by line terminals on the concentration matrix. If on the other hand it offers too much traffic to be treated as a line terminal then it must be served by trunks. A PBX can have its own completely separate numbering scheme for intra-PBX calls or it can be part of an integrated numbering plan.

1.2.5.2. Equipment Provided as Part of the Local Exchange

Use of equipment provided as part of the local exchange is called CO Centrex in the U.S.A. In this case the customers' lines are brought into the local exchange in the same way as individual residential lines. However, the lines are identified as serving a single customer by a class of service and are provided the same features as if they were served by a separate PABX. The attendants' consoles also terminate as individual lines. The provision of this service is made easier by the large number of classes of service and features which stored-program-controlled systems can offer. This service is particularly useful when an administration has to provide service for many different tenants in a large office building. The services required will vary because of initiation, expansion and termination of service and it easier to accomplish such changes by data changes with a minimum of distribution frame wiring than to remove equipment from the customers' premises.

1.2.6. Operator Services

Administrations are providing all possible services on an automatic basis with all the control actions performed by the customers. However there are still services which have to be provided by operators. These services are necessary when a customer wants a long-distance call directed to a specific person or when assistance or information requiring manual intervention is necessary. Such services are obtained by dialing special codes which automatically route the call to an operator.

In countries which have single-fee automatic coin boxes, an operator will have to control the establishment of multifee calls from a coin box. This service may be provided by multifee coin boxes which automatically assess and indicate the cost of the call per unit of time and only allow the call to proceed if sufficient coins have been deposited to cover the cost of the call.

Modern administrative methods require the centralizing of such services which are provided by stored-program-control operator consoles and centralized directory assistance positions also using stored-program control. The Bell System traffic service position system (TSPS)[2] and directory assistance services[3] are examples of the provision of such services.

1.3. Numbering Plans

The switching network requires an addressing structure, or numbering plan for routing calls. This numbering plan must identify all stations and services in the world so that there is a unique telephone number for each main station, which can either be residential line or PBX. This identifies the customers' geographical location and in its complete form identifies the country, the area, the exchange, and the main station code. The basis for the international numbering plan is described in Chapter 2.

1.3.1. Number of Digits Dialed

The number of digits that have to be dialed to reach a telephone will vary depending on the relative locations of the two parties and the organization of the numbering plan.

1.3.2. Calling Devices

There are two methods of producing address information from a terminal (other than a verbal request), each of which requires a calling device. One of these methods uses a code which sends the information in the form of interruptions of the loop current in the calling line. The number of interruptions corresponds to the digit being sent. The other consists of using voice frequency signals, dual-tone multifrequency (DTMF). Combinations of two frequencies identify the digit being sent from the terminal.

1.3.3. Interpretation of the Numbering Plan

A switching center includes the means of interpreting and acting on the address information necessary to identify the destination of the call. These digits can represent access codes, service codes, area codes, or exchange codes. It may be necessary to analyze the combination of a number of digits before a final routing determination can be made.

1.4. Wire Communications

The preceding description is based on the existing telecommunications equipment using wire communications. There are other telecommunication methods (radio communication for example), but what this book is primarily concerned with is not modes and media but rather functions. Where and how the functions are provided in a network is largely a matter of economics and technology. The need for switching systems could diminish with developing technology but there are likely to be switching systems in use for many decades and the following chapters we hope will permit an understanding of their requirements.

1.5. Meeting the Needs of Public Telephone Service

A switching center is designed to meet the increasing demand for public telephone service to more and more points at less and less cost to the customer. It must meet the need to accommodate new features and services to the customer and operating administration as they emerge. New systems are designed to permit the elimination (or minimization) of switching system interfaces with digital transmission equipment. A switching system provides a concentration of the customers' traffic at a convenient point in the geographical area that it serves. This reduces the cost of service by using shared rather than dedicated transmission equipment since calls can be carried by a common group of trunks to the serving local exchange.

1.5.1. Commonality of Switching System Application

Any new switching center design should be applicable to both local and trunk exchanges. When it serves as a local exchange it should accept relatively small quantities of traffic from a large number of terminals and either distribute it directly to local terminals or transfer it to other exchanges. When it serves as a trunk exchange, it should accept large quantities of traffic from a number of exchanges, both local and trunk, and distribute it to local or trunk exchanges.

1.5.2. Application of Switching System

1.5.2.1. General

A telecommunications switching system is a part of a telecommunications network and must be designed to meet the constraints (Figure 1.4) of its location in that network. It must be able to connect and interwork with the existing and projected terminals and transmission equipment. It must be capable of interpreting and acting on the signals associated with terminals and existing switching centers. It must provide services. It must fit into the environment. These are some of the constraints on its design which require it to be flexible and adaptable to varying requirements.

The alternative of designing a different system for each application is not viable. A switching system provides service at a cost that allows economical telephone service and provides a return on the investment required to provide it. The need to interface with the existing equipment and the long life expectancy of the equipment has dictated that system development be evolutionary. Technology allows the application of new techniques in controlling calls and in the

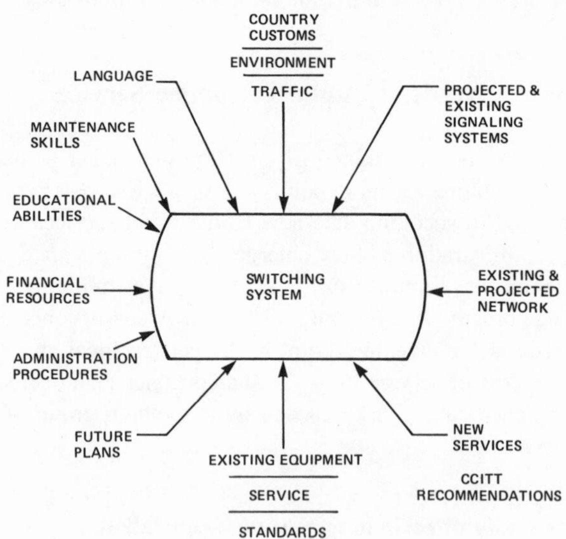

Figure 1.4. Basic system constraints.

combination of the transmission and switching functions, but the success or failure of a system design lies in its ability to interface with equipment which reflects the technology of many decades. A switching system is manufactured to a specification which states the exact needs of the user.

1.5.2.2. Acquisition of a Switching System

The first step in the provision of a switching center is the successful response to the user's request for a proposal. This request describes the user's specification and represents an attempt to find the lowest cost to provide this service. There may be heavy supporting costs in the provision of buildings to house equipment, and in air conditioning and heating equipment to maintain the required ambient temperature. There are also on-going costs—annual costs—in regard to the cost of the space used to house the equipment, the cost of the power used to operate the system, and the cost of the labor needed to maintain it, etc. The computation of the actual cost of the system must take into account all of the above factors and relate it in a meaningful way to the ability of the system to perform its task in a cost effective manner.

Switching centers handle telephone traffic, i.e., they switch calls of varying duration to varying destinations (with many calls in being at any time). The administration must recover the costs of calls from revenue. Hence there are limits on the amount that can be invested to purchase, install, and operate the system. Hence switching systems must meet such financial constraints to be viable.

1.5.2.3. Basis for Assessing the Cost of Service

Cost per line. One method used in assessing the cost of the equipment is to divide the initial cost of the system by the number of lines that it serves. While this gives a way of comparing systems made to meet a particular specification, it does not offer a good indication of the cost of providing the service in relation to revenue.

Cost per call. Another way of assessing the cost of service is to relate it to the amount of calls that the system must handle, since it is calls per line—traffic—that in the end determines the rates charged for telephone service. Therefore, any assessment of the ability of the system to meet its financial constraints must relate to the cost of handling individual calls. A convenient way of assessing this is on the cost per switched unit of traffic. Such a cost must include all of the supporting cost of maintenance, administration, servicing, providing services, etc. This evaluation based on the cost of switching traffic is the best way to compare competing systems.

The telephone traffic unit most widely used is the Erlang, which can equal either one call for one hour or the summation of all the calls occurring in an hour multiplied by their average duration. Traffic is usually computed from the average of the busiest hours of a number of days. Telephone equipment is only used intermittently, and it may be idle during the greater part of the day. The art of providing telephone equipment is to strike an acceptable compromise between the equipment needed for the idlest hours and the equipment needed for the busiest hours.

1.5.2.4. Operating Costs

The basis for establishing the operating costs of the switching center is shown in Figure 1.5.

Direct costs. There are parts of the telecommunications switching network which are dedicated to a terminal, such as the terminal itself and, usually, the permanently assigned local transmission channel and its interface with the switching center. The costs for these parts may be recovered by the basic monthly cost to the customer.

Indirect costs transmission path. There are parts of the telecommunication

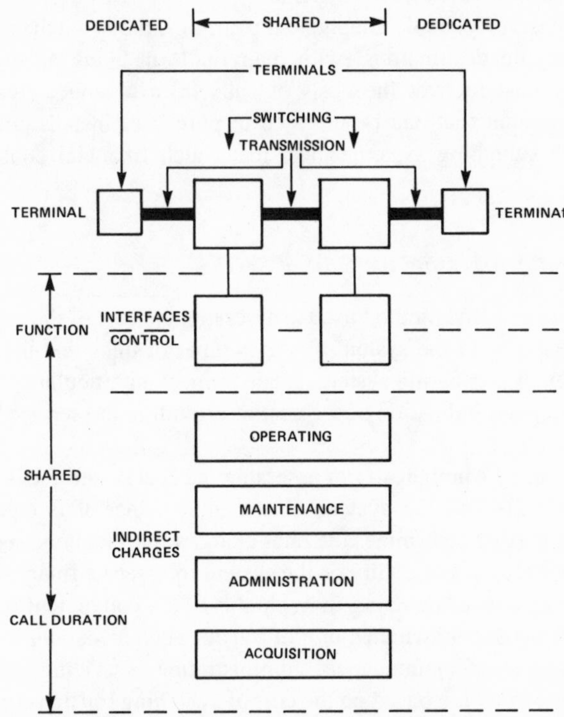

Figure 1.5. Basis of charging for calls.

switching network which are associated with a telephone call for its duration such as the transmission path used for the call after the first concentration stage. This includes the path through each of the switching centers used in the connection, their interfaces, and the trunk transmission equipment. The usage of the traffic-carrying units on a call can be obtained by monitoring and scanning the transmission path.

Other indirect costs. There are parts of a switching system which are only associated with the call to perform the necessary control operations and which are not held for the duration of the call. These costs can be recovered by assigning a percentage of them to each call. There are acquisition, operating, maintenance, and administrative costs, which have to be established and apportioned by the administration.

1.5.2.5. Computing Call Charges

An accounting system must be used to identify the cost of calls to distant terminals. These are usually called trunk (or toll calls) and may appear as separate charges on telephone bills. Trunk calls may be subject to compensation from the other telephone administrations involved in routing the call. This compensation is called separation of charges and it has an important bearing on the revenues of the telephone system in relation to mileage charges for the trunk transmission equipment.

The cost effectiveness of the operation can be determined by relating the revenue per switched Erlang to the cost per switched Erlang of calls to local area and long-distance destinations.

Call-charging methods. The methods by which the administration recovers its charges may vary. In addition to the basic charges for local telephone service, there will be additional charges for toll calls. These charges may be based on duration and distance in all cases or only apply to calls terminating outside an extended service area.

The manipulation of the data involved in determining the point of origin and the point of termination and the duration of the call should be an inherent part of the system operation. However, the format of the information and the way it is presented to the customer will be a matter of local choice. This may involve the bulk billing of all call charges. In this system each call is assessed in charging units. The number of units charged to a call will increase depending both on its duration and on its destination. Another method may involve the provision of individual bills in which each of the trunk calls charges are listed separately.

Any method used will almost certainly involve the use of centralized data-processing equipment which is external to the switching system for computing and preparing bills for service to the customer.

1.5.2.6. Environmental Constraints

Environmental constraints relate to the geographical location of the equipment. This may require special strengthening of equipment frames, etc., if the equipment is in an area which is subject to earthquakes. The equipment may require special precautions if the area has a heavy incidence of lightning. It will require special temperature and humidity controls depending on its location in either temperate, semitropic, or tropic zones.

The terrain has an influence on the system due to possibilities of variations in ground potential. This sometimes imposes constraints on the types of trunk transmission equipment that may be used. It may present problems in the case of line-of-sight systems such as microwave or in the installation of cables, etc., and in the organization of the network. This in turn may affect the number of switching systems interconnected in a call. The standards and working codes of the country and the availability of labor having the relevant skills in the country also may restrict or minimize the use of a particular technology.

1.5.2.7. Local Manufacture

There may be constraints due to the desire of a country to be either completely or partly self-sufficient in its manufacture of telecommunication equipment. This may require special attention in the design of the system in order that it be adaptable in part or completely to local manufacture in the country of application.

1.5.2.8. Compatibility

The most compelling constraint is the necessity to work with whatever switching equipment line and trunk transmission equipment exists in the country. This equipment may be equipment manufactured by a number of different suppliers. It may be equipment which has been in service for many years. It may involve equipment which is directly set by the calling device or indirectly controlled by intermediate storage devices (registers). It will have various methods of signaling between terminals and exchanges. Some of these methods will relate to the existence of transmission equipment which cannot be retired and some may relate to newly provided transmission equipment. Hence there may be a requirement for the equipment to interwork with exchanges which will be replaced gradually during the life of the various types of equipment and to interface with both old and new transmission requirements.

There are also other problems which are due to the provision of equipment by different manufacturers. In general in any interworking with other manufacturers' equipment, the work of the contractor providing the new switching center should be restricted to equipment manufactured by him. To attempt to intervene in other manufacturers' equipment presents all kinds of complications. Hence in providing equipment in an existing network, a separate arrangement may have to be made with the administration for the necessary adaption of the existing exchanges as part of the overall contract.

There are also constraints which are due to the standards necessary for interworking internationally. The need for overall methods of routing telephone calls between countries has resulted in recommendations for many of the parameters relating to factors which impact on the destination and originating countries involved in a connection. The organization concerned with recommending such standards is the Comite Consultatif International Telegraphique et Telephonique (CCITT),[4] and while in the past these standards were mainly for the purpose of rationalizing interworking between developed countries, the CCITT and its associated organizations are increasingly recommending switching systems organizations and networks for developing countries.

1.5.2.9. Interfacing (Figure 1.6)

The installation of a telephone switching system represents filling a gap in the network and may be considered akin to fitting a part of a jigsaw puzzle into its place. The parameters of the system, like the jigsaw puzzle part, are exactly defined by its place in the network, and the system must be configured to fit exactly into its place. This requires interfacing.

Interfacing requires matching the interconnecting and associated transmission facilities with the switching center. The positions of the interfaces are largely at the inputs and outputs of the switching system and may also be between portions of the switching system. Interfaces may be necessary for the following reasons.

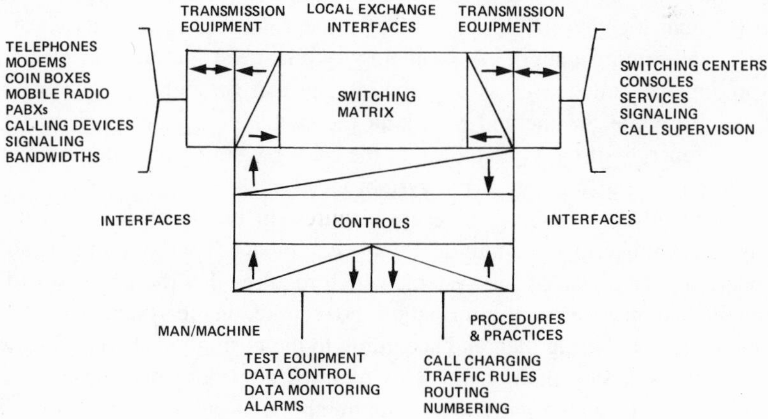

Figure 1.6. Scope of interfacing a local exchange.

Speed differences. Information received at one speed may have to be temporarily stored and then retransmitted at the new speed.

Level differences. Interfaces are also used for converting signal levels received at a low voltage level to a higher voltage level and vice versa or converting from low current to higher current and vice versa.

Physical differences. Some interfaces require transducers, such as those which convert sound energy to electrical energy or vice versa.

Transmission mode differences. Interfaces are sometimes needed to convert from one transmission mode to another—for instance, between a transmission mode which is analog in nature and a transmission mode which is digital in nature and vice versa.

Transmission means differences. Interfaces can also convert from unidirectional send and receive transmission paths to a bilateral transmission path (this is normally expressed as from four wire to two wire) and vice versa.

Cabling. There are interfaces between the outside plant, which constitute the transmission media, and the switching equipment. This involves the connection of cables suitable for burying underground to the cables suitable for use in connecting to switching systems.

Buffers. These are interfaces where a request for action is recorded and held until it can be acted upon.

Signal detectors. These are interfaces which permit the local switching system to respond to the calling devices in order to determine the destination address. This category also includes interfaces which detect control signals.

Local numbering plan. Interfacing also involves meeting the numbering plan requirements of the local administration.

The numbering plan for a given switching system location depends on the geographical position of the location in relation to the world numbering plan. This numbering plan uses a code based on world zones and (except for world zone 1) country codes. Each country has a local numbering plan based on the national call-routing pattern. A switching system must include the means for interpreting the called address and routing the call through the system to its destination. While the method by which the switching system performs these tasks is a function of the system design, the rules governing the interpretation of the addresses are a function of the system location.

Control of data base. There are also system interfaces that provide the means of administering and maintaining the equipment. These require the ability to update the data base of the system, which requires that the addresses of the terminals, and other relevant information, be recorded in the system according to the telephone numbering plan and according to the position of the interface with the local transmission equipment. This requires a specific data base for each switching system, which will vary as the number and types of customers' lines and services vary and as the locations of the customers vary. Hence a means of updating the data store to reflect any data changes relating to the terminals is necessary (a data base control). It also requires that as additional trunk transmission equipment is added to the system arrangements be available to include the trunk groups in the numbering and routing plans.

The data base for a modern switching system will be written in a program in memory. The data base has a number of separate portions. These include the operating program, the address program, and the diagnostic program. The operating program controls the actions of the switching center; it reflects the techniques used in the system and should remain unchanged. Hence the data base control cannot change it. The diagnostic program should also remain unchanged. The address program is specific to a single exchange and hence has to be updated as discussed above.

Means will have to be available so that the operating administration can train its own programmers in maintaining the data base. This points to the need for a common programming language for telecommunications switching. The CCITT is working on it, but since there are a lot of programs already written, changing could be quite a task!

Dimensioning. The amount of equipment provided must satisfy the number of calls which are being offered to the switching center at the required quality of service. It should be borne in mind that a switching system allows calls to fail. This is a matter of economy which allows the switching system to conform to acceptable costs. In order to determine the quality of service provided by the system, it is necessary to monitor the way in which the system is handling the telephone calls (''telephone traffic'').[(5)] This requires the collection of data at regular intervals and procedures to analyze the data and compare them to the forecasted and expected results.

Maintenance procedures. The maintenance of the system requires that there be diagnostic routines that constantly check that the system is operating satisfactorily. They must be capable of identifying the fault area to a small number of modules of equipment, so that maintenance of a service becomes a matter of replacing one package with another. The information that identifies the location of a fault must be available in a form to meet the requirements of the administration. The man–machine interfaces may require the use of a language different from that usually provided.

1.5.2.10. Repair of Equipment

When a fault is indicated in the system, it is categorized in terms of its impact on the system operation, and signals are transmitted to the point where personnel are available to deal with it. The personnel may have to be transported to the system to effect the necessary repair. Generally, such a repair consists of replacing one package with another. This requires the provision of sufficient spare parts to take care of repairs on a forecast basis. The forecasting of repairs on such a basis requires a statement of the number of faults that are expected to occur in a system owing to the random failure of components, which can be predicted, and in the meantime to repair such faults. The faulty module may either be modified on site (a procedure which becomes increasingly complex with modern technology) or sent to a central source for return and repair. The extent to which such repairs can be effected locally may have an impact on the acceptability of the system by the customer.

1.5.2.11. Maintenance of Transmission Equipment

Maintenance of the transmission facilities, both those that connect to the terminals that serve customers' telephones and those that connect to other exchanges and services, is necessary. This requires the provision of specified test routines which are either invoked on an automatic basis or applied manually. These check that the parameters are within the acceptable limits of the system. In any case tests may be applied on a manual basis in order to deal with a specific case where trouble has been encountered.

Detailed information must be provided on the system and the rules that govern its operation. This may have to be in the language of the user country.

1.5.2.12. Need for Switching Data

There is an increased requirement in modern switching centers to handle calls relating to machine data in addition to voice communication.

Data interfaces with present systems are effected through the use of modems which convert the data into analog signals. However, as data speeds increase there will be a requirement for switching at a higher bit rate than that possible with the bandwidth of the voice channel. This bandwidth is such that it allows the transmission of 4 kHz, but in practice it only utilizes from about 300 to about 3000 Hz. Digital switching systems will have to provide for interfacing at the level in the switching system related to the bit rate being switched.

1.5.3. Adaptability of Design

A switching system must be of modular design so that it may be adaptable to various quantities of terminals, trunks, and various services. It must (i) interface with many types of transmission equipment, (ii) work with various signaling systems, (iii) meet the national requirements for network management and numbering, and (iv) be adaptable to the specifications of the country of application and to the relevant international recommendations.

The structure of the switching center, especially the number and code translators, must be flexible so that it can deal with the many types of numbering plans in use in the world including some which are called ''open ended'' in which the number of digits dialed is not constant for a given exchange or area but varies. Its basic design must recognize the constraints which make its design a compromise.

1.6. System Documentation

1.6.1. General

It is necessary to provide adequate descriptions of switching systems for maintenance and evaluation purposes. The documentation must be easily understood and should not be restricted to written descriptions but rather based on universally understood symbols. The CCITT has been active in evolving a functional specification and descriptive language (SDL). Its objectives and recommendations are published by the CCITT.[6] A summary of them follows.

Narrative and numerical information is not standardized but a formalized presentation method has been proposed. The recommendation is a graphical method, based on state transition diagrams, using the symbols and rules of the SDL described below. Whenever appropriate the symbols of the SDL have been taken from the ISO standard for flow charts.[7]

1.6.2. General Objectives

The objectives of the SDL are to provide a standardized method of presentation

- that is easy to learn, to use, and to interpret in relation to the needs of operating organizations;
- that provides unambiguous specifications and/or descriptions for tendering and ordering;
- that provides the capability of meaningful comparisons between competitive types of Stored Program Control (SPC) telephone exchanges;
- that is open ended to be extended to cover new developments.

1.6.3. Areas of Applications

The main area of application covers all types of stored-program-controlled telephone switching systems. Within these systems the following functions are included among others:

- call processing (e.g., call handling, routing, signaling, metering, etc.);
- maintenance and fault treatment (e.g., alarms, automatic fault clearing, configuration control, routing tests, etc.);
- system control (e.g., overload control, modification and extension procedures, etc.).

1.6.4. Basis for SDL

The SDL is applicable both to the specification of what the user requires and the description of a particular system. Hence the description applies to actions performed by a system in providing telephone service. These actions are described in terms of inputs, states, decisions, tasks, and outputs. Functional blocks which are used to illustrate processes or parts of processes are broken down into a number of levels of increasing detail. The governing factor is that each process should have a well-defined boundary across which signals pass.

Basic Concepts

The SDL method for graphical presentation is based on the following definitions.

Signals. A signal is a flow of data conveying information to a process. A signal may be either in hardware or in software form. It should be noted that if the information flow is from a process described by a block to a process described by another block it is an external signal. If the flow is between processes described by the same block it is an internal signal.

Inputs. An input which may be external or internal is an incoming signal which is recognized by a process. (It is not to be confused with input as applied to normal data processing.)

States. A state is a condition in which the action of a process is suspended awaiting an input.

Transitions. A transition is a sequence of actions which occurs when a process changes from one state to another in response to an input. A process can be either in one of its states or in a transition at any one instant.

Outputs. An output is an action within a transition which generates a signal which in turn acts as an input elsewhere. (It is not to be confused with output as applied to normal data processing.) An output can be either internal or external.

Decisions. A decision is an action within a transition which asks a question to which the answer can be obtained at that instant and chooses one of several paths to continue the transition.

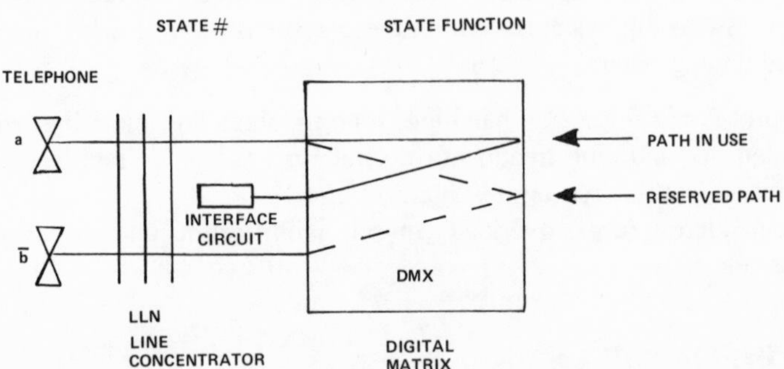

Figure 1.7. Digital switching system state function symbols. a = handset off hook, $\overline{b}$ = handset on hook. A prime is often used, as in b′, to indicate handset hook flash.

Tasks. A task is any action within a transition which is neither a decision nor an output.

1.6.5. Symbols

Figure 1.7 shows the symbols proposed for functional diagrams. Each part of the functional diagram contains a state number and a state function. Symbols shown include line idle (handset on hook) and line in use (handset off hook). There are symbols to show the connection paths (both those in use and those reserved for future use in the call). There are also symbols for interface circuits, networks, and so on.

Figure 1.8 shows the application of the diagrams to the ringing state and its transition to a talking state. The path for the talking state is shown as reserved in the talking state.

Figure 1.9 shows the same functions using a flow chart approach (based on the flow chart symbols shown in Figure 1.10). It can be concluded that the state

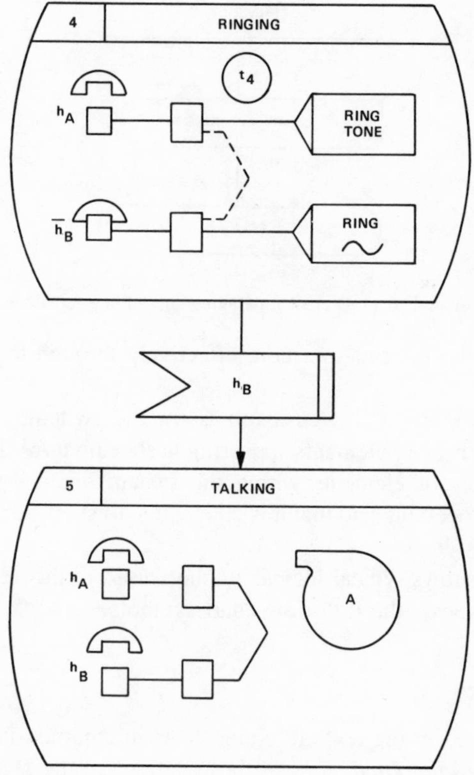

Figure 1.8. State diagram for ringing and talking functions.

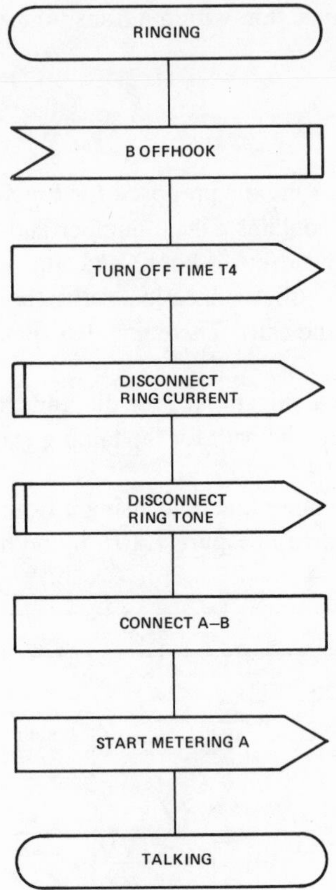

Figure 1.9. Flow chart for ringing and talking functions.

diagrams present the information more effectively through the use of pictorial symbols.

Table 1 shows the suggested symbols for the switching, signaling, call-charging, and supervisory elements appearing in state pictures. It should be noted that only the states of elements within the system boundary can be directly changed by call processing and that it is convenient to show terminals outside the system boundary.

Figure 1.11 shows typical logical symbols used in this text.

Figure 1.12 shows the ISO flow chart symbols.

1.6.6. Vocabulary

The terms used in the text are either those in common use and defined in *Webster's International Dictionary* or those defined in the *IEEE Standard Dictionary of Electrical Terms*.[8]

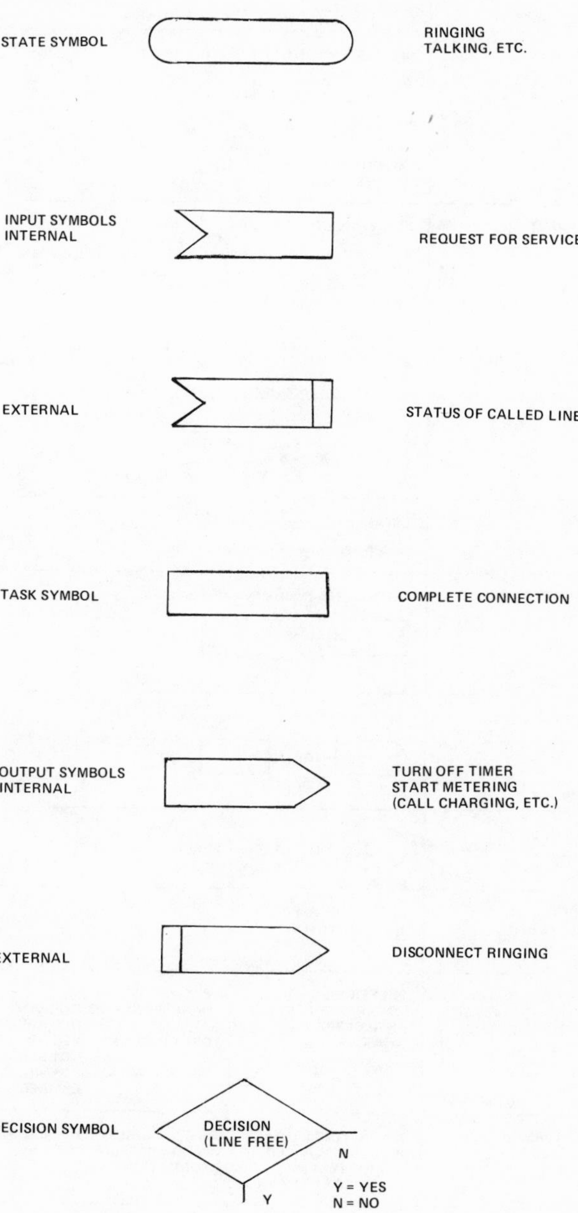

Figure 1.10. Proposed CCITT flow chart symbols.

Table 1. Suggested CCITT Symbols

CONCEPT OF ELEMENT	SYMBOL	
SYSTEM BOUNDARIES	a) SYSTEM PORT b) SYSTEM BOUNDARY OUTSIDE SYSTEM	
TERMINAL EQUIPMENT	a) HANDSETS ON-HOOK OR $\bar{b}$ OFF-HOOK OR b b) OTHER TERMINAL EQUIPMENT GENERAL MANUAL SWITCHBOARD	
SWITCHING PATHS	ESTABLISHED RESERVED	
GENERALIZED SIGNALLING RECEIVERS	(WITH LABEL INSIDE IN EACH CASE) OR OR	
GENERALIZED SIGNALLING SENDERS	(WITH LABEL INSIDE IN EACH CASE) OR OR	
TIMERS SUPERVISING A PROCESS	WITH LABEL INSIDE TO IDENTIFY TIMER	
CHARGING IN PROGRESS (AND WHICH CUSTOMER IS BEING CHARGED)	A IF A PARTY IS TO BE CHARGED	
INPUT VARIABLES	PREFERABLY USING LOWERCASE LETTERS AS ALGEBRAIC SYMBOLS	E.G. HANDSET STATES: ON-HOOK OFF-HOOK DIGIT RECOGN: RECOGNIZED NOT RECOGNIZED TIMER STATES: RUNNING EXPIRED
SUBSCRIBER (OR TERMINAL) CATEGORY AND IDENTITY INFORMATION	A = CALLING PARTY B = (FIRST) CALLED PARTY C = THIRD PARTY, ETC.	USE RECTANGLES TO CONTAIN CATEGORY INFORMATION E.G. BAR TRUNK ACCESS
INCLUSIVE CASE SYMBOL: A SUBSTITUTE FOR DELIBERATELY UNDEFINED MICROSTATE INFORMATION THAT IS SHOWN UNAMBIGUOUSLY IN OTHER STATE PICTURES.	USE OF ASTERISK: E.G. i) * HANDSET EITHER ON-HOOK $\bar{h}$ OR OFF-HOOK h	ii) * TRUNK ACCESS SUBSCR. CATEGORY EITHER "BAR TRUNK ACCESS" OR NOT, IN THIS STATE OF THE PROCESS.

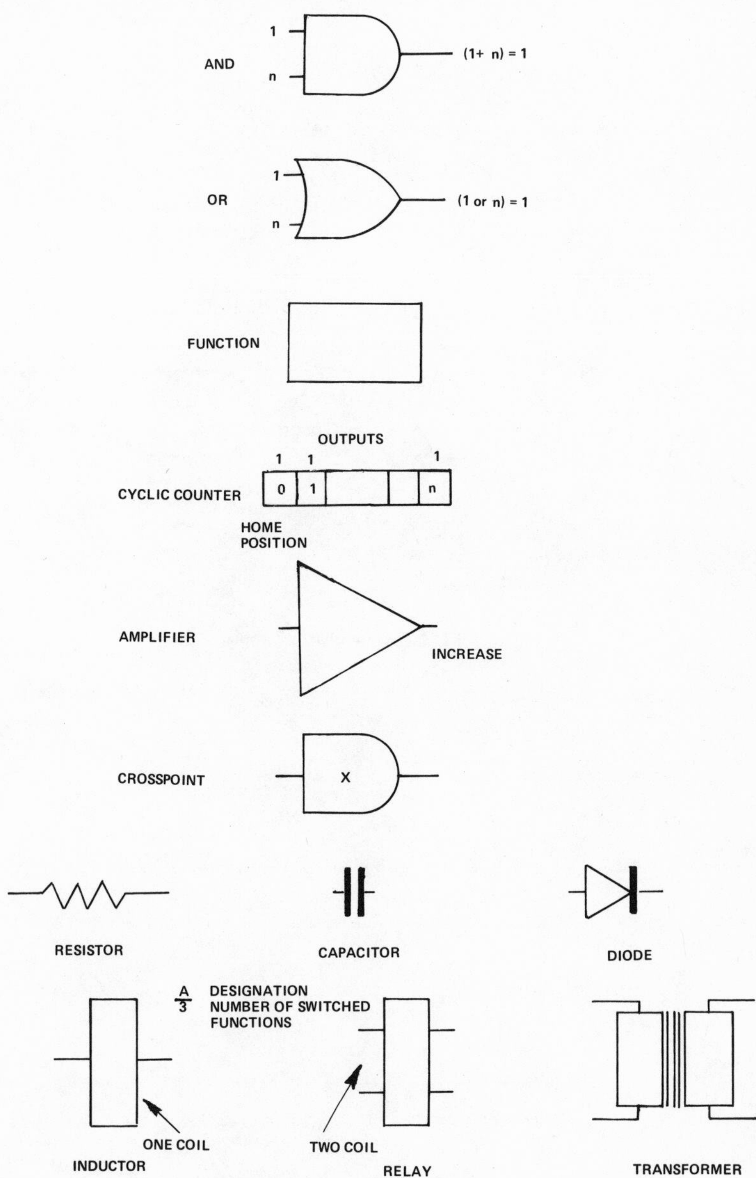

Figure 1.11. Logic symbols used in the text.

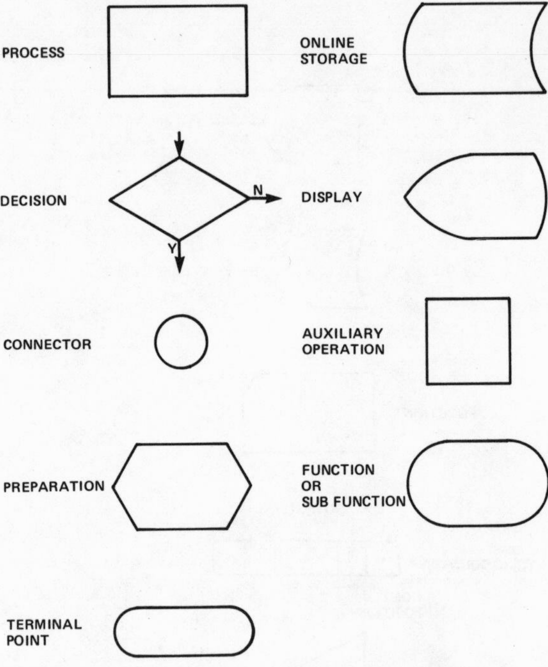

Figure 1.12. ISO flow chart symbols.

2

System Organization

2.1. Introduction

A telephone switching center is part of a telecommunications network. The structure and organization of a switching center must therefore be responsive to the needs of the network. A switching system is a means of routing traffic from the various concentrations of population to destinations which may be anywhere in the world. Since the telephone traffic from a telephone line is small, telephone lines are connected to a switching center which concentrates the traffic and distributes it to its destinations.

The organization of a switching network requires that the local exchanges which collect and distribute the telephone traffic be located in areas of relatively high population density. This is so that the common equipment such as power equipment which is used to operate the system and the maintenance and other facilities can be concentrated and shared among groups of users.

Local exchanges are connected to trunk exchanges by trunk routes which carry the concentrated traffic towards its destination. The routing depends on the number of telephone lines per area, the size of the country, the geographical position of the centers of population, and other factors such as terrain. The trunk routes funnel the outgoing traffic into large groups. These connect to trunk exchanges which serve to route the trunk traffic through the network and, if necessary into different areas and to other countries.

2.2. Organization of National Networks

Each country has an organization of switching centers which provides for routing traffic through its hierarchy of exchanges. If the country is very large then there may be many levels in the switching network hierarchy, as in the case of the Bell System here in the United States of America.

2.2.1. Hierarchy of Bell System Network

Figure 2.1 shows the basic structure of the Bell System network.

The local exchange, which connects subscribers' line transmission equipment to and from trunk transmission equipment, is the fifth level in the system and in AT&T terminology is called an end (class 5) office. The end offices connect to the fourth level of trunk switching offices which are (class 4) trunk exchanges. These are termed toll centers and toll switching points and can provide operator assistance. The toll centers connect to other (class four) exchanges and also to the next highest level exchange in the network, which is a primary center (class 3). Primary centers connect to the next exchange classification, which is a sectional center (class 2). Sectional centers connect to (class 1) regional centers. Regional centers connect to other regional centers and to international exchanges. This provides all the necessary interconnection for routing traffic in the North American numbering plan area.

When there is a sufficient volume of calls between areas then direct routes may be used. These occur at various levels in the system.

The operation of the Bell System network is described in the "Blue Book."[8]

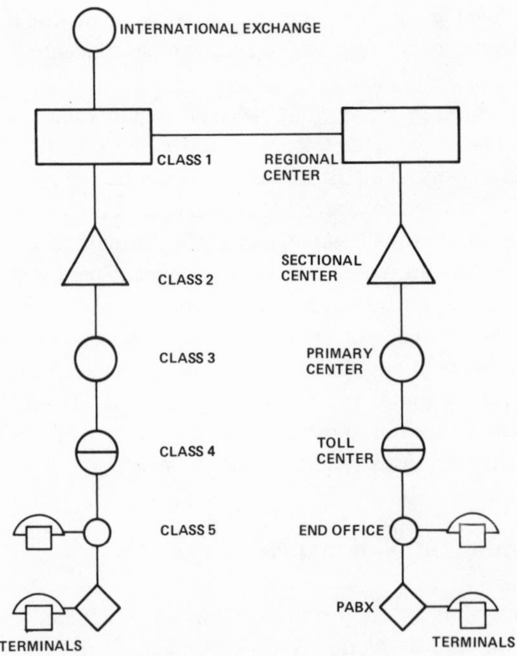

Figure 2.1. National network organization (U.S.A.).

2.2.2. Typical Hierarchies of National Networks

A country having fewer telephones than the U.S.A. may use an hierarchy of exchanges with a smaller number of levels.

A country may have as many levels in its hierarchy but use different terms for the exchanges; examples are: terminal center, group center, zone center, regional center, national center or local exchange, tandem exchange, primary exchange, secondary exchange, tertiary exchange.

However, the same basic routing principles outlined above always apply.

2.2.3. Network Evolution

The Bell System network of switching centers has grown with the evolution of telephony and the development of the country and is restricted to much of the existing trunk transmission equipment. Hence it is basically an analog network. The rapid application of stored-program control has provided a means of controlling the network by separating the signaling functions from the transmission path. Since address information does not traverse the same path as the telephone message, a separate signaling network is superimposed on the existing transmission network. This system, which only routes signal information, is described in more detail in Chapter 10. The application of trunk exchanges[(9)] which employ digitally coded voice transmission and the extensive use of digital carrier[(10)] offer possibilities of integrating the switching system and the trunk transmission equipment. This promises both greater efficiency in operating and a reduction in the quantity of equipment.

Existing national networks of all countries have developed in an evolutionary network with larger networks in the more developed countries. Digital switching and transmission is increasingly being applied in these countries and stored-program control is being used increasingly. In many countries the absence of an elaborate trunk network allows these techniques to be applied more freely; in others, with large existing networks, its introduction may have constraints.

2.3. International Routing

In addition to the necessity to provide a hierarchy of switching centers to serve the needs of a country it is necessary to provide the necessary routes for international calls. This involves the use of a hierarchy of levels of international exchanges, the number of levels depending on the amount of traffic that a particular country generates and receives. There are a number of such gateway exchanges in the U.S.A. interfacing with the Far East or with Europe, etc. They interface at level 1 (CT1). Figure 2.2 shows the general routing pattern from the AT&T network. There may be less than three levels of international exchanges

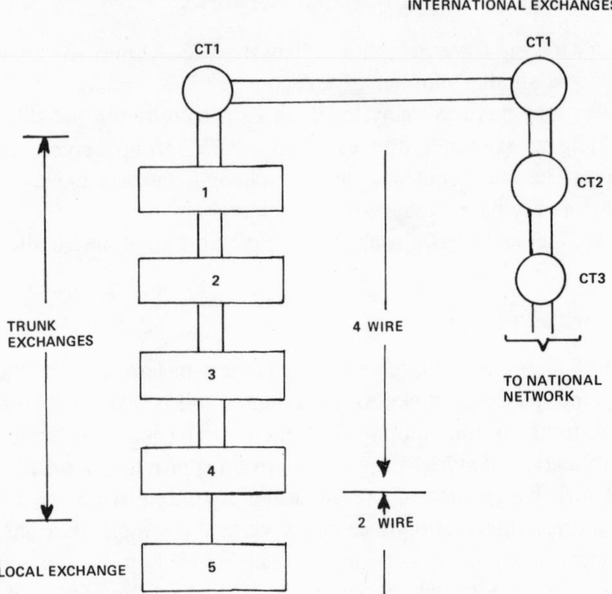

Figure 2.2. International network organization.

depending on the number of telephones in the country. The maximum number of international exchanges allowed in an international call is six.

Trunk Routes

The CCITT recommendations for international routing are to connect CT1 exchanges together by final routes, to connect CT2 exchanges together by low-loss-probability routes, and to provide high-usage groups between CT2, CT1, and CT3 category transit exchanges as required. Routing is accomplished by high-usage or overflow routes.

High-usage routing. High-usage routes, direct routes, are based on the most economical cost of routing which includes the cost of the terminations in the exchanges as well as the transmission paths.

Alternate routes. These are used to carry the traffic which overflows from high-usage routes. In the case of an alternate route the cost of switching (i.e., the cost incurred at the intermediate exchange) must be included. In all cases the costs are expressed on an average incremental annual basis.

Economic factor. The economic factor is determined by the ratio of the cost of an alternate route (AR) to the cost of direct route (DR). The greatest efficiency is obtained when the ratio

$$\frac{\text{AR}}{\text{DR}} = \frac{\text{the usage of an additional alternate route}}{\text{the usage of the last circuit in the direct route}}$$

equals T1/T2, where T1 is the traffic in Erlangs carried by an incremental alternate route and T2 is the optimum traffic in Erlangs carried economically by the last trunk in the direct route.

2.4. Transmission Plan

As pointed out earlier, transmission losses have to be assigned to each portion of the network. This is the basis for a transmission plan.

2.4.1. Transmission Network

A telecommunications network involves the routing of calls over long distances. A transmission plan has to be devised to ensure that losses in the levels of the voice and control signals be held to an acceptable level. In addition attenuation of signals of differing frequency must be minimized and precautions taken against overhearing other calls through signal leakage, etc. This requires that each portion of the network be assigned limits on the losses, etc., that it can introduce. The recovery of the signal level requires amplification to replace losses. The possibilities for amplification on a bilateral two-wire path are quite limited, hence separate paths have to be provided for each direction of transmission to allow amplification. In general all telephone lines and most local exchanges of the pre-1977 vintage are two wire. This is an economical compromise with the technologies of that era. Trunk exchanges are predominately four wire. A two-wire/four-wire interface is necessary to convert from two wire to four wire; this is called a hybrid. Unfortunately, hybrids require balancing the two-wire circuit, to prevent recirculation of signals, by canceling them out magnetically and a transformer or other ac coupling. This introduces its own losses. Balancing circuit tolerances limit the amount of amplification that can be applied.

Figure 2.3 shows the allocation of losses in the transmission plan for the Bell System analog network. Emerging transmission and switching equipment will ultimately result in a revised plan.

Bell System VNL concept. The VNL (via net loss) concept allows a maximum loss after amplification at each of the levels in a routing pattern.

CCITT *concept.* The losses assigned according to CCITT recommendations can be found in the *Green Book.*[11]

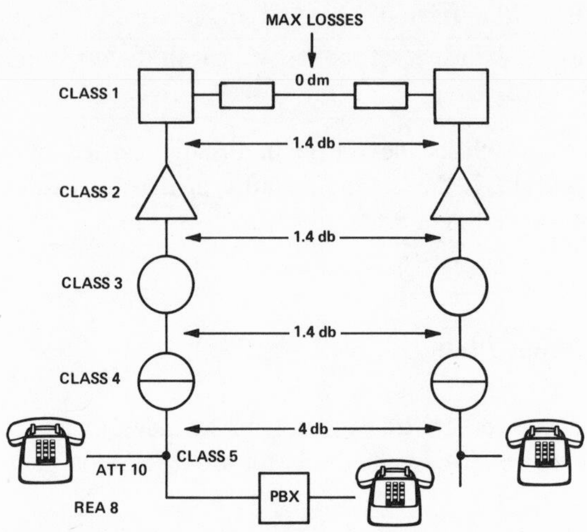

Figure 2.3. Transmission plan for Bell System.

2.4.2. Basic Transmission Modes

Figure 2.4 shows some basic transmission modes.

2.4.2.1. Interconnecting Equipment

The interconnecting of terminals and switching centers by transmission means requires the provision of physical interfaces with the outside transmission media. An interface with a metallic line is of the simplest form since it only has to connect from the outside cable to the inside cable. This is done in order to connect the lines to distribution frames in an orderly way. The transmission means to telephones are usually two-wire lines without amplification. This sets limits (due to transmission and signaling needs) on the length of the connections between the station equipment and the local exchange that serves it. The use of two-wire lines means that the receiving and transmission functions of the telephone have to be separated at the telephone and makes for a more complicated telephone.

2.4.2.2. Four-Wire Transmission

A four-wire transmission system had in its original concept two separate pairs of wires: one for one direction of transmission and one for the return direction of transmission. The provision of four physical wires to a telephone is uneconomical and hence four wires are normally used for a trunk transmission equipment. However, the four-wire transmission channel may be a derived channel (that is, one of a group of channels associated either through the allocation of separate frequency bands in a frequency division carrier system, or as a separate time divided path in a time-divided system).

It is usually necessary to convert from four wire to two wire for interconnection to the switching center. This involves the use of hybrid transformers, instead

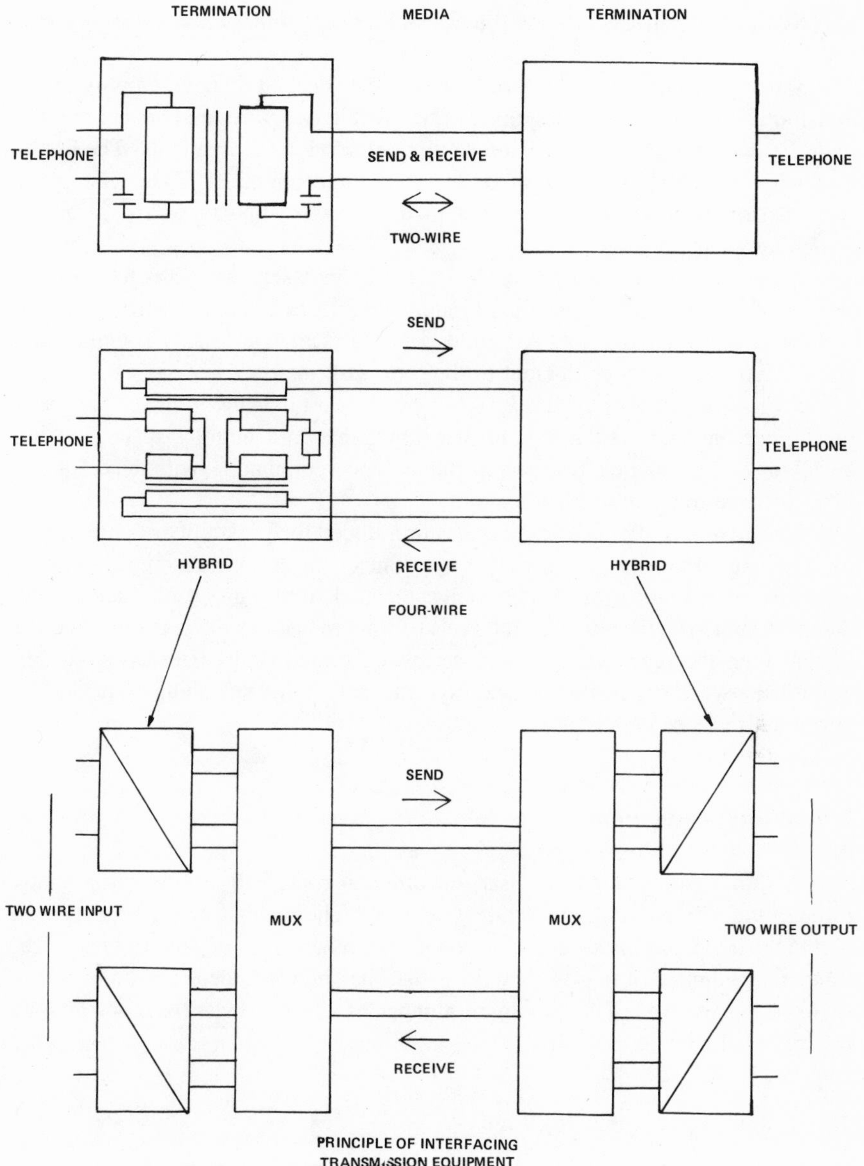

Figure 2.4. Transmission modes.

of the usual simple line transformers, as shown in Figure 2.4, to properly terminate the four-wire transmission facility.

2.4.2.3. Switching Matrices

Switching matrices provide the transmission path through a switching system.

Space division switching matrices. Most local exchanges employ space division and use metallic crosspoints. They switch two wires and are compatible with the two-wire line transmission equipment serving the terminals. The use of four-wire switching to a terminal is usually uneconomical. Four-wire trunk transmission equipment is converted to a two wire transmission path at the switching center.

Time division switching matrices. These have separate paths for each direction of transmission. Hence they require interfaces to adapt the transmission path with the outside plant. A hybrid per interface is necessary for matching two-wire transmission equipment to the four-wire matrix.

Switching matrices for trunk exchanges. Most existing trunk exchanges at the higher hierachies of a national switching network employ space division switching. They switch four wires but are not compatible with most of the long-distance transmission equipment.

The advent of digital switching has introduced the possibility of compatibility with digital carrier systems. If, for instance, the trunk transmission system allocates intervals of time to a particular channel from a group of channels and the switching systems allocates intervals of time to each of the paths through the system then the interface is greatly simplified. Such direct interfacing is only possible between digital transmission systems and digital switching systems. It is described in more detail later.

2.5. Numbering Plan

A numbering plan is necessary in order to route calls to their destination through the national and international network. The numbering plan identifies a particular telephone in a particular country in relation to its geographical location. The number of digits used in a national network depends on how the network has evolved. The maximum number of address digits that constitute an international address is 12.

2.5.1. Basis for Numbering Plan

A numbering plan center consists of the following portions.

2.5.1.1. World Zone Codes

A world zone number is based on the geographical location of the country. The world zone number is a number which has been allocated as a recommendation of the CCITT and it is usually the first digit of a country code. In the world numbering plan this number identifies one of the larger geographical areas into which the world was arranged in order to permit the routing of international calls. Figure 2.5 shows the geographical allocation of the world zone numbers.

Allocation of world zone codes. The allocation of the world zone code digits is as follows:

Code 1 is allocated to North America, which includes areas operating within the regional numbering plan and hence includes the U.S.A., Canada, Mexico, Bermuda, Puerto Rico, some Caribbean islands, Alaska, and Hawaii.

Code 2 identifies the continent of Africa.

Codes 3 and 4 identify Europe, which has been allocated two zones because of the high telephone density and the large number of countries in it.

Code 5 serves South America, Cuba, Central America, and part of Mexico.

Code 6 serves the South Pacific.

Code 7 serves the Union of Soviet Socialistic Republics.

Code 8 serves the North Pacific and Eastern Asia.

Code 9 serves the Far and Middle East.

Code 0 has been left spare and may be used ultimately, if a local world zone requires a greater number of stations than that possible with the present number-

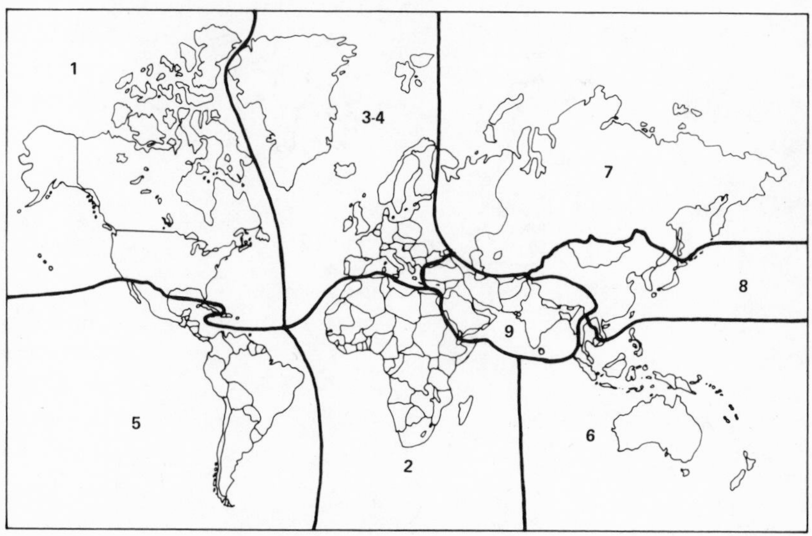

Figure 2.5. World numbering zones.

ing scheme. However, the nonallocation of digit 0 to a world zone allows its use for shorter codes, such as operator access.

2.5.1.2. Country Codes

A country code is required to identify the geographical area within the world zone; the initial digit of the country code is always the world zone number.

Allocation of country codes. A country code usually consists of the World zone code plus digits (depending on the number of telephones in the country). Hence a country code can consist of one digit (as in the unique case of world zone 1), two digits (as in the typical cases of France or the United Kingdom), or three digits as in the typical cases of Nigeria or Guatemala. This country code iden-

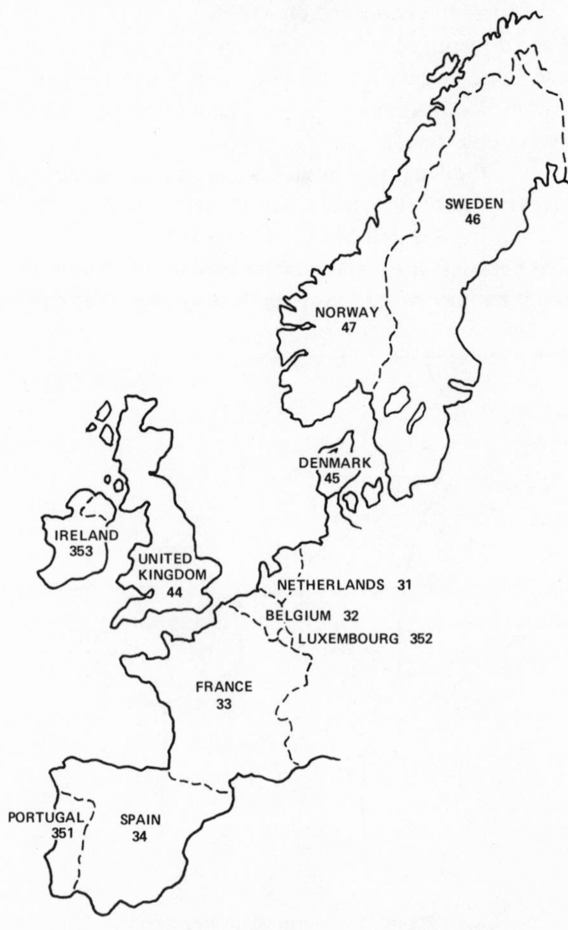

Figure 2.6. Country codes used in Western Europe.

tifies an integrated numbering plan area. Its sole purpose is to direct traffic to its destination and therefore the allocation of a country code does not always relate to political boundaries (for example, as in the case of Canada, which is included in world zone 1).

Identification of countries in world zone. In world zone 1 the numbering plan areas identify the geographical areas, e.g., 403 the Province of Alberta Canada, 406 the State of Montana, 809 Bermuda and some of the Caribbean islands, 716 the northwest portion of New York State. There are no separate country codes.

The allocation of the numbering plan areas in world zone 1 is specified in the Blue Book.[8]

Country codes used in Europe. Figure 2.6 shows the country codes used in Western Europe. Two world zone codes are involved, digit 3 or digit 4. Countries with large numbers of telephones are identified by two digits such as the United Kingdom (44) and France (33), whereas Luxembourg (352) and Portugal (351) have three digits.

2.5.1.3. Area Codes

An area code is necessary to identify a numbering plan area. It follows after the country identifying code. An area code may consist of one, two, or three digit numbers and generally relates to a geographical subdivision of a country or territory that is covered by a separate national or integrated numbering plan.

2.5.1.4. Exchange Codes

The exchange code identifies the local exchange to which the required destination terminal is connected. An exchange code may consist of one, two, or three digits.

2.5.1.5. Main Station Code

A main station code identifies a terminal, and since sometimes there is more than one customer served by the same line transmission equipment, then more than one station code may be assignable to a terminal. It is possible that, since a four-digit station code serves a maximum of 10,000 main station codes, more than one exchange code may have to be provided for very large switching centers. There are numbering schemes in which the number of digits needed to identify the main station in an exchange may be more or less than four.

2.5.2. Application of Numbering Plan

Countries vary in population size and number of telephones and in the numbers of geographical areas, hence it does not follow that each of the above

Table 2. Typical Numbering Plans

Country	Numbering plan, maximum 12 digits[a]				
	World zone	Country	Area	EXCH	Station
U.S.A.	1	None	N 0/1 X	NNX	XXXX
U.K.	4	4	Varies 7 to 9 digits		
Nigeria	2	34	None	NN or NNX	XXXX
Guatamala	5	02	NNX	N or NN	XXXX

[a] X = Any number 0 through 9; N = any number from 2 through 9; 0/1 = 0 or 1.

codes will appear as a separate identifiable entity in every numbering plan. This is illustrated by Table 2, which gives an example of four numbering plans.

2.5.2.1. North American Numbering Plan

The basic construction of a numbering plan depends on the geographical locations of the switching centers that it serves. The North American numbering plan is an example: Figure 2.7 shows the general arrangement. The codes used in this figure are shown in parentheses in the following paragraph.

Allocation of codes. Each terminal has a terminal (main station) code of four digits (6493). Each switching center has an office code (461) or codes, each of three digits. Each toll switching center area has a three-digit area code (716), and there is a common world zone code. In the case of North America this is world zone 1. The addressing of a telephone call in this network could involve dialing as many as 11 digits (if the call is to an area other than that of which the originating exchange forms part). It can consist of seven digits (if the switching center is within the same area). Calls to operators or services can be reached by dialing as few as three digits. International calls may require the dialing of an access code before the desired number.

2.5.2.2. Service Codes

In addition to addresses which relate to stations, other address information may be required from the calling terminal. For example, in certain conditions additional digits may be required for access to a particular area because of the type of equipment in use in the area and the types of switching centers already installed there.

Language codes. It may be necessary under certain circumstances, e.g., in the case of multilingual areas, to identify the language to be used by an operator called into the connection.

Operator codes. These are necessary to identify a call requiring the services of an operator. The majority of telephone calls are now dialed directly, so it is necessary to identify whether the call requires operator intervention for completion. For example, a call to a main station in which anyone who answers is an acceptable charge can be handled automatically. On the other hand a call to a destination for conversation with a specified individual requires, at the present state of the art, an intervention by an operator. This is necessary in order to determine that the appropriate person is available before the call is completed and a charge is made for it.

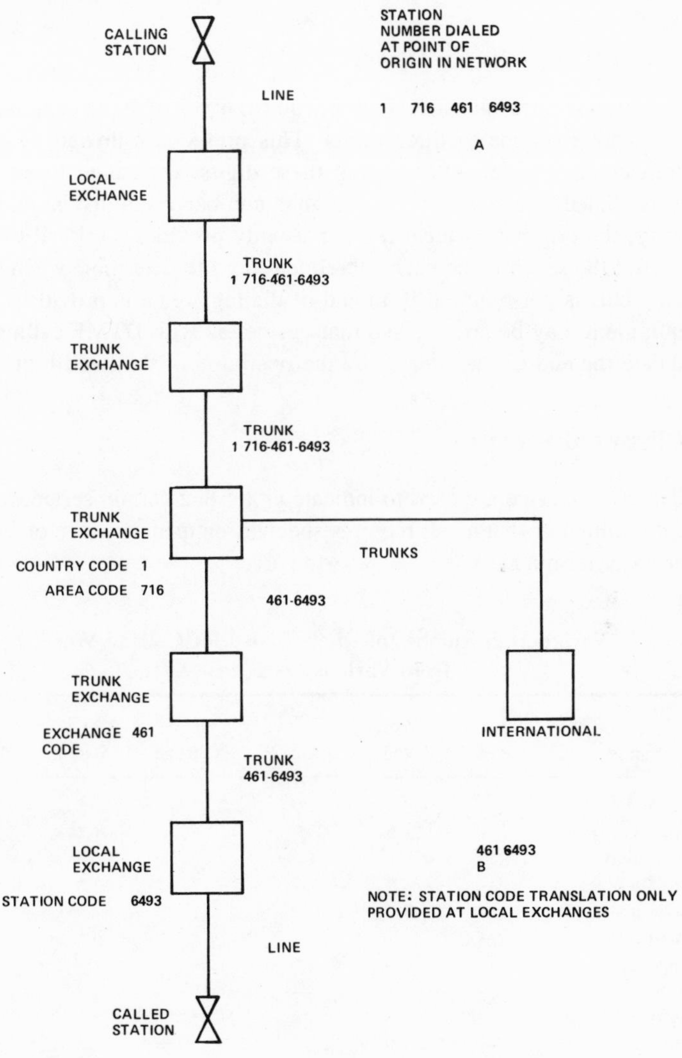

Figure 2.7. World zone 1 main station code address structure.

Emergency codes. There is also a need for codes for access to particular services. These may or may not require the dialing of a station code. An example of such a code is a code which is provided in order to route calls intended for emergency services such as fire, ambulance, police (911, 999, etc.).

2.5.3. Number of Digits Dialed

An example of the varying numbers of digits that can be dialed is given by the examples in Table 3.

International Direct Distance Dialing

International direct distance dialing requires up to 16 digits beginning with a prefix consisting of one to three digits. This prefix is followed by a one- to three-digit country code. After dialing these digits, the called listed directory number is dialed. Because of the variable number of digits in international numbering, the originating local register usually provides a critical time out of 3–5 sec after the seventh and each subsequent digit to determine when dialing is complete. This is not required if an end-of-dialing signal is provided. The register equipment may be arranged so that customers with DTMF calling devices can indicate the end of the address by the operation of the twelfth button (#).

2.6. Classes of Service

Classes of service are used to indicate either that certain services are available to a terminal or that a call requires special treatment because of its point of origination or termination.

Table 3. Variation in Number of Digits Dialed for Calls to World Zone 1 from Various Locations

From	Prefix	World zone	Area	Exchange	Station	Number of digits dialed
International	*[a]	*	*	*	*	Variable
World zone 1 (to other area)	—	*	*	*	*	11
Same area (to other exchange non-EAS)	—	*	—	*	*	8
Same area (to other exchange EAS)	—	—	—	*	*	7
Same exchange	—	—	—	*	*	7
Special services trunk	—	—	—	*	—	3
Special services line	—	—	—	*	*	7

[a] The asterisks indicate the number of digits dialed.

2.6.1. Identifying the Class of Service

The address of the called party is signaled by the caller, but the routing of calls may vary according to the type of service to which the calling and/or called party is entitled. Hence it is necessary to provide a means of identifying the classes of service which are available to a terminal for originating calls from or terminating calls to it. This is accomplished by identifying the main station number of the party. This is then referred to its memory area and the relevant class of service obtained for the calling or called parties.

2.6.2. Types of Classes of Service

Classes of service are used to produce a number of differing features. These features can be broken down into two main categories: for the operating administration or for the telephone customer. Operating administration classes of service have primarily to do with call routing, accounting, and network management. Customer services relate to features, usually tariffed separately, which are system options.

2.6.2.1. Call-Routing Classes of Service

Call-routing classes of service are used to indicate groups of terminals serving the same destination or to change the routing of calls or to override busy signals.

Trunk classes of service. These are used to identify trunk groups and identify the trunks in a group. This is necessary because while there is only one line which serves a residential main station there are many trunks in the relevant trunk group which are capable of routing the calls to the given destination. This follows since the traffic from many sources is combined into a stream of calls going toward a specific destination. It is necessary to identify each trunk in the trunk group by its interface equipment location to route the call to the called number via the trunk group. It is also necessary to indicate any special conditions which may apply to a trunk such as the type of signaling, whether it is an incoming or outgoing trunk, and so on. It may be necessary to prevent calls coming in on a particular trunk group to go out on that same trunk group. This procedure could allow the interconnecting in tandem of many trunks, in an illegal manner, on a given call. Access to certain kinds of trunks may be limited, for example official telecommunications services. In some networks it may be necessary to reconstitute some of the address information which was used to route the call from a serving local exchange. An appropriate class of service can provide this feature.

Private branch exchanges. Classes of service indicate the lines, or trunks, which either provide access to private branch exchanges or provide the services as part of the local exchange. They can be used to indicate a terminal number which allows a test to be made on each of the lines or trunks serving a PBX, before indicating that all lines to it are busy. It may also be used to permit direct dialing of the PBX lines if the PBX and local exchange share the same exchange code. Calls originated from the PBX may also be allocated special classes of service, for instance to restrict certain lines to certain destinations.

The class of service can also indicate lines reserved for outgoing calls only or for incoming calls only (in order to serve the business needs of the private branch exchange).

Internal call routing. Classes of service may be used to change the destination of a call (for example in case it was directed to an unallocated or temporarily out-of-service line). They may be used to permit line testing by overriding a busy signal and so on.

Classes of service for special terminals. Classes of service may also indicate special kinds of terminal equipment. An example of this is a coin box. Coin box calls require special treatment under various conditions. For example the coin box may be restricted to the dialing of a restricted or limited number of destinations (because of a limitation on the mechanical collection of coins to pay for the charge). Hence a call from a coin box which does not allow for automatic coin collection on a time and distance basis will require the intervention of an operator. The operator must know that the call is from a coin box in order that he or she may check for the deposit of the correct coins into the appropriate slot, or slots, of the coin box.

Classes of service for special accounting methods. There are other types of classes of service, for example, a destination for which no charge is made to the calling line when calls reach it. Such calls may be paid for by the administration or by the called party.

2.6.2.2. Customer Classes of Service

Customer classes of services can be tariff-related features which involve the paying of addition charges such as an abilıty to dial a shorter number of digits to identify a frequently called destination (abbreviated addressing).

They may provide an indication that a call to a certain destination, may under certain circumstances, be rerouted to another destination (call forwarding).

Another class of service is one that provides an indication, when the line is in use on a call and a second call is made to the line, that another person requires to reach the terminal. This is called *call waiting*.

There are many other classes of service which are possible especially in conjunction with modern signaling systems. These and the above are discussed in detail in Chapter 12.

2.7. Signaling

In order to transfer address and other information between terminals and switching centers it is necessary to encode the information into signals. There are three categories of signals: station signaling, which controls the line transmission path and provides address information to a register in a local switching system; line signaling, which controls the trunk transmission path; and register signaling, which provides the address information for routing calls to their destination and uses senders to signal to registers in other levels of the network.

Figure 2.8 shows where these occur in the network. Station signals from the calling device (‘‘ing’’ customers) are stored in a register (R) in the local exchange and sent as register signals by a sender (S) to the register in the terminating local exchange in the example. Line signaling is used between the two local exchanges to control the call.

2.7.1. Interpretation of Signals

The interpretation of the signals and the compiling of the necessary routing data, classes of service, etc. are obtained by using a translator. This receives address information from a register and produces data for call routing and other purposes. Routing information is sent to a sender. Figure 2.9 shows the association of the functional units.

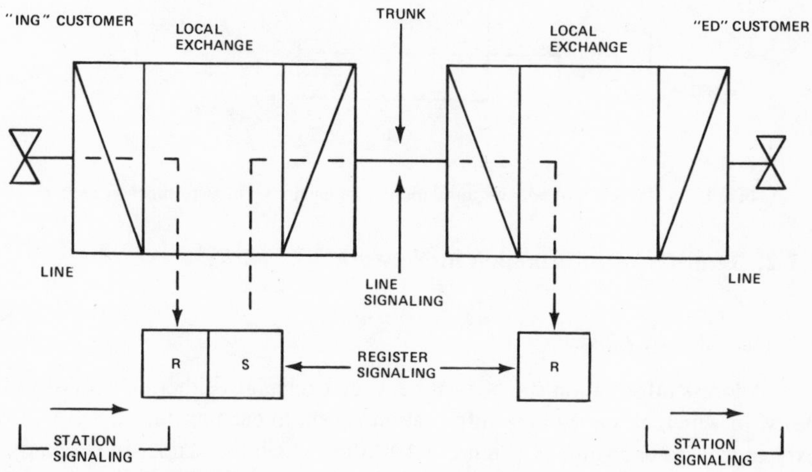

Figure 2.8. Classes of signaling. R = register, S = sender.

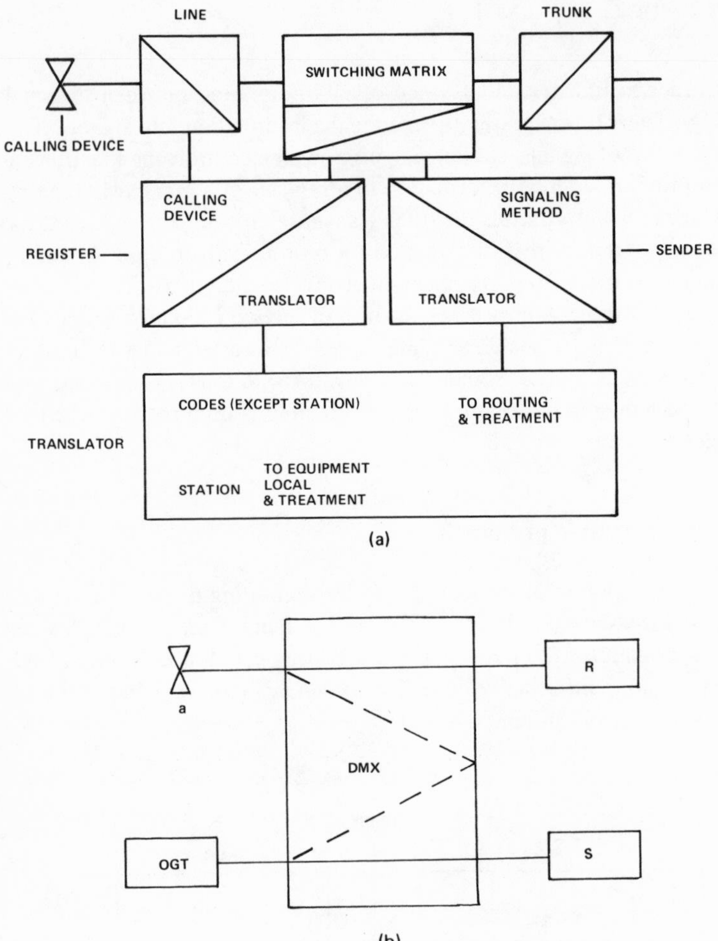

Figure 2.9. Principle of address translation. (a) Functions, (b) state function picture.

2.7.2. Transfer of Information in Network

2.7.2.1. Link by Link

Address information can be sent between exchanges either on a link-by-link basis, in which all the address information is sent to each exchange—e.g., from exchange A to exchange B (Figure 2.10) after which the sender in exchange A releases from the connection. Subsequently exchange B sends all the information to exchange C (Figure 2.11). Each exchange acts on the address information before sending to the next exchange.

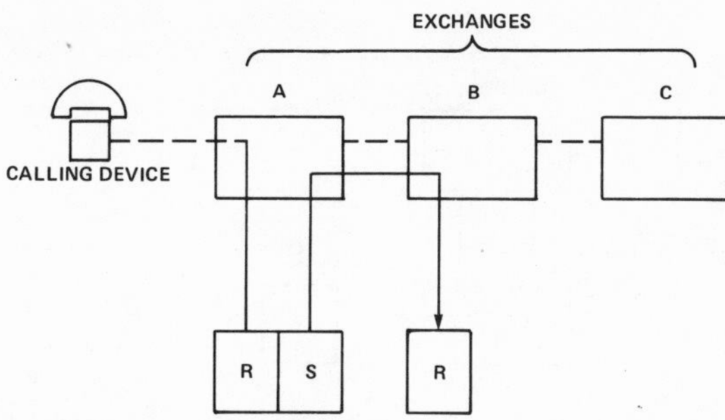

Figure 2.10. Link-by-link signaling exchange A to exchange B.

2.7.2.2. End to End

The alternative method is end-to-end signaling, in which the original register (R) remains in the connection and the original sender (S) sends only enough information to route the call to the next exchange (Figure 2.12). Exchange A then sends information to C, Figure 2.13. (The register in exchange B released from the connection after completing the routes through B to C.)

2.7.3. Signaling Methods

Signaling is either sent without acknowledgment from the distant end (pulse signaling) or with acknowledgment (compelled signaling).

Pulse signaling. In this method the signals are sent as a series. The beginning of the series is controlled by a signal from the distant register and it may be interrupted by the controlling register.

Compelled signaling. In this method each signal persists until it is acknowledged by the receiving register. It does not require a start or stop signal from a register as does pulse signaling.

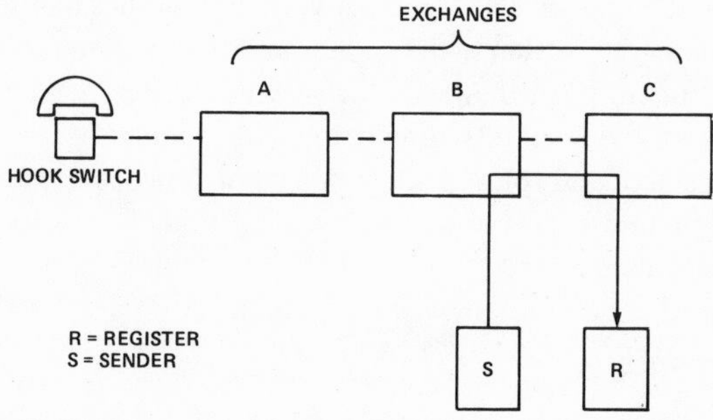

Figure 2.11. Link-by-link signaling exchange B to exchange C.

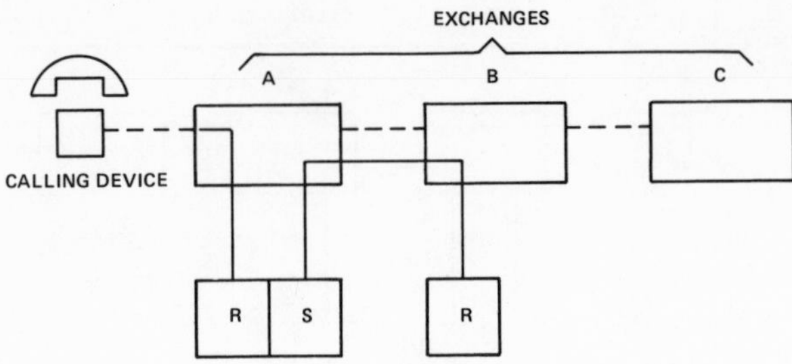

Figure 2.12. End-to-end signaling exchange A to exchange B.

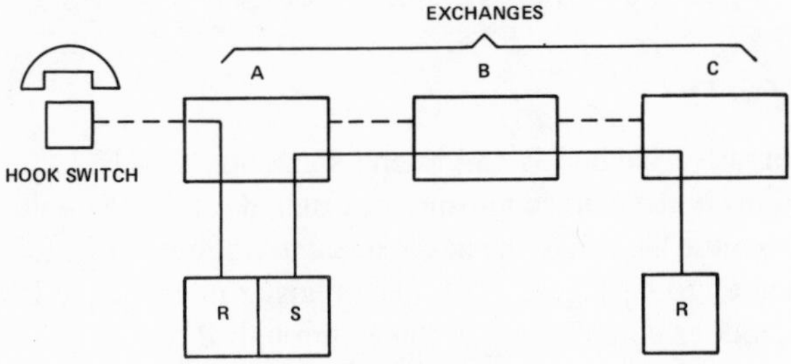

Figure 2.13. End-to-end signaling exchange A to exchange C.

2.7.4. Signaling Systems

There are a variety of signaling systems in use in the world varying from simple direct current loop signals to complex computer-controlled voice frequency or digital systems. Modern switching systems must be capable of interworking with all of them. They are described in detail in Chapters 9 and 10.

2.8. Administration and Maintenance Plans

A telecommunications switching network needs plans to operate the system. It needs an administrative plan with routines and practices to collect revenues and update the system (to meet demands for services to new customers). This plan must also set out the methods to monitor the quality of service, to provide network management and routing strategies. It also needs a maintenance plan to ensure that the system is performing its tasks adequately and to provide routines for fault correction.

2.8.1. Maintenance

Early switching systems used relatively simple controls and relatively complex mechanical components and techniques but demanded manual dexterity on the part of the maintenance personnel. Modern day systems have simpler mechanical components but more intricate controls and hence need different skills for their maintenance from earlier systems. Maintenance has changed from correcting relatively large numbers of simple faults, with a limited impact on the system operation, to correcting relatively small numbers of difficult faults, with often a major impact on the system operation. This has required a different approach to maintenance in which the facilities are often centralized to make better use of personnel.

2.8.1.1. Maintenance Centers

Figure 2.14 shows the general arrangements of maintenance centers. A group of switching centers is maintained from a first-level maintenance center. This is staffed by personnel capable of dealing with routine faults. Faults which cannot be handled at level 1 are referred to a second-level maintenance center serving a number of level 2 centers. This permits the economic provision of elaborate diagnostic routines driven by computers and the concentration of skilled personnel.

2.8.1.2. Local Maintenance

Maintenance at the exchange level is necessary to correct faults requiring the physical replacement of equipment modules and to maintain line and trunk transmission equipment.

Figure 2.15 shows the association of this equipment with a local exchange. The maintenance and administration equipment has functional units for providing

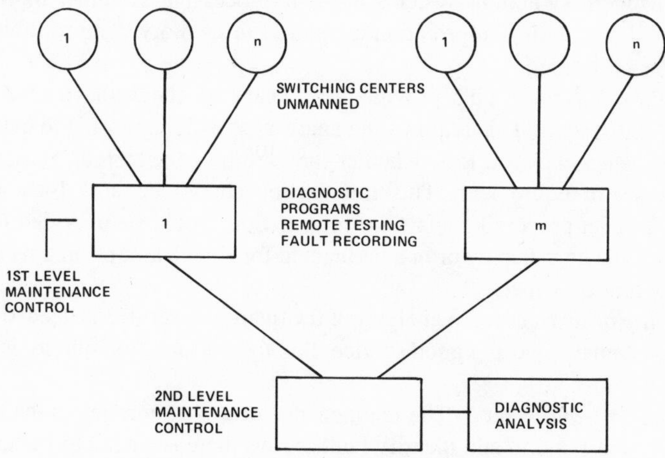

Figure 2.14. Centralized maintenance principle.

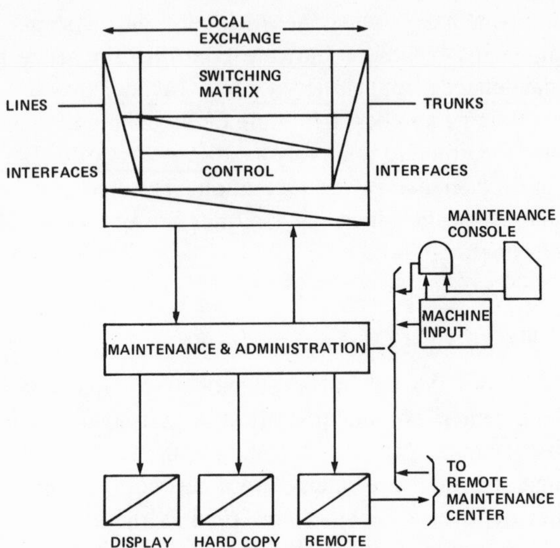

Figure 2.15. Local maintenance principle.

access to and from the exchange. These last units provide information to and from the exchange.

2.8.1.3. Man–Machine Interfaces

Man–machine interfaces are needed so that the personnel responsible for operation or administration of the telephone exchange can perform their tasks. They provide a means of inserting data into the system or extracting data from the system. Modern switching systems are data-processing systems, in which the address and the routing information are stored in memory. The following tasks have to be performed.

Traffic analysis. This provides the means to check the traffic carrying ability of the system. This requires the analysis of data in relation to calls and an indication when calls are lost, whether they are lost due to fault reasons or for insufficiency of equipment. This information should be in a form which is suitable for data-processing analysis. Information should be provided locally or remotely either as a display or in a permanent form such as a printed recording or printed output on paper.

Quality of service. An analysis of the quality of service is needed to check that the customer is getting good service, that the transmission quality is satisfactory, etc.

Repair of equipment. The maintenance of a system requires the resources and skills to respond to and identify faults. Thus there is a need to indicate to the operating staff where the fault is believed to be located. This information should indicate the suspected location of the module or modules involved in the fault.

Hence the maintenance personnel are provided with man–machine interfaces which generate messages to be displayed locally or transmitted to centralized maintenance centers. A standard telephone data link and data modem will perform this task provided that the system output is encoded in the appropriate way to drive the electrical typewriter which provides the information at the local, or distant, maintenance center.

The man–machine interface provides diagnostic routines in response to instructions from maintenance personnel. This is necessary in order to locate the fault when the machine output cannot relate the fault to a particular group of packages. A man–machine interface for a stored program control should be capable of extracting all the data or part of the data in the machine relating to a fault and of transmitting it for analysis to the central maintenance centers, which have more powerful analytical equipment and the correspondingly trained personnel to operate them.

2.8.2. Administration

Many of the administrative practices have to do with managing people and facilities using the same principles as any other business. But the administration of a switching network differs in its methods of collecting revenue.

2.8.2.1. Call Charging

Call charging is the principal means of collecting revenues for the switching system. It requires a means for assessing the costs of calls (on a time and duration of call basis) recording them and billing the charges. This requires data relating to the point of origin, the point of termination, the class of service of the calling or called party, and the duration of the call. There should be a means for assessing charges due to other telecommunication carriers who may be involved in the connection. These data are now being provided in the form of magnetic tape recordings rather than by electromechanical recorders which were provided on a per line basis. There are a number of ways of call charging; the following are those most frequently encountered.

Flat-rate service. The monthly bill is a fixed amount which covers all local calls in the base rate area. This is normally provided in most of world zone 1. Trunk calls are billed separately (toll ticketed).

Message-rate service. This service has a fixed charge for the station equipment and sometimes a fixed number of local calls. Charges based upon usage are applied to local calls. In the past these charges were accumulated by a counter which was connected to the line terminal. These charges were periodically recorded photographically. Modern systems store this information in memory. Trunk calls are billed separately.

Local message-rate service. Local message-rate service provides detailed billing information for local calls. This includes the calling station identity, the called station identity, date, and duration of each chargeable local call. At the conclusion of such a call, the data so accumulated are stored in a dedicated memory area. A calling class of service is used to provide a detailed listing of all chargeable local calls originated from any given station. Usually an alternative nondetailed billing service which provides accumulated charges for local calls is offered as a standard service. Trunk calls are billed separately.

Bulk billing. Bulk billing provides a means of metering calls, both local and trunk, using time and zone metering with a single pulse repeated at variable time intervals.

The charge is based upon the distance to the called zone from answer until disconnect. The greater the distance the shorter will be the interval between unit charges.

No provision is made for the detailed listing of any chargeable calls; hence customers wishing to monitor their call charges may subscribe to a service to provide remote metering at their premises.

Remote meter control. Remote message-rate meter control provides for advancing a message rate meter or other type of counter at the customer's premises in step with the accumulation of charges recorded in the local exchange. This is accomplished by the application of a tone pulse to the customer's line for each unit recorded at the local exchange. A filter is provided at the local exchange to prevent the tone from being applied toward the called end and a filter at the customer's end of the line prevents the tone from reaching him or her.

Karlsson metering. The Karlsson metering principle provides for charging local calls at a fixed rate from a common pulse source. This starts with the first pulse which occurs after answer and it continues at the fixed rate until the call is disconnected.

A variation of the Karlsson metering principle provides for a charge immediately upon answer followed by pulses from the common pulse source. This ensures at least one count for each chargeable call regardless of the point at which it is answered in the common pulse cycle.

Trunk calls are billed separately.

2.8.2.2. Methods of Accumulating Charges

The methods of accumulating charges have evolved as switching systems and networks have developed. There are two basic methods. One is to apply a special metering signal through the switching matrix. The other is to use internal system signals to dedicated memory areas.

The latter system produces data which can easily be processed by modern business computers and is rapidly becoming the preferred method as stored program systems are being installed in increasing quantities.

Single-pulse metering. A single unit is applied for each answered chargeable originating call made in the base rate area regardless of its duration.

Zone metering. In this arrangement the local area is devided into zones, typically from four to six. A calling station is assessed one to four (or one to six) units upon the zone of the called station, in relation to the calling station, regardless of its duration, for each chargeable originating call.

Time and zone metering. Time and zone metering assess units to a call, based upon both the duration and the destination zone of the called station; charges start on the answer of a chargeable call and continue until either party disconnects. There are two methods of time and zone metering. The first is based on fixed time intervals (usually three minutes) with a variable number of pulses per time interval. The second is based on a single unit which is repeated at variable time intervals (the interval depending inversely on the destination zone).

2.8.2.3. Automatic Message Accounting

A modern switching system must provide the charging data required for producing an individual billing record of any call switched by the system for which there is charge for service. This should cater to any service measured on a time and charge basis and produce an individual record of each chargeable call.

In addition to providing automatic billing, means must be available to deal with chargeable services requiring operator assistance in setting up the call. This type of service is provided by special operator positions for such calls as person to person, which cannot be fully automated.

The services that may be handled include:

- Station-to-station calls charged to a third party or the called party such as collect, credit card, advise time and charge, third party charge.
- Person-to-person calls such as normal billing, third party charge, credit card, collect.

Output. The output of the automatic message accounting system provides a permanent record of each completed chargeable call. This includes the number of the calling station, the number of the called station, the date, the time of answer, and the elapsed time of the call. This permanent record should be in a form that is suitable either for transmission over data links to distant points, or for operating a printer.

Immediate readout. In addition to the above an immediate readout feature which provides information relating to a specific call may be required. It should be immediately available after the call has been completed. It is usual to provide this service by means of an appropriate class of service. This information should be in a form suitable for transmission to a customer via his or her normal telephone, teletype, or by other means as specified by the customer.

2.8.2.4. Implementation of Automatic Message Accounting

The equipment needed to prepare the charges for calls can be localized (LAMA) or centralized (CAMA). In either case a means of identifying the calling line is necessary; this is called automatic number identification.

Automatic number identification. The switching system must provide the means for automatically recording the directory number and either storing it for use locally or transmitting it to a centralized facility.

Local automatic message accounting. The term ''local automatic message accounting'' is usually applied to systems which provide itemized bills for toll calls. Charges for local calls are computed separately by one of the methods described earlier. The equipment for collating the relevant data, and sometimes also processing it, is located at the local exchange.

Centralized automatic message accounting. This system, which also only deals with toll charges, serves as a central facility for processing call charges. Figure 2.16 shows the principle of operation. A call originates in the local exchange where the calling line identification equipment obtains the calling number. The address information is also stored in this exchange. Since the call is to a distant destination it is routed through the exchange which has the centralized automatic message accounting equipment. The called address is transmitted to this exchange in order to route the call. A signal from this exchange at the appropriate stage of the call results in the transmission of the calling number, usually over the transmission path. The whole of the data to produce a charge is now available in the centralized facility where it can be processed by a rate and charge computer.

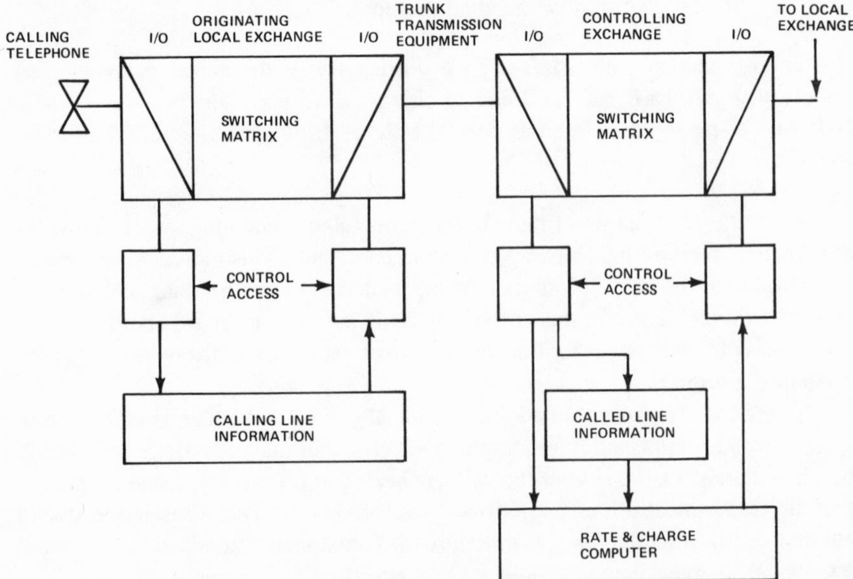

Figure 2.16. Centralized call charging (CAMA).

3

Switching System Organization

3.1. Basic System Organization

The telecommunications system organization used in the text is shown in Figure 3.1. It is divided into three separate functional areas. These are (i) transmission (traffic-carrying units), which includes the terminals, the line transmission equipment, the line interface, the switching matrix, the trunk interface, the trunk transmission equipment; (ii) the switching system control interfaces; and (iii) the switching system controls.

3.1.1. Transmission

For the present purpose of describing a switching system the switching center boundary is considered to extend to the interface between the system and the line or trunk transmission equipment. Later chapters will show an extension of the switching system into the transmission equipment, but initially we will consider applications in which the switching system and the line or trunk transmission equipment have clearly defined boundaries. This necessitates interfaces to change from one mode or another and from one means to another. The transmission equipment in the switching system is used for the duration of a telephone call.

3.1.2. Interfaces

The second portion of the switching center is the functional units which interface equipment having dissimilar signaling or dissimilar holding times (for example line and trunk interfaces with controls).

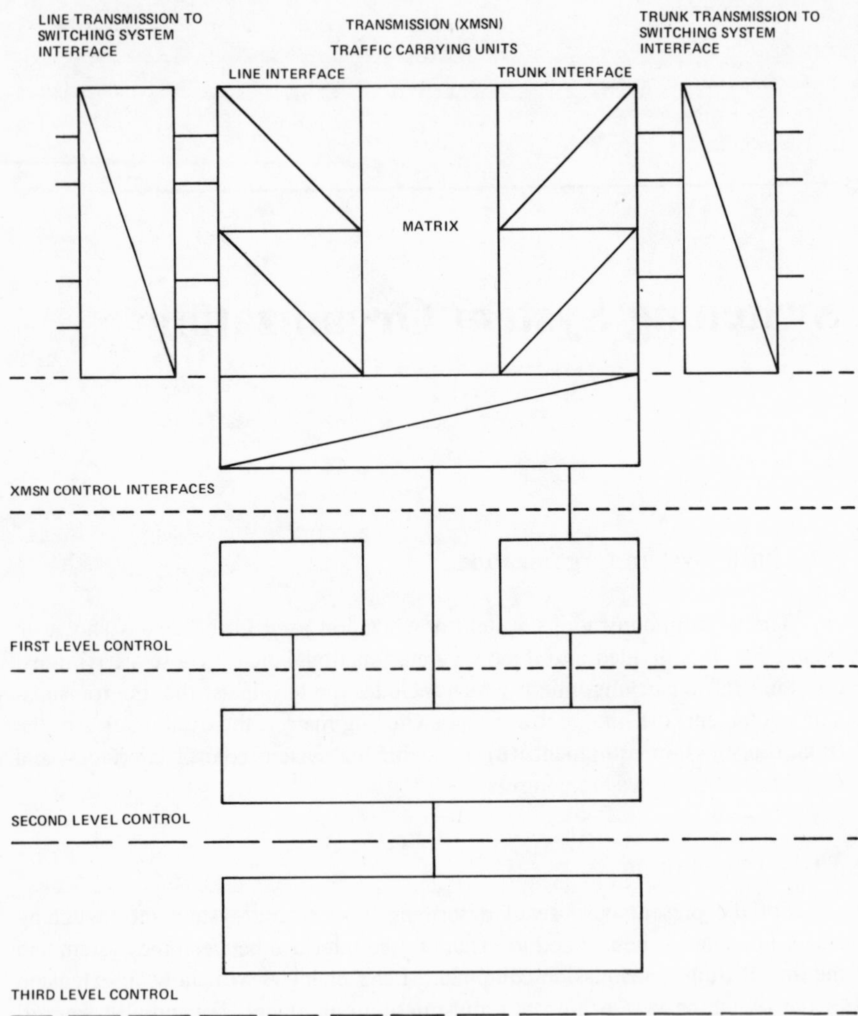

Figure 3.1. Switching system organization.

3.1.3. Controls

The third portion of the switching center is the functional units providing the control function. The control function has been broken down into three levels for the purpose of discussion.

3.1.3.1. First-Level Controls

The first-level control is that portion of the control which has a holding time which is controlled by a system input or output, and hence the holding time may not be definable. Such an interface provides a means of relating the address information to the call-routing pattern used in the system. In this case the equipment must be associated with the call during the time the calling customer is using the calling device. This time interval can vary widely depending on the destination of a call and the calling habits of the customer.

3.1.3.2. Second-Level Controls

The second level of control is that portion of the control which only receives complete tasks as inputs and therefore has a definable holding time. These tasks include the establishment of transmission paths through the system and the association of traffic-carrying units and functional devices as needed to handle the call.

3.1.3.3. Third-Level Controls

The third level of control involves those functions which are auxiliary to the operation of the system, such as diagnostic routines and other information which does not have to be present for the system to function.

3.2. Need for Interfacing

Interfaces with the switching system are necessary in order for it to work with terminals, line transmission equipment, and trunk transmission equipment. The interfaces between the line transmission equipment and the system are necessary to convert from one means of transmission to another. They are necessary where a number of telephone lines or trunks are multiplexed onto a common carrier system, each line or trunk being allocated a separate channel. In this case the individual channel will have to appear as either an inlet or an outlet to the switching system at a bandwidth which the switching system can handle, which is normally 4 kHz. The interface between the switching system and the carrier system is necessary so that the trunk interface can be associated with the relevant channel in the carrier system. It is also necessary so that the signals sent from the exchange can be converted into signals which the trunk transmission equipment is capable of carrying and vice versa. Such transmission equipment is normally used on interexchange trunks. There are, however, cases where the line transmission equipment is also a carrier-type system. In this case an appropriate interface will have to be provided for each line appearance.

3.2.1. Interfaces with Telephones

Interfaces with telephones are required to connect the line transmission equipment to the switching center and to interpret and apply station signals.

3.2.1.1. *Station Signaling*

Station signaling has two distinct sets of signals. The first are those required when originating a call from the telephone, which are the following:

- The request-for-service signal
- The address information signals
- The recall signal
- The release signal

These signals are sent to the switching center which responds by sending progress tones or recorded messages such as

- Dial tone
- Called line busy signal
- Called line being rung
- Unused code or line (usually a recorded message) signal
- Receiver-off-hook signal

The second set of signals are those required when a call terminates on a line, which are the following:

- Ringing
- Intrusion tone and other special service progress tones

3.2.1.2. *Line Transmission*

The line transmission can be a wire pair or a derived circuit.

Wire line. In the majority of cases in the existing telecommunication network terminals will be connected into the switching center by means of wire paths. In these cases the interface may have to convert from the line transmission mode to the exchange transmission mode. The exchange transmission means can be a matrix made up of crosspoints which have similar parameters to the wire paths. If this is the case then a simple interface is all that is necessary. This interface serves to acknowledge the request for service and subsequently transfer the control point for the call to some other portion of the system.

Derived channel. On the other hand if the transmission mode in the exchange differs from the line transmission means then the interface will become more elaborate since all the functions shown in Figure 3.2 have to be accomplished at the line circuit.

Interfaces at the telephone. These are required for coupling the various telephone elements to the line transmission equipment.

3.2.2. Interfacing Lines with the Switching Center

Interfaces between lines and the switching center provide the following: a means of detecting a request for service and of responding to the calling device

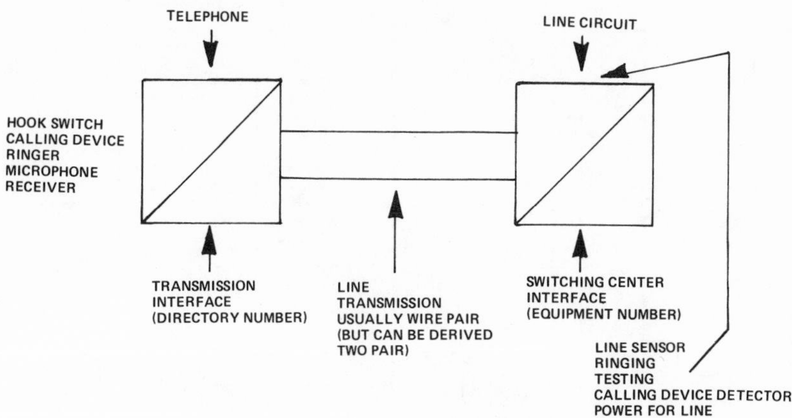

Figure 3.2. Telephone interfacing.

on originated calls, and a means to extend ringing on terminated calls. The line circuit must also provide the power source for the telephone transmitter and DTMF key pad.

3.2.2.1. Ringing

Figure 3.3 shows the principles involved in the line circuit ringing function. This figure is based on the following assumptions:

- The line transmission equipment and switching matrix are not compatible (being wire and electronic, respectively).
- A voltage of 50 V is necessary to power the telephone.
- A voltage of 5 V is necessary to drive the electronic circuitry used for the functional unit.
- The first-level system control is electronic.

3.2.2.2. Description of System Interface

The control used in the system operates at a high speed, hence its instruction to ring a line has to be stored until it can be acted on. This is accomplished by providing a storage device at the line circuit. This is shown as STORE in Figure 3.3 and it is activated from the control via a coordinate marking which enables the control gate momentarily. The control gate records the activation in the store. The store actuates a driver which operates the relay R. (It is assumed that the relay is not of the latching type.) A circuit associated with the ringing function detects when the called party has answered and causes the resetting of the store and hence the release of relay R and the disconnection of the ringing.

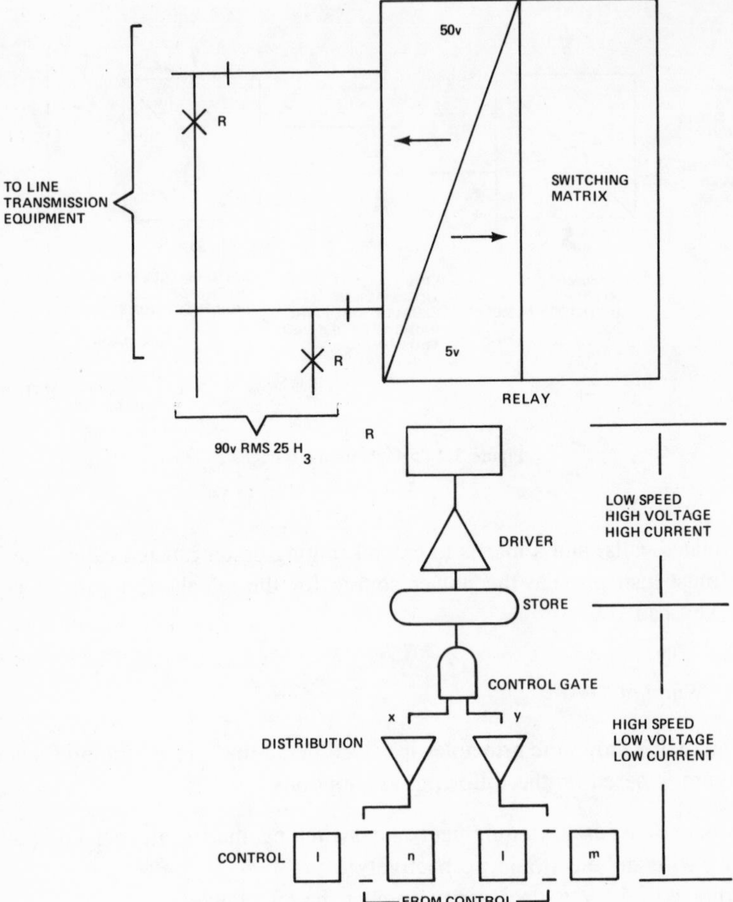

Figure 3.3. Line circuit interfacing.

3.2.3. Interfacing with Trunks

An interface has to be provided between the trunks and switching system. The functions provided depend on the needs of the trunk transmission equipment and the signaling system.

3.2.3.1. Wire Trunks

Most existing switching centers connect to other switching centers using trunk connecting equipment of the wire cable type. This occurs in most instances in long-established telecommunication networks. Telecommunications equipment is only rarely replaced since it has a long useful life. Thus much of this wire equipment will be in service for long periods of time. Even though the trunk transmission equipment and the switching center transmission path may be compatible there is still a need for an interface. This is because switching centers of the metallic crosspoint space division type require control leads (in addition to those providing the transmission path). These control leads are needed to maintain the transmission path for the duration of a call. Hence, since the trunk transmission equipment only provides the transmission path, other interface functions are necessary to provide for the control leads.

3.2.3.2. Derived Trunks

In the case of derived channels interfaces are necessary for the conversion of means and modes. Trunk interfaces are needed to provide and respond to interexchange (line) signals.

3.2.4. Interfacing Trunks with Switching Centers

The switching center also has to provide means of interfacing the slow-speed external signals from trunk transmission equipment with high-speed internal signals used to control the operation of the switching center. This requires the use of speed buffers where information at a low speed is accumulated and stored. When all the necessary information has been stored it is transferred at the high speed into the control for analysis and execution.

3.2.4.1. Interfacing Digital Trunks

Figure 3.4 shows the principle of operation when trunk transmission equipment interfaces with a digital switching matrix. The following assumptions are made:

- The trunk transmission equipment has a speed of S1 and a format which is word organized (i.e., A1 to A8 . . . n1 to n8).
- The digital switching matrix has a speed of S2 and a format which is bit organized (i.e., A1 to n1 . . . A8 to n8).

The operation is as follows: The incoming information is stored as a complete frame, then processed and stored in the new format. The information is then transmitted at the new speed. There is a pair of incoming stores which are used alternately. One store is accumulating information and the other is dispersing it. There is a pair of outgoing stores operating in the same manner.

3.2.4.2. One-Way and Two-Way Trunks

Trunk transmission equipment (trunks) are often classified by originated traffic direction (Figure 3.5). There are two directions for trunks: one in which

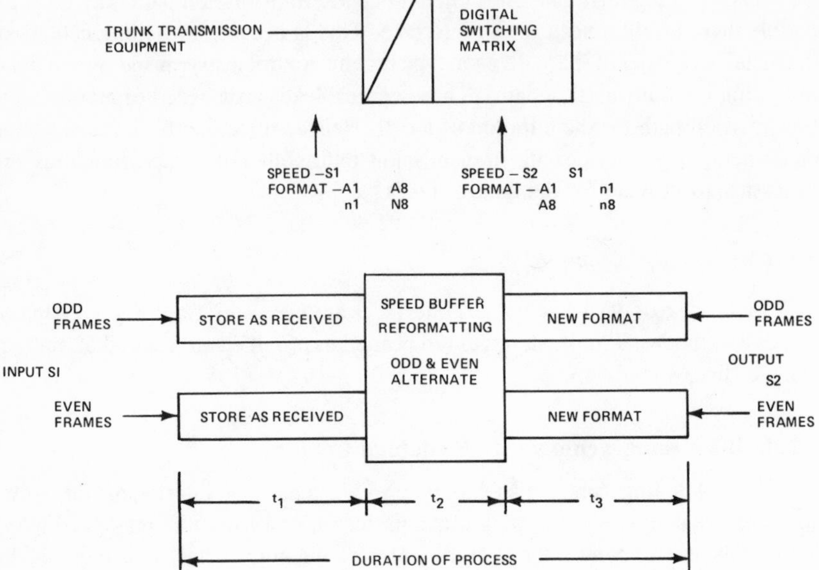

Figure 3.4. Digital trunk interfacing.

the calls always originate in one of the exchanges and terminate in the other, even though the transmission path is bilateral, and one in which the call may originate and terminate at either end of the connection. The first kind of trunk is called a one-way trunk and each end only requires access to or the provision of the functions associated with either an originating or terminating call. In the case of a two-way circuit originating and terminating functions must be available at both ends, since either end can serve as an origination or termination.

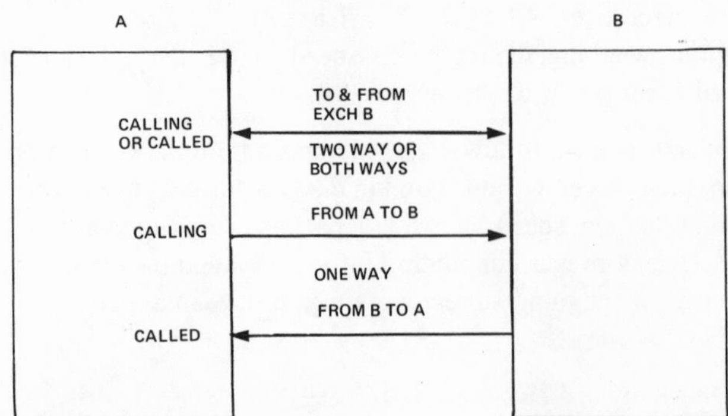

Figure 3.5. Classification of trunks. Note: One-way or two-way trunks can be two wire (combined send and receive) or four wire (separate send and receive).

3.2.4.3. Unidirectional and Bidirectional Trunks

Trunks are also classified by transmission direction. There are two directions for transmission over trunks.

Bidirectional (two-wire trunks). Bidirectional trunks are trunks in which a single pair of wires is used for transmission for both directions. This combined send and receive transmission path is called a two-wire path and employs bidirectional transmission.

Unidirectional (four-wire trunks). This second category of trunks has separate send and receive paths.

An example of such a trunk is a carrier system in which derived trunks provide unidirectional transmission and have separate send and receive paths. Such trunks are called four-wire trunks (even though the actual transmission path may be derived from a carrier system). There will still be four wires to carry the separate transmission and return paths at an interface with a switching center of the metallic crosspoint type.

3.3. Switching Matrix

A transmission matrix provides access from inlets to outlets and provides sufficient simultaneous connections through the system to handle the traffic. There must be one connection per call through the system at any given time. The path for the call can be by means of a matrix of interconnected crosspoints arranged in stages. The call is routed through the system from its point of origination (inlet) to its point of termination (outlet). The number of stages in the system corresponds to a degree to the number of inlets and outlets which it serves. In the case of terminals serving individual lines there must be a dedicated combined inlet and outlet for each path. In the case of trunk transmission equipment over which traffic to many terminals may be routed there are a number of paths and hence inlets and outlets that may be used. Any one of a group of trunks will serve to route the call to the destination. The amount of traffic carried by trunks is greater than the amount of traffic offered to and received from the lines. This is because lines serve single sources and trunks serve many sources. Switching system matrices are designed to concentrate the traffic from the line matrices into the trunk matrices.

3.3.1. Metallic Transmission Path

A transmission matrix for a switching system may be of the metallic type, which uses electromechanical crosspoints for the transmission path. The switching matrix consists of an arrangement of such controlled paths in stages which permits the steering of the traffic through the system to its destination (space division). Such a transmission medium is compatible with the wire line subscriber transmission equipment, and a system which uses this technique can apply the functions at whatever stage in the matrix is the most appropriate.

A typical arrangement for such a system is shown in Figure 3.6. This interconnects the line interfaces (line circuits) with the selection portion via a space division concentrator. The concentrator interconnects the inlets to a smaller number of outlets. These outlets are interconnected via a selection matrix. This provides access to service circuits, trunk interfaces, etc. The system provides access to its control via an auxiliary access matrix. In such a system the functional units can be provided on a traffic basis (that is according to the number of calls on a system) and routed to, or included in the call path by the control equipment.

Such a system is the basis for many existing systems. It has the advantage of simple interface circuits (especially with the line connecting equipment) but the disadvantage of large quantities of crosspoint contacts. Such crosspoint contacts can be sealed or open to the atmosphere but the net result is that the system occupies a great deal of space.

It is interesting to note that such a switching matrix arrangement forms the basis of a number of major existing switching systems even though these were originated independently by different manufacturers in different countries. Other arrangements of the switching matrix are possible (such as an arrangement in

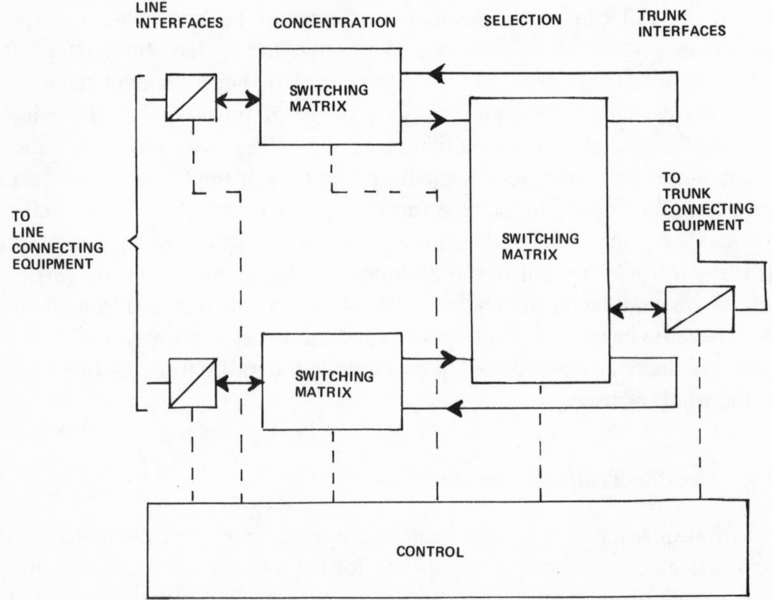

Figure 3.6. Traffic routing in local exchange. Note: Arrows show traffic from point of origin.

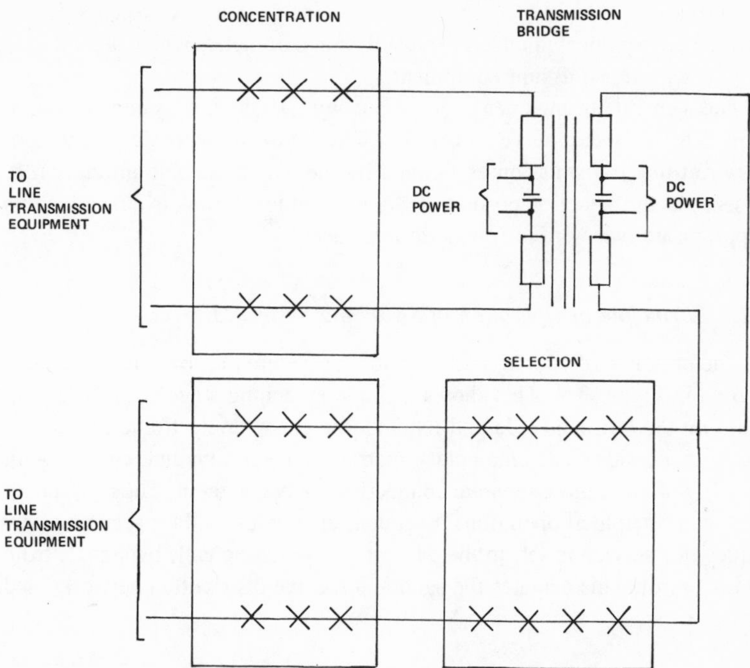

Figure 3.7. Two-wire line-to-line switching (wire-compatible crosspoints).

which the lines and trunks appear on the same side of the matrix). The equivalent of the trunk connecting stage then provides connections between inlets and outlets via links at one side of the matrix. This type of switching matrix will be described later.

3.3.1.1. Two-Wire Matrix

A transmission switching matrix may switch four wires (two undirectional paths) or two wires (bidirectional). Two-wire switching is usually based on metallic crosspoints. A typical transmission path between two lines is shown in Figure 3.7. This figure shows three transmission stages of concentration to transmission bridges, which provides the means for driving the telephone transmitter and for supervising the call, and four stages of selection, which provide access to the various destinations.

3.3.1.2. Four-Wire Matrix

A four-wire switching matrix can be based on metallic crosspoints. There are many such trunk switching systems in use. Modern electronic technology now makes four-wire switching economical for general use. Such systems use electronic crosspoints which have transmission parameters which are not compatible with wire transmission equipment.

The general arrangement for a four-wire switching system is shown in Figure 3.8. This can serve either two-wire trunk transmission equipment or four-wire trunk transmission equipment by the use of suitable interfaces. It can also serve as the selection portion of a combined local/trunk exchange, again via an appropriate two-wire to four-wire interface.

3.3.1.3. Principle of Operation of Switching System Matrix

The principle of operation of a switching system matrix using space division is shown in Figure 3.9. This shows a single switching stage with the crosspoint symbol on the left-hand side and its embodiment with electronic crosspoints on the right-hand side. The elementary matrix connects four line circuits with the possibility of up to three separate connections between them. The right-hand side shows the principle of operation. A sequential counter used for scanning lines for a request for service is set, in the case of a terminating call, by signals from the system control. This actuates the send and receive distribution networks to drive the matrix gates.

3.3.2. Time-Divided Transmission Path

A switching system in which the transmission matrix is not compatible with the outside trunk or lines requires a more complex interface which must provide all the signal and control functions. Such a switching system greatly reduces the amount of equipment involved in the switching matrix by using time-shared connections (time division multiplex) but at the expense of more elaborate interface circuits.

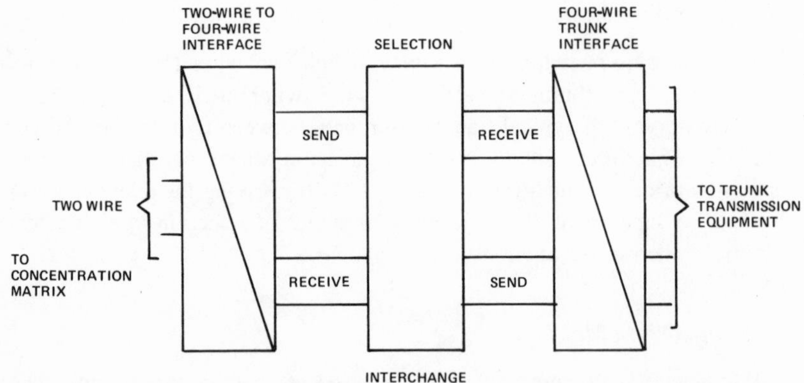

Figure 3.8. Four-wire connection (noncompatible crosspoints).

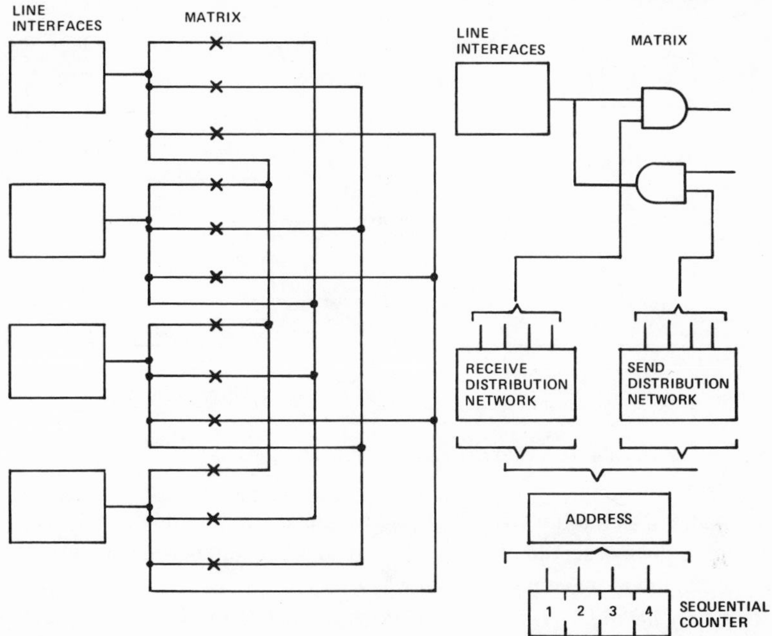

Figure 3.9. Switching system matrix principle.

3.3.2.1. Principle of Operation

A time division multiplex switching matrix is one in which intervals of time are allocated to each call. The intervals of time occur at a recurring frame rate which is rapid enough to effect the necessary transfer of the speech and other information. (This frame rate can be shown to be twice the speed of the maximum frequency to be carried.[(12)]) The intervals of time are called *time slots*. A number of such time slots, which occur sequentially on a recurring basis, are carried by a *highway*. Time slots in the highway are allocated on a traffic basis for transmission paths.

The principle of operation of a highway is shown in Figure 3.10. This shows three time slots and hence can carry three connection paths. These three time slots may be used to connect any of the n inlets to any of the m outlets via coders and decoders which convert the speech to samples suitable for transmission by the time slots and vice versa. Each time slot has a portion for use as transmission and a portion (guard time) to allow the signal on the highway to decay before applying the next time slot. (The guard time prevents or minimizes interference between successive time slots.)

Switched highways. The routing of calls requires a number of highways in series. The interconnection between the highways is provided by means of a

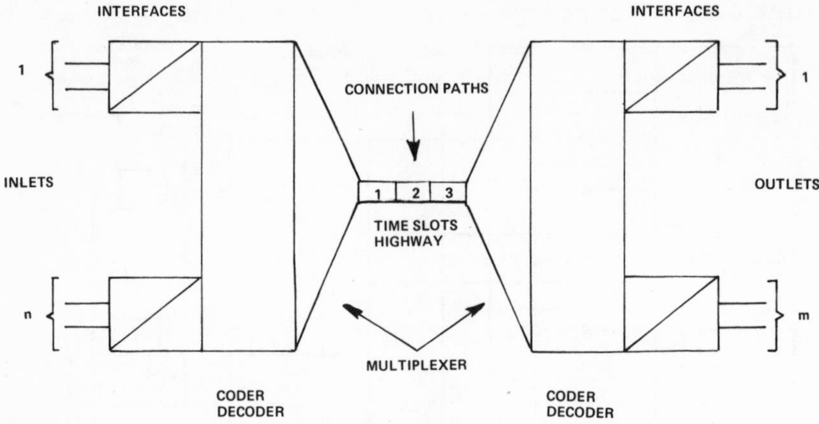

Figure 3.10. Principle of TDM transmission matrix.

spatial matrix which allows any highway to connect to any other highway. The number of highways served can be proportional to the number of stages of spatial "and" gates interconnecting the highways.

Capacity of highway. The number of time slots on the highway depends on the format of the system being used. However, the number of time slots is restricted because of the increasing difficulties (due to the reduced guard time) of providing an adequate transmission means if amplitude modulation is used. Digital transmission is not affected in this way.

Bit rate of highway. Time division multiplexing of speech requires as a minimum that the frame of time slots recurs at 8 kHz. (Information theory developed by Nyquist of Bell Telephone Laboratories[12] teaches that a 4-kHz voice band can be transmitted with adequate fidelity if sampled at twice the voice band, i.e., 8 kHz.)

In the case of digitally coded time division multiplex the bit rate of the highway is a function of twice the maximum frequency (which equals 8000 Hz) multiplied by the 8 bits that are necessary to encode the information in a digital form.[13] This equals 64,000 bits per second. Hence digitally encoded voice signals require 64,000 bits per second for adequate transmission.

The actual bit rate of a highway depends on the number of time slots that it carries. 32 time slots is a typical number. This would produce a bit rate of 2,048,000. (This is the bit rate of one of the CCITT primary multiplex systems.[14])

Bit rate of super highway. A number of such highways may be served by a common highway operating at a proportionately higher bit rate.

If four such highways were combined the bit rate would be approximately 8 MHz. This then would be the highest speed at which the system switches data and the space division interhighway "and" gates must be capable of responding to such speeds. A call through such a matrix has a dedicated time slot at the calling and called inlets and outlets but these time slots need not be the same since a memory can be introduced into the center of the system which stores the information from one highway and reformates it and transfers it to its destination in other highways in a suitable time slot in the highway used. This function is termed a *time slot interchange*.

3.3.2.2. Conversion from Analog to Digital

In order to convert from analog signals to digital signals it is necessary to analyze the value of the signals before encoding them according to the desired law. This requires a stage of amplitude modulation in which the analog signal is sampled. The samples have appropriate voltage amplitudes. A switching system can operate without converting to intermediate digital transmission. Some of the early electronic switching systems used this technology as do some small private branch exchanges. However, there are advantages to the use of digital transmission. This is because it is necessary to preserve an amplitude signal accurately as it is switched through the system (this imposes severe requirements on component parameters). On the other hand in order to reconstruct an analog signal in a digital system it is only necessary to indicate if an amplitude signal is greater or less in value than each of a series of voltage levels and code it as a 1 or 0. The conversion back to analog is made at the outlet from the system where the reconstitutions of the voltage pulses are used to reconstruct the analog signal.

3.3.2.3. Use of Wire-Compatible Concentrator

A switching system may use a space division wire-compatible concentrator in addition to a time division matrix for the selection stage. This is because a space-division-compatible crosspoint concentrator reduces the complexity of the wire line transmission interfaces. This technique is only effective if there are many wire lines and the cost of the simpler line circuit and the space division concentrator is less that the corresponding items in the time divided concentrator. It results in a smaller number of digital to analog traffic carrying devices in the selection network. Access to the control is made at the line interface for the request for service information, and at the selection matrix, for the addressing information. There are a number of modern telecommunications switching systems utilizing this technique.

3.4. System Controls

The controls which establish the connections through the matrix operate according to a program. This may be a program stored in the circuitry of interconnected components (wired program control) sometimes also called firmware. On the other hand it may operate according to a program stored in memory and acted on by a processor (stored-program control).

3.4.1. Level-1 Control

In general, wired programs may be used for repetitive well-established system functions, especially those which consume real time. At this point an explanation of the expression "real time" is necessary: When a system control performs its function it is operating according to a process which has a finite operating time and hence can process a finite number of instructions within a given hour. The number of instructions to be processed could be impacted if there were delays in transmitting the instructions to the processor or in acting on the instructions from the processor. In either case the time to carry out the instructions would be increased thus reducing the number of instructions which can be dealt with. Maximum efficiency in terms of processing is effected if there is no appreciable delay in the transfer of instructions to or from the control. This requires the use of the intermediate level (level-1) controls for time-consuming functions. These controls provide the means to store information from calling devices, etc. for later transfer to level-2 controls.

3.4.2. Level-2 Control

In general, stored-program control is used for call processing. The stored-program technique provides a very flexible method of handling the data necessary to process calls. These actions include processing the following: the addresses of the terminals and trunks; the organization of the numbering plan; services available; etc. Stored-program-control systems are adaptable to varying system applications (since many of the adaptions can be made by software changes only), that is, by changes in the program of the level-2 control processor.

3.4.3. Level-3 Control

Level-3 control should always use a stored program. A switching system requires a third level of control for fault location and other administrative functions. While it is possible to include third-level control functions in level-2 controls, this has disadvantages. This is because security of operation requires that controls (other than level-3 controls) be at least duplicated. This requires that all the data to control the system be stored in both the controls. However, only one of the controls drives the system at any time. If diagnostics were included in these data then the size of the program would be larger. It would also be necessary to check the satisfactory operation of the diagnostics as well as the operation of the call processing program in order to ensure satisfactory operation of the system.

The diagnostic and other infrequently used programs are best stored in a level-3 control (which does not need to be duplicated since the system can operate without it). Then, not only is the amount of memory in level-3 control reduced, but also the program is simplified; the system can operate even if the diagnostic program is inoperative.

3.4.4. Memory Organization

A stored-program system operates from a program stored in memory and manipulates data stored in memory. This requires that all the call processing data and diagnostic and operating data be assigned to addressable memory area. Hence a system which operates in stored-program mode has to be initialized. This requires writing in of the relevant data for it to be operable. The first stage of bringing such a system into service is to write the program into one of the level-2 controls. The second state is to transfer the program to the other control and then the system is operative. This process involves instructing and dimensioning and configuring the system so that it will provide all the intended services. In addition, this portion of the program relates to the basic operating parameters of the system, its modes of operation, available services, classes of services, switching matrix organization, etc.

Use of temporary memory. The operating program can be stored in an electrically alterable memory (if there is a separate, readily accessible source for rewriting the data in the event of faults). This method has replaced earlier systems which used permanent memories which required mechanical or electrical methods to initialize and change them.

The data concerned with the establishment of individual calls cannot be stored in a permanent form since it varies from call to call and such information is only stored in electrically alterable memories.

3.4.5. Control Functions

A stored-program control responds to data and issues data in the form of instructions to all control levels of the system. In order for the system to perform its tasks there are basic functions that must be accomplished in the gathering of data for the instructions and the executing of the instructions. These include the following.

3.4.5.1. Locating Requests for Service

Locating requests for service involves line scanning in which all line terminals are interrogated at regular intervals from a level-1 control to determine the status of the line. This status includes the quiescent idle condition; the request for service condition; the request for service acknowledged condition; the call progress condition, etc. It also provides for the special treatment of lines, as for example in the case of a faulty line, which has been taken out of service. In this case a request for service will not be honored until the faulty line has been repaired (line lockout).

Figure 3.11 shows the functional units used in response to a request for service. The control (a level-1 function) detects the request and associates equipment capable of responding to the calling device with the calling line, in this case via the switching matrix.

Figure 3.12 shows the control functions used in responding to a request for service. These functions may be performed in different parts of the signal and involve level-1 and level-2 controls. The coordinates of the line interface x and y are converted to the control code, in this case binary-coded-decimal code, and correspond to the physical location of the line interface. Hence the full equipment number of the interface can be obtained. This information is applied to the memory (number translator) and the directory number and class of service (COS) obtained. The storage of the class of service and the directory number results in the acknowledgement of the service request and the connection of dial tone. The information is stored in a register area and the system then stores the address signals in another area of the same register memory. This provides the association of the calling and called party information. Thus all the relevant information relating to the call can be obtained by reference to the relevant portion of the system memory.

Association of register. The result of the request for service is the associa-

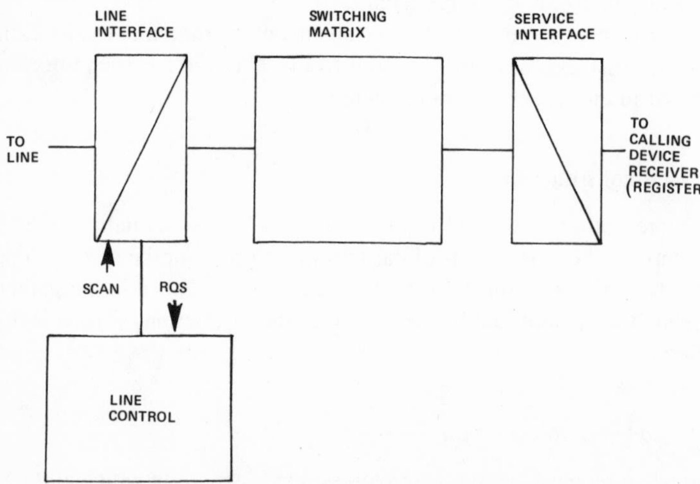

Figure 3.11. Functional units used in request for service.

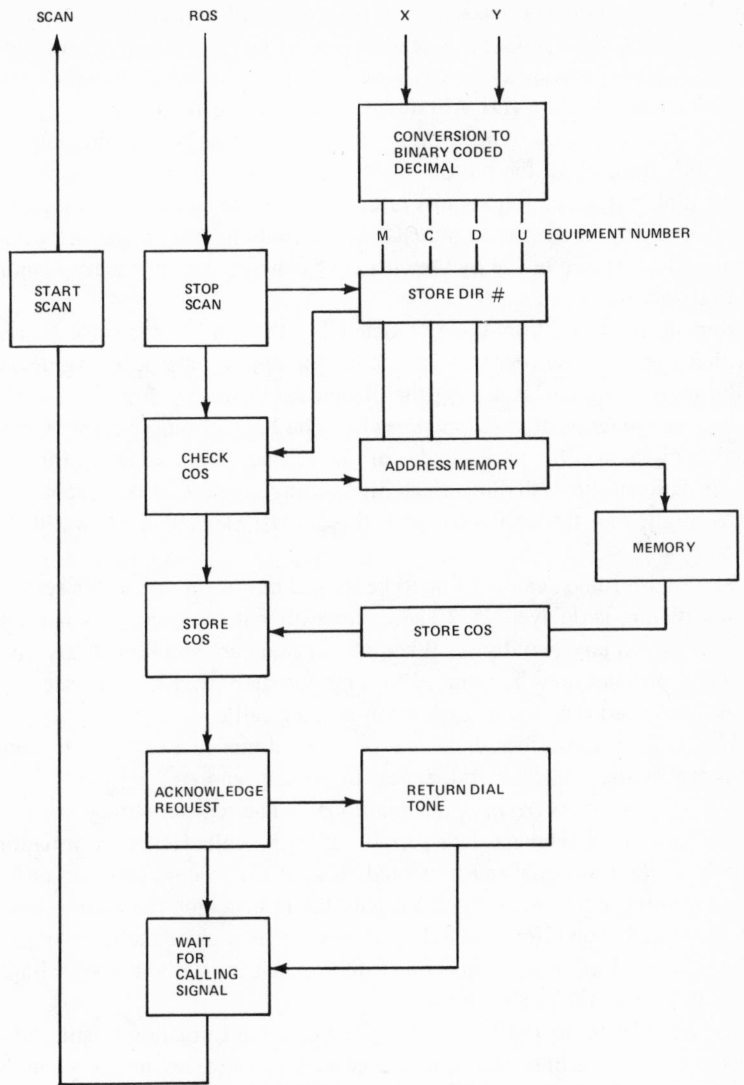

Figure 3.12. Response to request for service.

tion of a register capable of responding to the caller's calling device. The signals to the register may be restricted to the following types of calling devices: The rotary dial, which produces a train of interruptions of a direct current circuit (the number of interruptions corresponding to the digits on the dial which has been chosen and rotated). It should be noted that such a calling device transmits the digits when it returns, under its own control, to its normal condition. (This return of the dial provides the necessary speed control usually either 10 or 20 pulses per second.) Such a dialing system is relatively slow and it is rapidly being replaced by tone signals produced by a series of push buttons called dual-tone-multifrequency signaling (DTMF) and having many marketing terms applied by different manufacturers. This calling device produces a combination of two frequencies for each of the buttons depressed. It is obviously much more rapid than the dial, but, since it transmits information in the speech band, it requires a different receiver to decode its signals. The switching system can determine the type of calling device in use by the customer from its memory when responding to the request for service.

Registers serving trunks are associated with a call in response to a signal from that trunk. These registers must have the appropriate decoding devices to suit the register signals sent from the distant exchange.

Register functions (called numbers). The register must accept as many as 13 dialed digits plus the prefix codes of one to three digits required for national and international direct dialing. Each digit-storage area must be capable of storing any digit, one through zero, plus the DTMF eleventh and twelfth button signals.

The register must cause a line to be locked out of service if it is seized by a line, but dialing is delayed 25–30 sec, either after it has been associated with a line, or between any two digits. When all registers are simultaneously busy the timing out interval may be reduced to approximately 8, 12, or 15 sec. After a register has timed out it is available for another call.

If a calling subscriber dials one or more digits in excess of the number anticipated by the translator, the excess digits are ignored.

Register functions (routing information). The register storage areas maintain a record of all relevant data pertaining to the call. Hence information obtained from the translator can be stored in it, if the system operates on a bulk transfer of information, or it can store only the information required to process a stage of the call. The information thus obtained may include the following: digits to be outpulsed; location of trunks on switching matrix; method of signaling; type of call; next action address; and so on.

Register functions (calling line). Part of the information relating to a call relates to the calling line. Hence it is necessary to store the directory number of the calling line. This will be used for billing purposes and also to obtain the class of service.

3.4.5.2. Registering Address Information

The second basic control function is the address receiving and sending function. This is the means whereby the signals from the customer's calling device are stored and assembled in a form suitable for analysis by the system. Arrangements are subsequently made for sending address information to the other switching centers involved in routing the call to its destination. The principle of operation is shown in Figure 3.13. This shows the line interface (which had responded to the request for service) the switching matrix, and the register sender interface to the register sender. The register sender connects to a translator which performs the necessary analysis of the address. This determines whether the call terminates in this exchange or whether it has to be routed to some other exchange. It should be noted that the blocks do not represent functional hardware but only functions and that in a system operating with noncompatible crosspoints the register sender interface may be at the line interface and not be part of the switching matrix. Translators provide data on lines and trunks for the national numbering plan, special terminals, coin boxes, direct inward dialing to PBXs, station numbering, service codes, etc.

Number translation. Number translation is necessary in order to determine the equipment number, class or service, etc. of the calling and called numbers.

The directory number identifies a telephone by its location in the numbering plan area to which it belongs by the terminal which is connected to a particular exchange. In order that the system may route calls to terminals it must identify the interface address (the equipment number) for the terminal. The interface for the terminal will not have the same address as the called directory number. This is because the lines which serve to connect terminals to exchanges are parts of telephone cables which are buried under the streets of towns and cities. These cables terminate in the exchange in their order of appearance. They are then cross-connected into the switching center interfaces so that the terminal that they serve may obtain the required service (since different kinds of terminals may require different kinds of interfaces). This interface location (the equipment

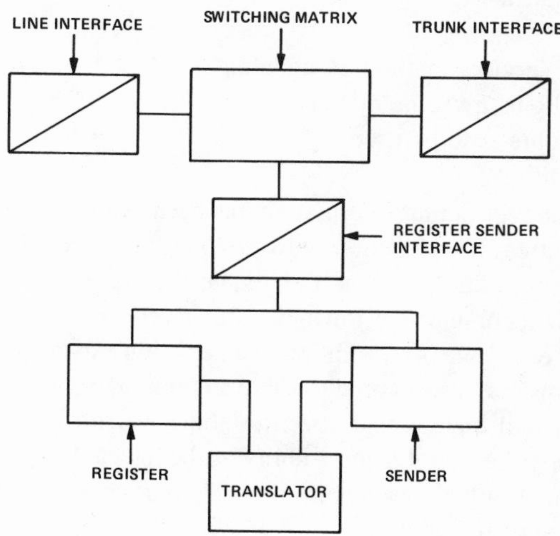

Figure 3.13. Functional units used in address receiving and sending.

number) identifies the physical position of the interface which is necessary for routing calls to it.

The line translation function relates equipment numbers to directory numbers for originating calls and vice versa for terminating calls.

Functions of number translator. The basic function of the number translator is to convert the calling equipment number whenever the switching system requires it for routing purposes. This action will require the conversion of the equipment number to the directory number in order to identify in memory the relevant class of service information pertaining to the calling line. This information is necessary in routing calls which require special treatment. A terminating call must be routed to a particular line interface in order to reach the required terminal, hence the called address must be converted to the corresponding equipment position.

The number translator stores the information required for each line in its memory area including its directory number, equipment position number, and class of service. Its data can be entered or altered from a local or remote point.

Code translation. The code translator stores information relating to each code and prefix required for call routing. This information includes, where applicable, call destination (local or outgoing), trunk group, trunk class of service, alternate routes in sequence, type of interoffice signaling, etc. Its data can be entered and altered from a local or remote control.

Translator functions. The address information is analyzed after each digit has been stored to determine if there is sufficient information to determine the destination.

A modern system translator function has the capability of translating the first one to six digits into routing instructions. The following information is provided by a translator:

- Class of service, method of signaling
- Which digits are to be outpulsed
- Which route is to be used

When the above routing information has been provided, no further translation will be required. The function must provide the translation required by the national numbering plan and to accommodate the present interim and ultimate numbering plans recommended for international working.[15]

If the local exchange is an addition to an existing exchange, the existing and the added systems may be required to share a common exchange code. Since the existing portion will have already been established with its numbering assignments, it will be necessary for the addition to be adapted to fit into the existing numbering plan. Available groups of directory numbers should be assigned for the exclusive use of the addition.

3.4.5.3. Path Selection

The response to the address signals requires a means of choosing a path to the desired destination (shown on a flow chart as "invoke selection routine"). Figure 3.14 shows the general arrangements. The information relating to the calling and called numbers is obtained from the first-level control. The input relating to the status of the calling and called lines together with the status of the possible connection paths (links) through the concentration and selection stages is obtained. The second-level control utilizes these data in conjunction with the path-selecting algorithm for the switching system to select the path. The path data are then used to complete the connection by using the relevant first-level control functional units.

Principle of outlet selection. Figure 3.15 (1) shows the principle of selection from an inlet to one of *n* outlets. Outlets 1 to *n* are tested sequentially until either one is found to be free or all have been found to be busy. The figure shows a fan-type network having only one path from an inlet to any of the outlets. If the network has more than one stage then there may be no way to reach a free outlet. This condition can be obviated if, prior to testing the outlets, the system identifies all free and accessible outlets. The selection of an outlet results in the

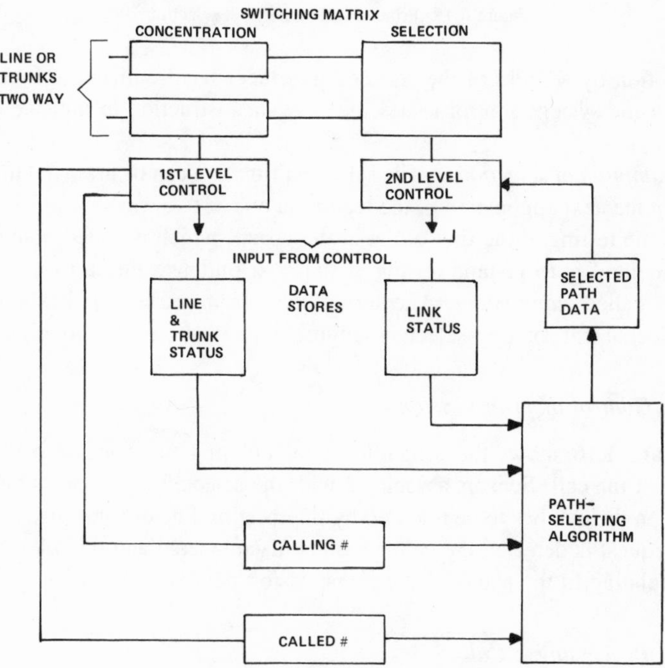

Figure 3.14. Principle of path selection.

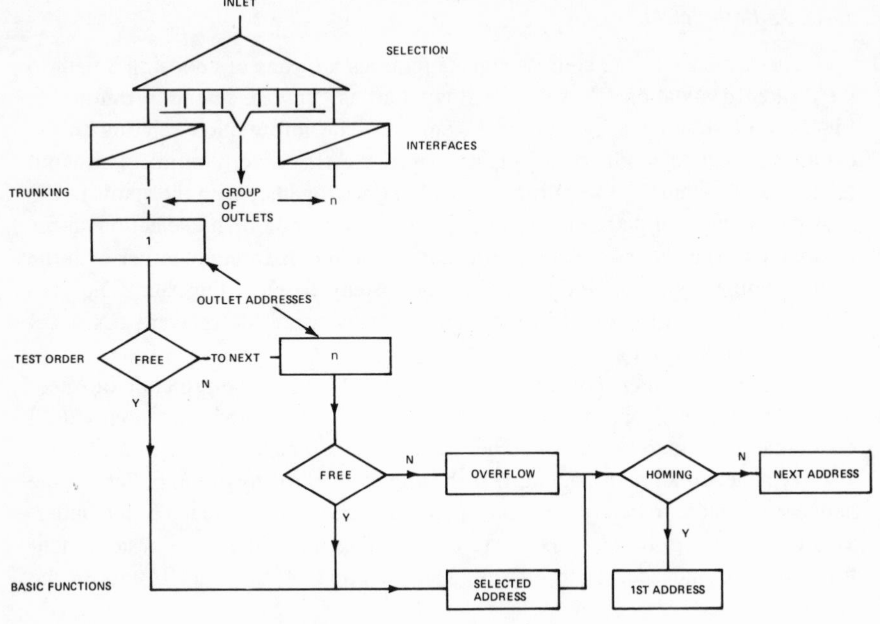

Figure 3.15. Principle of path outlet selection.

identification by address of the selected interface. If all outlets are found to be busy then the system control issues the relevant instruction to indicate congestion.

Resumption of selection. The selection function can be arranged to resume testing on the next application of the feature in one of two ways. The first method is to resume testing at the first outlet in the group, which is called *homing*. The second method is to resume testing at the next outlet to the last one selected, which is called *nonhoming*. The terms stem from earlier applications using electromechanical rotating selectors. Figure 3.15[11] shows the principle.

3.4.5.4. Control of Transmission Path

Figure 3.16 shows the principle of maintaining the selected path for the duration of the call. Sensors associated with the selected input and output determine when the call has to be released by interpreting line or trunk signals. When such a request is detected the connection is disconnected and the relevant information relating to the status of the transmission path is updated.

3.4.5.5. Processing a Call

The establishment of a call follows the sequence shown in Figure 3.17 for a local call and Figure 3.18 for a trunk call. It must be realized that both of these

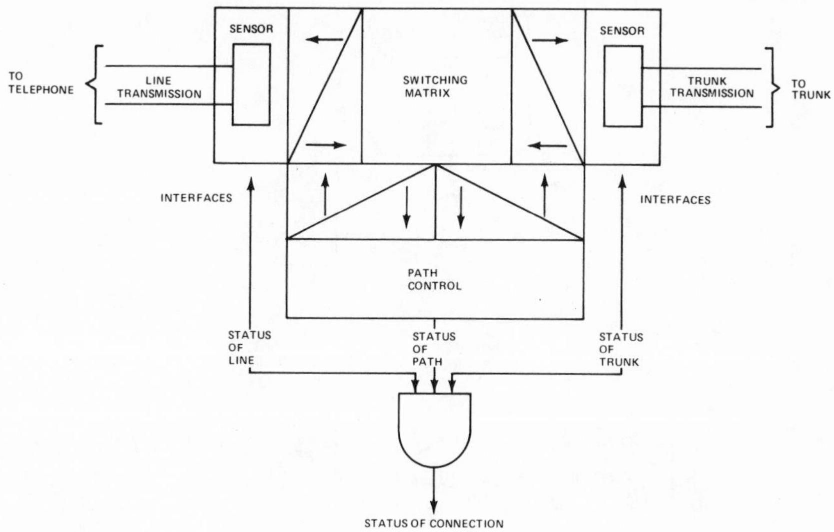

Figure 3.16. Principle of path control.

flow charts are overly simplified and that in practice there will be many other steps. However, the sequence will be the same even though the number of steps will depend on the system architecture. Most of the defined functions are self-explanatory, but the following explanation is necessary for the term *map in memory*.

Check map in memory. A system which employs map in memory holds details of all established calls and the paths through the system that they are using. Hence the existence of free paths and the status of the inputs and outputs to the system can be determined by reference to the system data banks.

3.4.5.6. Release of Connection

Release of connection can occur at any time during a call or call attempt. Typical cases of call release follow.

Local-to-local call. If the calling party disconnects first, the calling line will be made free following a release time test interval of 180 msec. The called line will be locked out and marked busy until disconnection by the called station occurs. When the called party disconnects first, no release action normally takes place immediately. If the calling party fails to disconnect after an interval of approximately 30 sec, the called line is automatically released and made free. Lifting the handset of the called line before the timing period has elapsed will reconnect it to the calling line and time-out will be canceled. After time-out, the calling line is locked out of service.

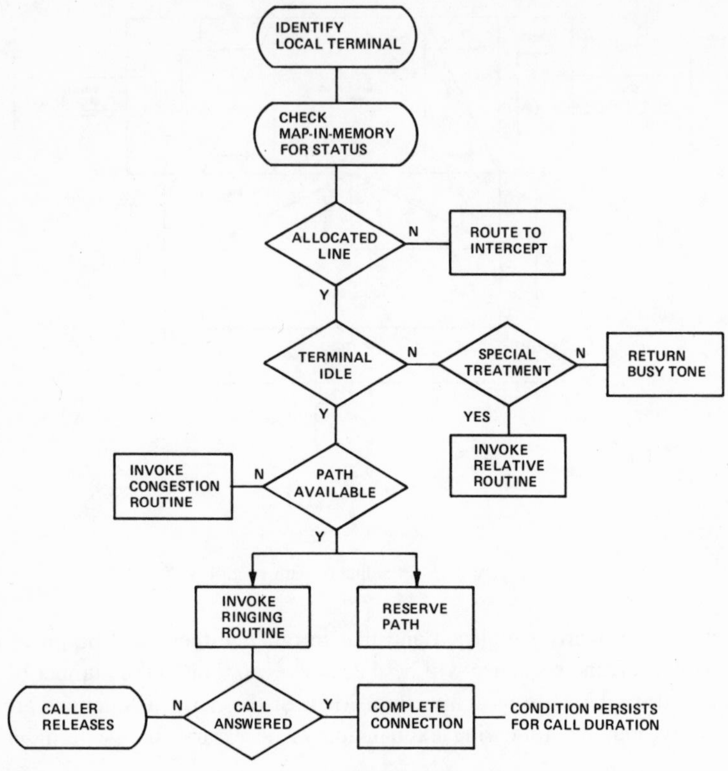

Figure 3.17. Outline of local call.

Local to special service. When a special service is called by a local line, the calling party normally controls the release of the call. A disconnect by the local calling station initiates release at the far end. This releases the local switching equipment and frees the calling line. The local end of the outgoing trunk is not made free until the distant end is free and ready to receive a call. Normally a disconnect by the called party first has no effect except to return an on-hook signal to the originating exchange.

Local to operator. The release of a call from a local line to an operator is normally under the control of both the calling local party and the operator. Often the control of release is transferred to the operator after the call has been answered.

Incoming calls. If a local station is called over a trunk from another office and the local station disconnects first, the release action is the same as for a local-to-local call except that, on calls from the operator to coin boxes, the connection will be held if required until the operator disconnects. This permits re-ringing the terminal and the exercise of coin-control functions.

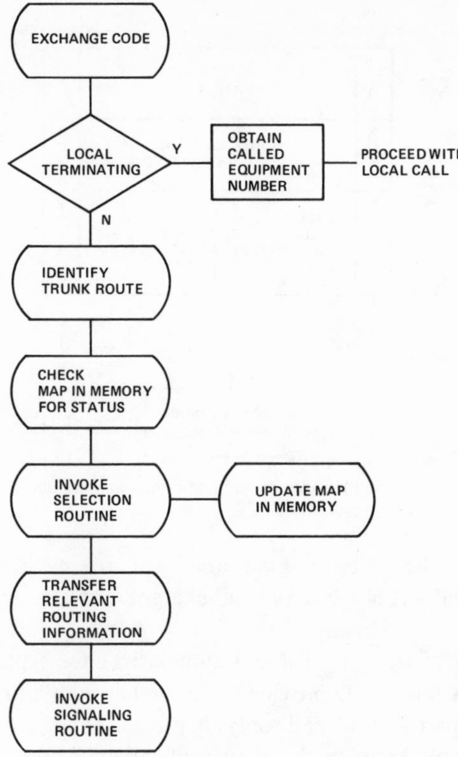

Figure 3.18. Outline of trunk call.

3.4.5.7. Charging for a Call (Toll Ticketing)

The data necessary to provide billing information are accumulated as the call progresses. The principle of operation is shown in Figure 3.19. The calling number is obtained during the serving of a request for service by the calling line identification equipment. The called number is obtained during the storing of the address information. The start of charging occurs when the called party answers and it ceases when the calling party disconnects. The routing information, the time of day, and the date are supplied to the rate and charge computer, which computes them and formats them for the billing output.

Basis for call charging. Figure 3.20 shows the general arrangements for assessing the cost of a call. There are portions of the system with dedicated equipment: line interfaces, wire pairs, and telephones which are associated with a main station. There are portions which are in use only when calls are in progress. These are allocated to users according to the tariffs of the operating administration. As a general principle calls are charged on a time and distance basis. Thus

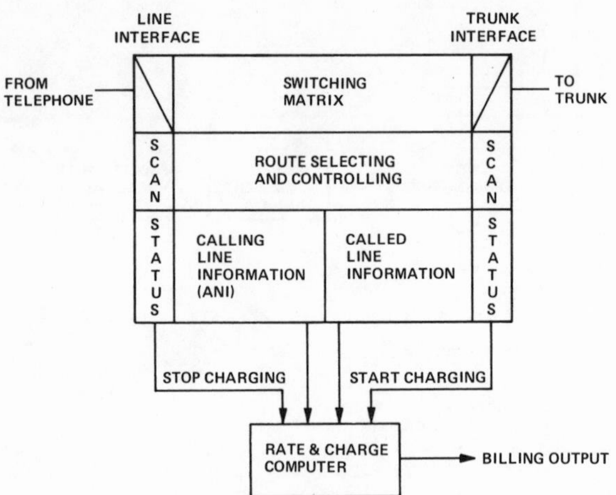

Figure 3.19. Principle of automatic message accounting.

costs increase from the lowest cost call in a local area, which may not be timed (or indeed be the subject of a cost per call charge) to relatively high-cost international calls.

Types of call charging. Table 4 summarizes the types of call charging methods. Types A through D provide the customer with a printed bill which is fully (all calls) or partly (trunk calls only) itemized. Type E, which is called *bulk billing,* provides a bill based on the number of call units used for all calls. This is analogous to a bill produced to indicate the consumption of electricity.

3.4.5.8. Charging for a Call (Bulk Billing)

In the bulk billing method, no tickets are provided and no details of the breakdown of the bill are available.

The equipment used operates in a way similar to that which produces the toll tickets: it assesses the charge in terms of a single pulse repeated at a variable interval (the longer the distance the shorter the interval). The rate and charge computer applies these unit charges to a dedicated memory area corresponding to the calling number.

Since no breakdown of charges for individual calls is provided it is necessary to provide an alternative way to determine the charges for individual calls. This takes the form of a meter at the customer's premises which is actuated every time a unit charge occurs from a line interface.

The principle of operation is described in detail in Chapter 4.

3.4.5.9. Line Testing

Line testing is necessary for maintenance of a switching center. This requires checking for loop leakage insulation, line continuity, etc. This involves the

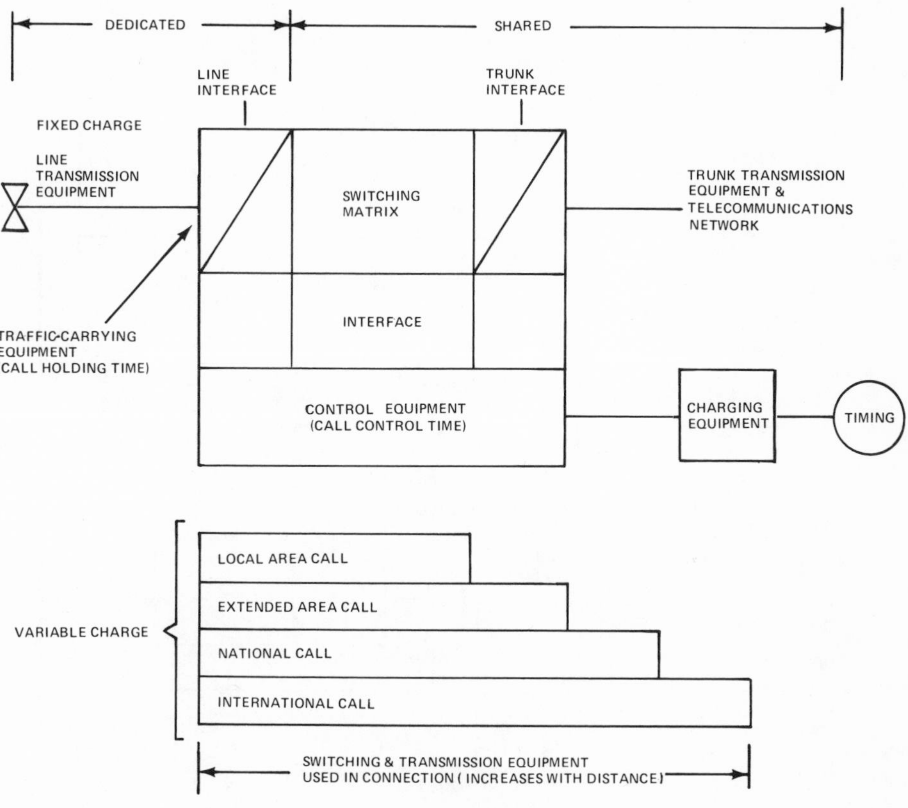

Figure 3.20. Basis for call charging.

connection of test equipment to the line either through a compatible switching matrix or directly from the line interface. The first case uses the transmission matrix. The second case requires a special test access arrangement. In each case the testing equipment is called a *test desk*. The general arrangement is shown in Figure 3.21. A manual means of line testing via test jacks, which are provided per line, should also be provided.

Table 4. Types of Call Charging

Type	Local area	Extended area	National	International
A	Flat rate	Flat rate	Detailed ticket	Detailed ticket
B	LMRS and detailed	LMRS and detailed	Detailed ticket	Detailed ticket
C	Single fee not timed	Multifee not timed	Detailed ticket	Detailed ticket
D	Single fee repetitive timing	Multifee repetitive timing	Detailed ticket	Detailed ticket
E	Single fee variable repetition	Single fee variable repetition	Single fee variable repetition	Single fee variable repetition

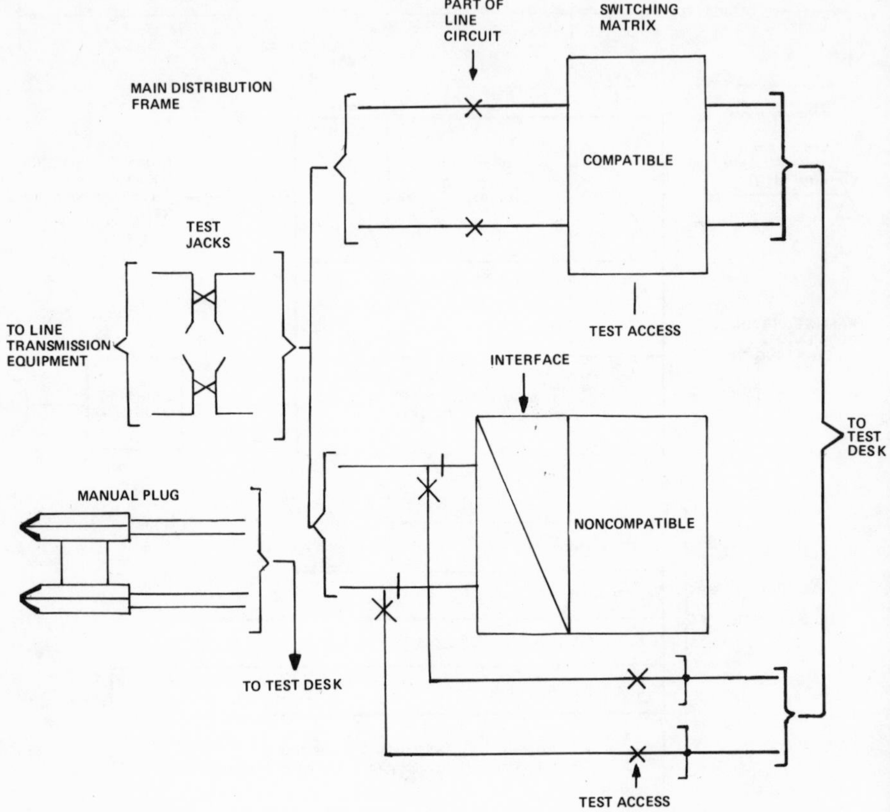

Figure 3.21. Principle of line testing.

4

Basic Signaling

4.1. General

Signaling is the means used to establish and control telephone calls. It includes signals from terminals (station signaling), to and from switching centers (register signaling), and also signals between switching centers (line signaling). Signals are carried by the line and trunk transmission equipment and the switching centers. Signaling also includes the tones and recorded messages which are used to indicate the progress of calls through the telecommunications network.

There are also signals internal to a switching system which convey information and commands to and from the various parts of the system and operating and administration signals.

The modes of signaling which are summarized in Figure 4.1 include the following.

4.1.1. Direct Current Signaling

In the direct current signaling method (loop-disconnect signaling), a signaling code is derived from the duration and direction of the current flowing through a loop. This loop includes the customer's telephone and the line transmission equipment and its switching center interface. Direct current loop signaling is also used with trunk transmission equipment of the wire type.

An alternative method of direct current signaling uses only one of the pair of wires and is called *leg signaling*.

4.1.2. Alternating Current Signaling

Alternating current signaling is based on signals of different frequency either in the same bandwidth (in band) as the speech transmission path (300 to 3400 Hz) or at a lower, <300 Hz, or at a higher, >3400 Hz, frequency (out band). Figure 4.2 shows the bandwidth allocations.

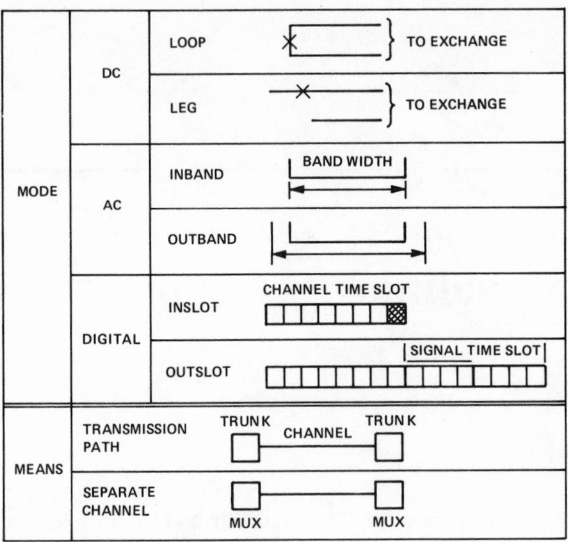

Figure 4.1. Signaling modes and means.

4.1.3. Digital Signals

Digital signals take the form of a series of successive pulses, binary signals, which are coded to produce a signaling format. Digital signals may occupy a portion of one of the time slots used for the transmission of speech (in slot) or may use a dedicated time slot (out slot).

4.1.4. Signaling Path

Signaling may use the same transmission path as the speech or data signals or it may use a separate channel. The first case is called *channel-associated signaling*, the second case *common channel signaling*.

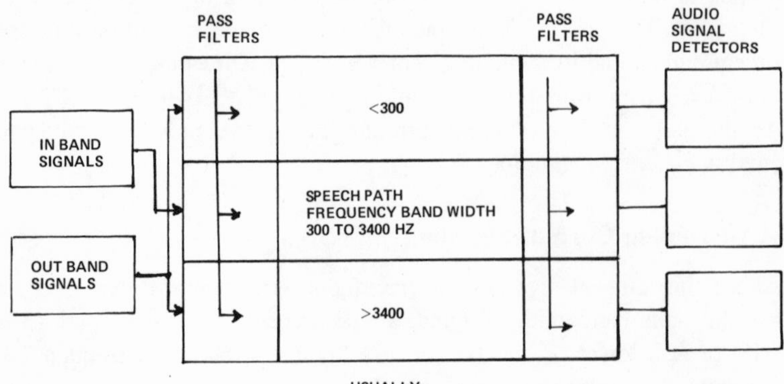

Figure 4.2. Audio frequency in-band–out-band signaling allocation.

Common channel signaling. Figure 4.3 shows the principle of operating for common channel signaling. The signaling function for both line and register signaling is removed from the path carrying the transmission signals. In lieu of this a series of messages indicating the originating and terminating trunk identities, together with the relevant address or line signals, is sent over a separate dedicated channel. This information is processed in the system controls at either end and decoded to produce the relevant actions. There are two such systems in use, or planned for use: CCITT system #6, which uses an analog signaling in band method by the use of data modems, and CCITT system #7, which uses pulse code modulation (PCM) coded signals operating at 64,000 bits per second. They are discussed in more detail in Chapter 10.

4.1.5. Signaling over a Carrier System

The signaling path may consist of wire conductors or it may be derived from a carrier system. Figure 4.4 shows the general arrangments of a carrier system. This provides a dedicated derived channel from one of the n inputs to one of the n outputs. It uses either frequency multiplexing or pulse code modulation to multiplex the channels.

4.1.5.1. Frequency Division Carrier

In the case of frequency division a number of channels are each allocated a portion of the bandwidth of the carrier system. Each portion is multiplexed with the carrier frequency by the multiplexing equipment at the send end. It is demul-

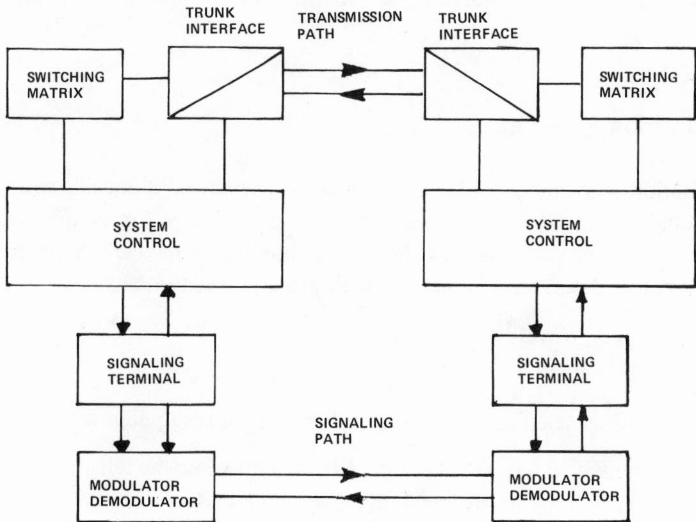

Figure 4.3. Common channel signaling.

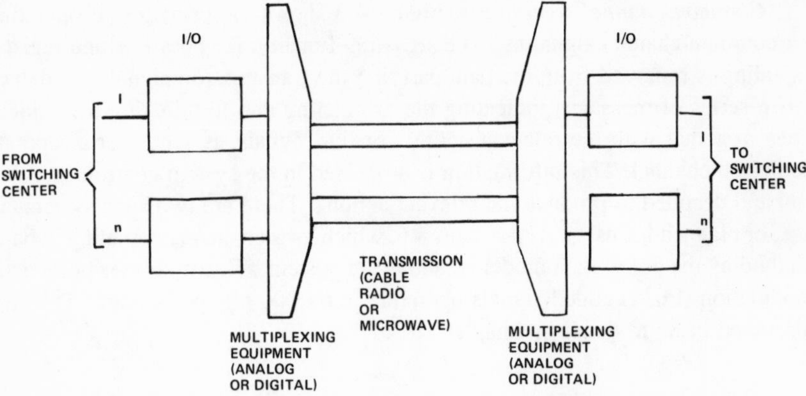

Figure 4.4. Multiplexing of carrier channels.

tiplexed and filtered to select the relevant channel at the receive end to restore the speech wave form.

4.1.5.2. Pulse Code Modulation Carrier

In the case of pulse code modulation a number of channels are each allocated a time slot which has a fixed recurring position in a frame of time slots. The analog input to each channel is sampled and coded at the send side and interleaved with the other time slots by the multiplexing equipment. The channel is extracted at the receive end and demultiplexed back to analog.

Signaling systems are designed for the various transmission modes and means.

4.2. Station Signaling

Station signaling controls the transmission path between the terminal and the exchange and provides address information. It is used for the duration of the call to control the call. Address signaling is used to set the call up and to initiate certain kinds of services, some of which are described in Chapter 12.

4.2.1. Signals from Telephone

Present-day telephones have the following signaling elements:

- A hook switch which completes the loop circuit when the telephone handset is lifted. This is used to originate or answer a call. Replacement of the handset opens the loop and terminates the call.

- A calling device, which is either a rotary dial or a key pad, to produce the address signals, hence the station signals address information to the switching center register using either dial pulses or DTMF pulses.
- An alerting device, usually a telephone bell, to indicate a call terminating on the line.

4.2.1.1. Rotary Dial

The calling device which is in most common use is the rotary dial. The rotary dial produces dial-pulse (loop–disconnect) signals.

This calling arrangement uses the standard telephone dial (Figure 4.5) rotary type, usually having a pulsing rate of ten impulses per second (IPS). However, 20 IPS dials are used in certain countries. Dial pulsing utilizes a train of one to ten pulses, each pulse consisting of a momentary opening of the signaling loop. The digit to be dialed is selected by rotating the dial manually from the selected digit to a finger stop; releasing the dial produces the requisite pulse train.

The loop disconnect signals in general use are shown in Table 5. These include switch hook or equivalent signals.

4.2.1.2. Loop–Disconnect Station Signaling

A calling line originates a call by a request for service signal which is produced when the telephone handset is removed. This signal is detected by the line circuit which associates a calling device detector and a register (Figure 4.6). Dial tone is returned to the caller, who then dials the address of the required party. The register stores the dialed digits and hence accumulates the address information. The line circuit register and calling device detector respond to the following loop–disconnect signals.

Closed-loop signals. These can indicate a request for service or the end of a train of digits. In the first use the signal must persist for a minimum of 10 msec. A signal occurring after request for service must persist for a minimum of 180

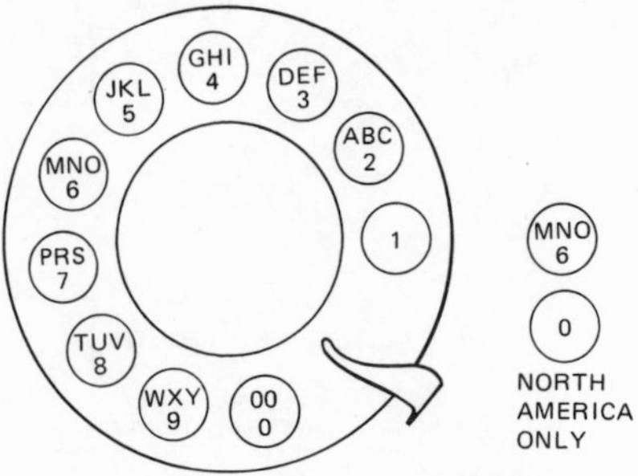

Figure 4.5. CCITT rotary dial faceplate.

Table 5. Loop-Disconnect Signals

Type of signal	Duration	Loop state
Request for service	180 msec	Closed
Call control		Closed
Answer		Reversal
Disconnect (regular)	180 msec	Open
Disconnect special (unanswered)	180 msec	Open
Disconnect special (answered)	2200 msec	Open
Dialing (10 IPS)	Speed 8 to 12 IPS 50% to 80%	Open
Dialing (20 IPS)	Speed 18 to 22 IPS 50% to 80%	Open
Interdigit	10 msec	Open
	10 msec	Closed
	180 msec	Closed

msec duration to be recognized as the end of a series of pulses corresponding to a digit (interdigit signal).

If the signal occurs after all digits have been received then it is a control signal.

Open-loop signals. An open-loop signal can either be the beginning of a dial pulse or the beginning of a disconnect or "hook-switch flash" signal.

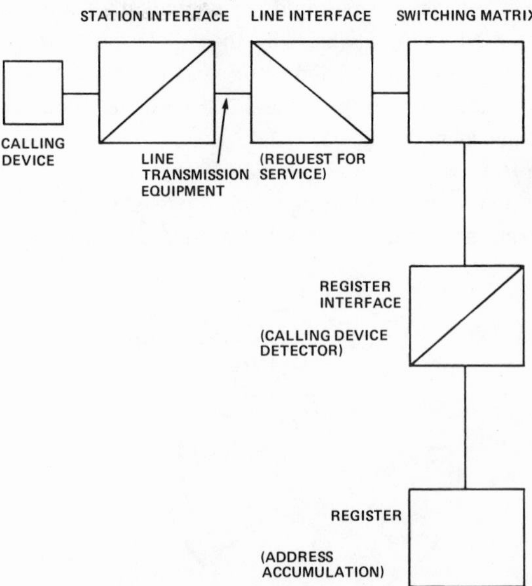

Figure 4.6. Station signaling to register.

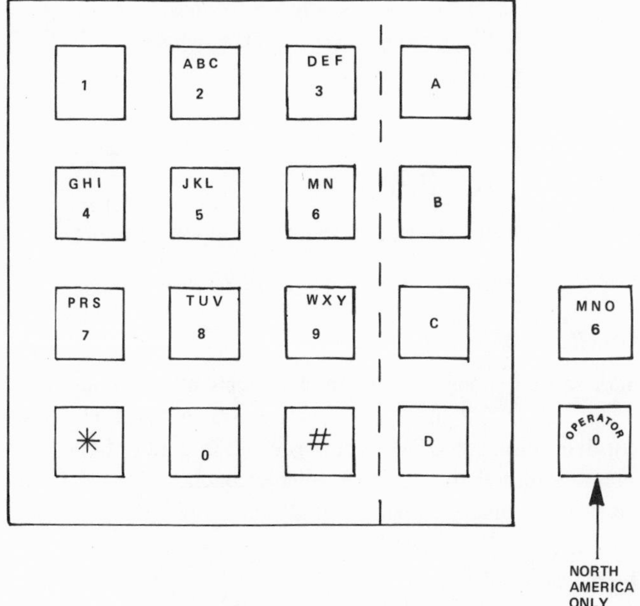

Figure 4.7. CCITT DTMF key pad faceplate.

If the signal is the beginning of a dial pulse it will be followed by a closed-loop condition within 180 msec. If the signal is the beginning of a "hook-switch flash" it will be followed by a closed loop condition within 1 or 2 sec. Services using "hook-switch flash" are those services which require the connection of functional units to an established call without disconnecting the established connection. Hence, depending on the class of service of the caller, a sustained open-loop condition which occurs after a closed-loop must last longer than 180 msec or 2 sec to be recognized as a disconnect signal.

4.2.1.3. Tone Signaling from Station

The advent of low-cost electronic components has made the use of key-controlled oscillators as calling devices feasible. The generic term is *dual tone multifrequency* (DTMF). This calling device has push-button keys on the telephone instruments which transmit two frequencies when a key is depressed. The arrangement of the key plate is shown on Figure 4.7; signals are those developed by the "Bell System" for Touch-Tone service, which is a registered trademark of Western Electric Co.

Figure 4.8 shows the frequency allocation. Key plates in normal use have only 10 or 12 of the 16 possible keys, but systems should be capable of responding to all 16 codes for future applications.

		HIGH			
		1209	1336	1477	1653
LOW	697	1	2	3	A
	770	4	5	6	B
	852	7	8	9	C
	941	(*)	0	(#)	D

○ SOME COUNTRIES MAY USE DIFFERENT SYMBOLS

Figure 4.8. Allocation of DTMF frequencies.

4.2.1.4. Association of Signal Detectors

Registers serving lines are designed to respond to either dial pulses or DTMF signals. Figure 4.9 shows the general arrangement. This uses separate processes for dial pulse and DTMF but the resulting digit (whether it was determined by counting pulses or decoding voice frequency signals) is coded in the system code (usually binary-coded decimal) and stored.

4.2.1.5. Binary-Coded Decimal

This coding method uses four bits per decimal digit. The first ten of the 15 possible combinations are used. The four bits have the most significant bit value

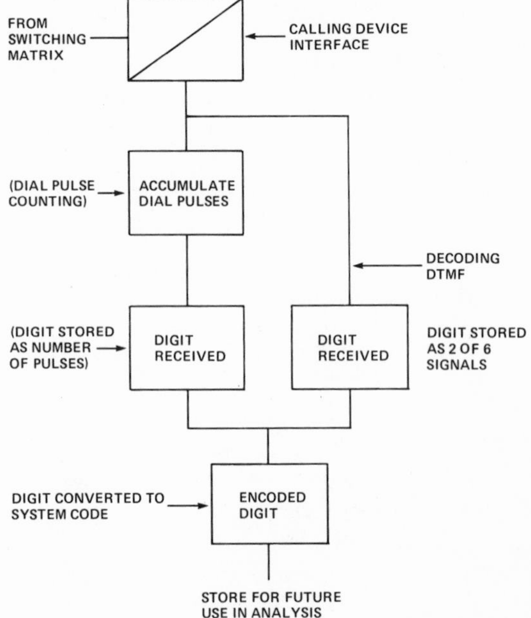

Figure 4.9. Association of signal detectors with line.

(eight) on the left-hand side and numbers are coded by summing the bit values; e.g., decimal value 1 is coded as 0001 and decimal value 10 is coded 1010. This method, which separately codes each decimal digit, has advantages in analyzing the digits dialed by the caller and relating them to the numbering plan. Table 6 shows its application.

4.2.2. Signals to Telephones

The signals to the telephone of the calling party consist of tone signals, or messages, to the caller. These direct the caller's actions or advise him or her of the progress of the call. The frequencies and cadences of these tones vary from country to country but the CCITT is recommending the use of standardized tones. This has become increasingly important with the introduction of direct international dialing. In addition to this there are signals which alert a called customer. Another signal informs a customer that the telephone handset has been left off the hook, without dialing, for a predetermined period (receiver off hook).

4.2.2.1. Call Progress Tones

Call progress tones provide information to customers to invite action, or to indicate what action is being taken by the system.

Typical tone signals are dial, line busy, congestion, ring, number unobtainable, intrusion, call interval, vacant line, information, and receiver off hook. Most tones are interrupted periodically.

Table 6. Binary-Coded Decimal Station Code

Value	8	4	2	1
0	0	0	0	0
1	0	0	0	1
2	0	0	1	0
3	0	0	1	1
4	0	1	0	0
5	0	1	0	1
6	0	1	1	0
7	0	1	1	1
8	1	0	0	0
9	1	0	0	1
10	1	0	1	0
11	1	0	1	1
12	1	1	0	0
13	1	1	0	1
14	1	1	1	0
15	1	1	1	1

Dial tone. Dial tone is used to tell the caller that the level-1 control in the originating local exchange is ready to receive the called address information from the calling line. It is normally returned in an average of 100 msec after the ''request for service signal'' is detected by the line sensor, normally when the subscriber lifts the handset of the telephone set. In times of traffic congestion, requests for service are handled on a randomly served delay basis with the return of dial tone delayed until one of the detection circuits has been assigned to the calling line. Dial tone must be removed during or immediately following the receipt of the first digit. This tone is 425 Hz unless otherwise specified and is continuous.

Line busy tone. Line busy tone is used to indicate that a called subscriber line is busy. It is returned from the nearest traffic-carrying device to the caller that the operation of the switching network permits. For example, if the interoffice line signaling system provides a signal from the remote office to indicate that the called line is busy, then the local switching system should release the trunk connection and connect the calling line to line busy tone at its local exchange, if the register signaling system includes a control signal to indicate this. In most cases a frequency of 400, 425, 435, 450, or 600 Hz is used.

Congestion tone. Congestion tone is used to indicate that a desired connection cannot be completed because there is no available path through the switching network to the required interface. A call on which congestion tone has been sent to the caller may be blocked from further progress until the originating party disconnects, or it may be rerouted, depending on the regular signaling system and the type of exchanges in use. This tone is also called ''paths busy'' or ''reorder'' tone. It uses the same frequency as busy tone but a different cadence. If the interoffice register signaling system provides a backward signal from the remote office when congestion is encountered, then the originating local exchange can release the trunk transmission equipment in the connection and either reroute via an alternate route or switch the calling line to congestion tone at the local exchange.

Ringing tone. Ringing tone informs the caller that the called line is being rung. It is applied at the local exchange which serves the called line. The tone is removed when the called party answers. Ringing tone is also returned on service calls (such as calls involving telephone operators). It is also applied on calls which are routed to intercept service. It is removed when the operator or service answers. This tone is usually 425 Hz.

Number unobtainable tone. Number unobtainable tone is applied toward a calling station when an unassigned directory number is called.

Intrusion tone. Intrusion tone is used as a warning on an established connection just before an operator intrudes on an established connection.

Call interval tone. Call interval indicating tone is provided at fixed intervals during a chargeable call—typically, three short bursts of tone sent 12 sec before the end of a chargeable time interval.

Call waiting tone. Call waiting tone informs a subscriber engaged on a call that there is another call, waiting to be answered, for his or her line.

Vacant code intercept tone. Vacant code intercept tone is applied toward a calling line when a vacant or unassigned code is dialed.

Reverting call tone. Reverting call tone is provided to a calling station calling another station on the same line.

Information tone. Information tone informs an international caller to obtain an operator in the country of origin for assistance if the recorded message on any other audible signal is incomprehensible (for example because of the language in which it is spoken). This tone consists of 950 Hz for 333 ± 50 msec, then 1400 Hz for 333 ± 50 msec, then 1800 Hz for 333 ± 50 msec followed by a silent period of 1000 ± 250 msec then 950 Hz and so on. This signal does not vary in application since it is an international standard recommended by the CCITT.

4.2.2.2. *Alerting Signals to Telephone*

A called line is signaled by ringing the telephone bell. Ringing consists of an interrupted alternating current low-frequency signal such as 25 Hz. The signal has a voltage level of 90 to 150 V. A series of alternate pulses superimposed on the transmission battery and transmission battery alone make up the ringing pattern. The ringing patterns commonly encountered are the same as those shown for ringing tones in Figure 4.10.

Types of ringing detectors. A ringing signal is automatically applied to a station line as soon as practical after a call is extended to it. The ringing voltage usually actuates a telephone bell (ringer). The ringer may be tuned or untuned, depending upon the type of line service. The audible signal at a telephone may also be produced by an electrical transducer which develops a tone signal in response to the application of ringing voltage.

A visual signal lamp may also be used at a telephone station in lieu of the audible signal or as a supplement to it. When special terminals such as key systems[16] are used as station equipment, the standard ringing signal from the switching equipment is used to produce visual signals at the key system telephones.

Repetition of ringing cycle. When the ringing cycle is greater than 3 sec it is usual to apply a signal immediately. This will be followed by the normal ringing cycle.

Tripping of ringing. Ringing must be immediately tripped, that is the switching equipment must remove the ringing signal, when the called party answers. This occurs at the local exchange on receipt of an answer supervision signal (usually looping the called line). Tripping must occur either during the ringing or silent period of the ringing cycle. The loop sensor and any other attachments to a called line must be disconnected from it before ringing is applied.

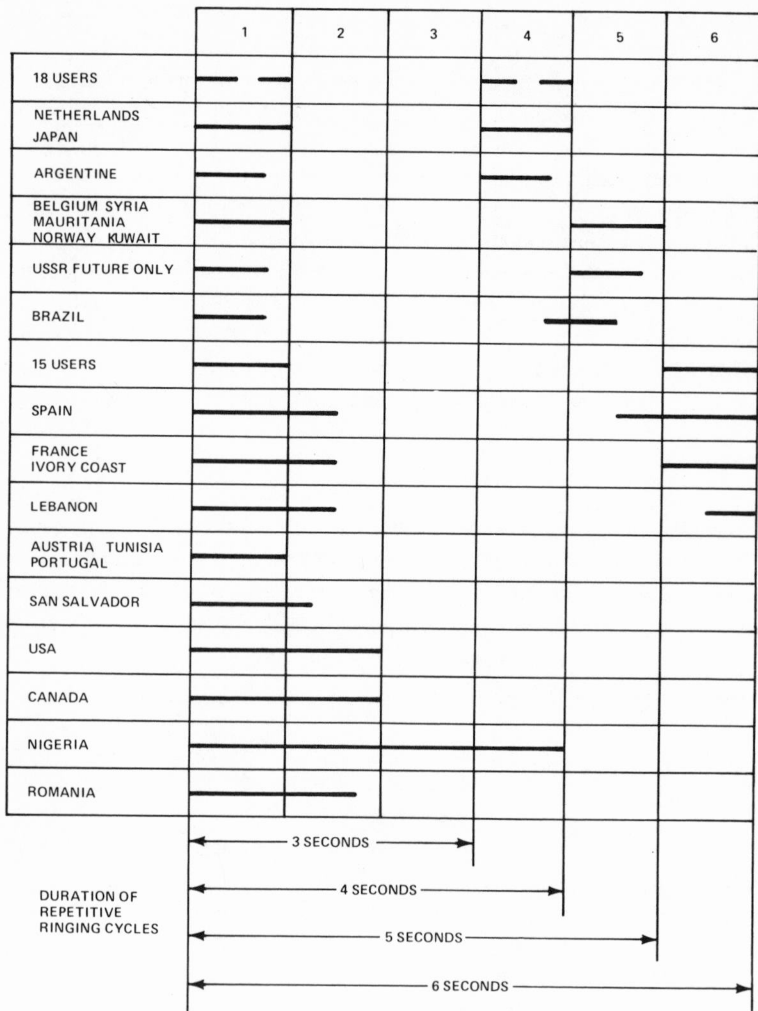

Figure 4.10. Typical ringing cadences.

4.2.2.3. Generation of Call Progress Tones

A switching system has to meet the requirements of the telephone administration for station signaling. The frequencies and cadences of the tones vary from country to country. Figure 4.10 shows a selection of ringing cadences and Figure 4.11 shows a selection of busy tone cadences.

The Bell System uses a series of precise tones based on four frequencies, 350, 440, 480, and 820 Hz, which have a plus, or minus, 0.5% percent frequency variation and plus, or minus, 3 dB in amplitude variation. Typical uses of the frequencies to produce audible tones are shown in Table 7.

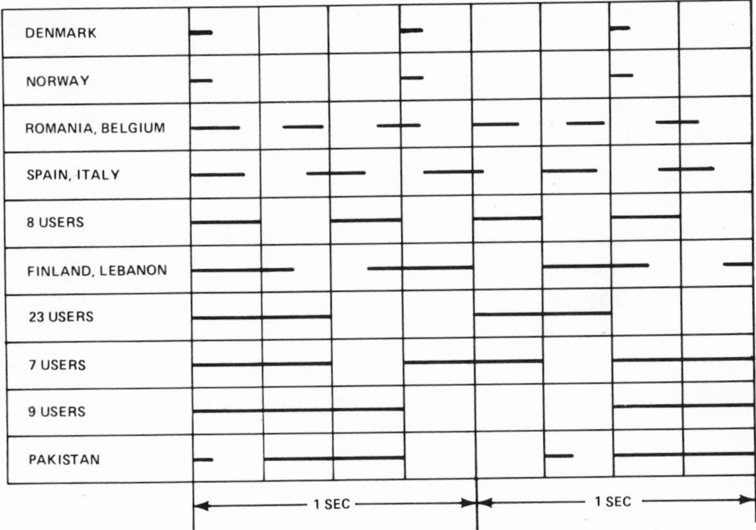

Figure 4.11. Typical busy tone cadences.

The progress tones may be generated as analog or digital signals depending on the transmission matrix and operating mode of the telephone exchange.

Analog signals are generated by tone generators, which are adaptable to the frequencies and cadences commonly encountered. These requirements can usually be met by using three four-frequency generators which should operate from the exchange battery.

Progress tone frequencies. The generator units should be adjustable to provide any of the following frequencies with a tolerance of ± 0.5%. The tone generators should produce frequencies in the following groupings: 120, 133, 200, and 350 Hz, 400, 425, 450, and 600 Hz and 900, 950, 1400, and 1800 Hz.

Ringing frequencies. Ringing requirements can usually be met by providing one of the following: 16⅔, 25, 33⅓, or 50 Hz.

Table 7. Some Precise Tones

Tone	350	440	480	620	Interruptions (sec)	Level per frequency (dB)
Dial	*[a]	*	—	—	None	−13
Line busy	—	—	*	*	0.5 on 0.5 off	−24
Ring	*	*	—	—	2000 on 400 off	−19
Congestion	—	—	*	*	0.25 on 0.25 off	−24

[a] Asterisks indicate frequencies used.

Output level. The output of each generator is amplified to supply the tone at an adjustable level suitable to produce an output at the point of application to a line or trunk interface of −5 dBmO to −15 dBmO. The output of the generators can be used to produce unmodulated tones or modulated tones.

Dual-frequency unmodulated tones. The tones are mixed and amplified to produce the above adjustable output level. The two frequencies making up a dual-frequency tone should not differ in levels by more than 3 dB.

Modulated tones. These are produced by modulating the output of one tone generator with the output of another and amplifying it to produce the above adjustable output level.

Tone addition. It may be necessary to produce a tone by adding a modulated tone made up of two frequencies, such as 450 modulated by 16⅔ Hz, to a third tone, such as 25 Hz, which is then amplified as above.

4.2.2.4. Generation of Interruption Cycle

The tone interruptions should provide adjustable interruption rates to provide the recommended CCITT cadences. These are shown in Table 8 for line busy and congestion tone cadences and Table 9 for ringing and ringing tone cadences.

Ringing tone is normally applied within one ringing period of called line seizure.

4.2.3. Signals to and from PABX

Signals to and from private automatic branch exchanges (PABXs) are used largely for accounting purposes and hence include signals to restrict, assess, or provide information on trunk and other chargeable calls.

4.2.3.1. Toll Restriction

Toll restriction limits direct dialed calls to some or all the dialable destinations. It separates the calls into categories in relation to the calling number, class of service, or identifying code and restricts outward dialing accordingly. This restriction signal usually consists of a loop battery reversal of 100 ± 50 msec to the PABX.

Table 8. Line Busy and Congestion Tone Cadences

	CCITT accepted (msec)	CCITT recommended (msec)
Tone on	100 to 660	120 to 660
Tone off	120 to 800	120 to 660
Total pulse	300 to 1100	300 to 1100
Ratio		0.67 to 1.5 (1.0 optimum)

Table 9. Ringing and Ringing Tone Cadences

	CCITT accepted (sec)	CCITT recommended (sec)
Tone on	0.67 to 2.5	0.67 to 1.5
Tone off	3.0 to 6.0	3.0 to 5.0
Total	3.67 to 8.5	3.67 to 6.5

Automatic number identification. A local exchange must be capable of requesting and storing the calling number, if the number is sent over the transmission path. It must also be capable of receiving the date over a separate channel if a system such as automatic outward identified signaling[17] is used. The calling number will be sent using the numerical codes of the register signaling system in use at the point of application. It will be sent in response to an off-hook signal or, in the case of the R2 signaling system, the appropriate code from the distant point.

4.2.4. Signals to and from Coin Boxes

Coin box signals must relate to the type of coin box in service, which may either provide only limited features and require operator intervention or be entirely automatic.

4.2.4.1. Signals for Automatic Operation for Local Calls

Coin boxes that provide signals for automatic operation for local calls can be entirely mechanical and require the operation of a button to complete a single fee call when the called party answers or be electromechanical with a signal from the exchange to collect calls.

Reverse battery coin collection. A reverse battery condition is applied from the local exchange when the called party answers. This results in the collection of the coin and the completion of the transmission circuit. If the call is not answered the coin is refunded automatically when the caller replaces the hand set.

Dc coin controls. In this system a coin is collected as a result of a 110-V positive potential to ground and refunded as a result of a 110-V negative potential to ground.

4.2.4.2. Signals for Trunk Calls

Signals for trunk calls are provided by an operator who assesses the charges for the call and requests the appropriate deposit. An audible signal is relayed to the operator as the coins strike one of a number of gongs. Each gong produces a distinctive sound and serves a slot for coins of a designated value. Any coins deposited in order to reach the operator are automatically refunded when the operator answers.

Ac coin control. This system has an interactive signaling system which automatically generates coin collect signals at intervals for the duration of the call. The coin box may include equipment which will automatically compute the total of coins deposited and will subtract from this in response to signals from the exchange. Alternating current signals commonly used are 12,000 Hz or 16,000 Hz applied for 50 msec or longer.

4.3. Register Signaling

Register signaling provides the address and other information for routing calls to their destination and treating them appropriately. It is used during the establishment of the call. The register accumulates information from the station which signals to the register in dial pulse, analog, or digital modes. The signals reach the register via the transmission path and register access arrangement from the station or via a separate signaling channel in lieu of this for sender to register.

The signaling that occurs after the register has obtained the signals from the caller's calling device and which occurs between a sender and a register is called register signaling. Figure 4.12 shows the general arrangements.

4.3.1. Register Sender Operation

The register and sender functions are related. A sender can only transfer information which was originally accumulated in a register. Hence a register function must always precede a sender function.

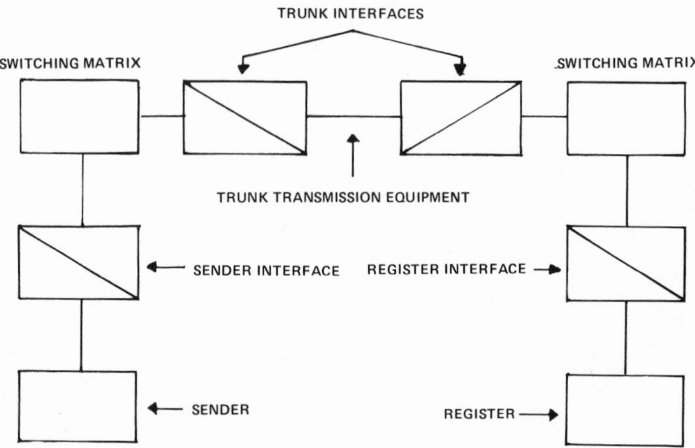

Figure 4.12. Register signaling.

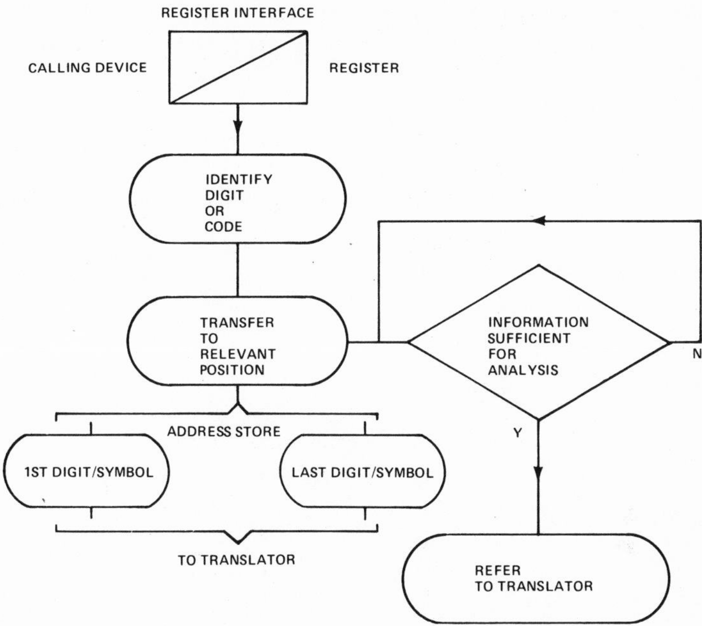

Figure 4.13. Called address registering.

4.3.1.1. Register

Figure 4.13 shows the principle of storing dialed digits. The digits sent by the calling device are decoded and transferred in the order of receipt to the address store. Each time that a digit is decoded it is offered to the translator with its preceding digit or digits, if any. When enough information has been received to identify the call destination, a sender function is invoked if the call is to another exchange.

4.3.1.2. Sender

Figure 4.14 shows the principle of sending digits. Digits are extracted, according to the routing given by the translator, one by one, from the address store. Each time a digit is extracted a check is made to determine whether any more digits have to be sent to complete the sending task.

Sender functions. Sender functions include the following:

- Sending address information in the signaling mode given by the code translator in its routing instruction.
- Receiving and sending information simultaneously in the same or differing signaling modes. This reduces the delay on connection which requires register signaling.

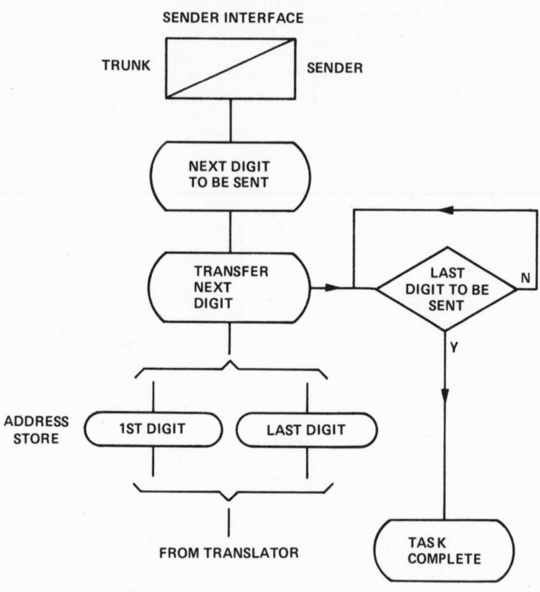

Figure 4.14. Called address sending.

- Delay sending when the remote exchange has not connected a register to the trunk and hence is not ready to receive digits. The signaling system will provide an indication when sending may be started or resumed.
- Sending loop/disconnect dial pulses at 10 ± 0.5 pps with a ratio of 39 ± 2% make and (61 ± 2%) break. The interdigital pause between digits should be 700 ± 35 msec.
- Sending in the various signaling modes and systems which are described in the chapters on signaling.

4.3.2. Types of Signaling

There are two basic types of signaling: pulse signaling and compelled signaling.

4.3.2.1. Pulse Signaling (Pulsing)

The signals are not acknowledged and have a timed duration. They start after the transmission of a line signal and the subsequent return of a "start-sending" signal from the register usually "link by link" as described in Chapter 2.

4.3.2.2. Compelled Signaling

All signals persist until they are either acknowledged or timed. The acknowledgement signal stops the signal which originated transmission and the removal of the acknowledgement signal starts the next originating signal, usually "end to end."

The principle of operation of compelled signaling is shown in Figure 4.15. A sender signals to a distant register over the trunk transmission means via the originating and terminating exchanges. Signals received are transmitted from the distant sender to a register in the originating exchange. The originating sender sends the signal until the return signal is received by its associated register. It removes the signal when the return signal is decoded. The removal of the signal causes the distant register to instruct the distant sender to remove the return signal. The originating sender applies the next signal when its associated register indicates that the signal from the distant sender has been removed. A signaling system operates in this mode for register signaling only. Station address signals are pulsed to the originating register.

4.4. Line Signaling

Line signaling controls the interexchange (and hence the intraexchange) transmission path.

It is used for establishing, maintaining, releasing, and monitoring the selected path for the duration of the call.

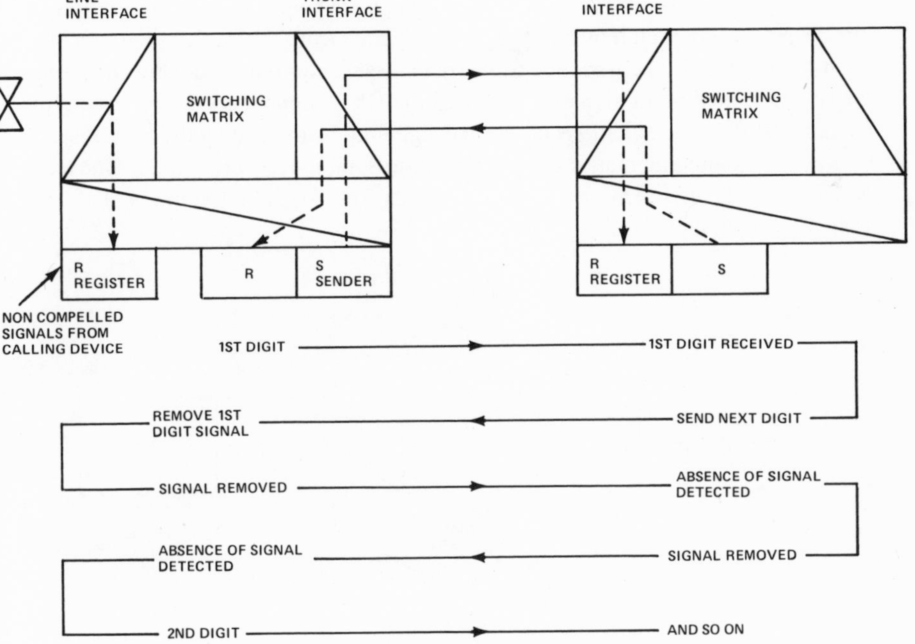

Figure 4.15. Principle of compelled signaling.

Figure 4.16 shows the principle of line signaling.

Line signaling may use direct current, digital, or alternating current signals.

4.4.1. Direct Current Signals

Direct current signals are used on wire trunk interfaces system. They may use a wire pair (loop signals) or a single wire (leg signals).

4.4.1.1. Loop Signals

Loop signals are based on opening and closing the loop to the distant exchange and reversing the direction of current from the distant end. A summary of loop signals is shown in Figure 4.17, which is self-explanatory.

Modern electronic switching systems and switching systems located in areas with varying ground potentials should use loop signals when direct current is being used. This is also the case with interfacing signaling systems such as the E&M system described below where a pair of wires replaces each of the single leads.

4.4.1.2. Leg Signals

Direct current signals are also used to produce the signals required for interfacing with derived channels. These signals are necessary since most signaling systems, other than loop signaling, are separate from the trunk equipment and are normally located between the trunk transmission equipment and the switching matrix. One method of signaling is called the E&M signaling system. It derives its name from the designations of the two signaling leads used by the Bell System between the switching matrix and the signaling equipment for the trunk. One lead is called the E lead. This lead carries signals from signaling equipment to the switching equipment.

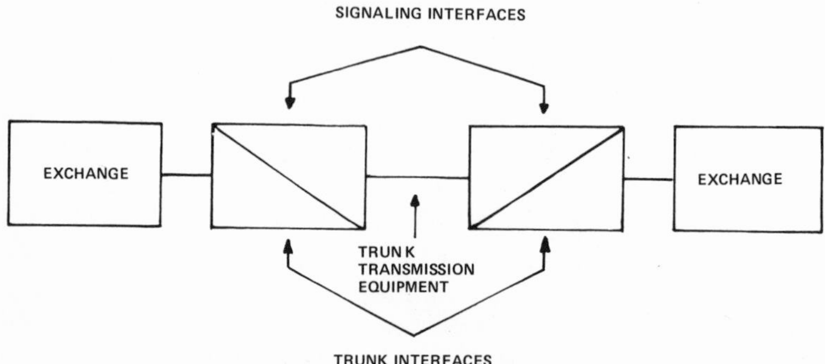

Figure 4.16. Line signaling.

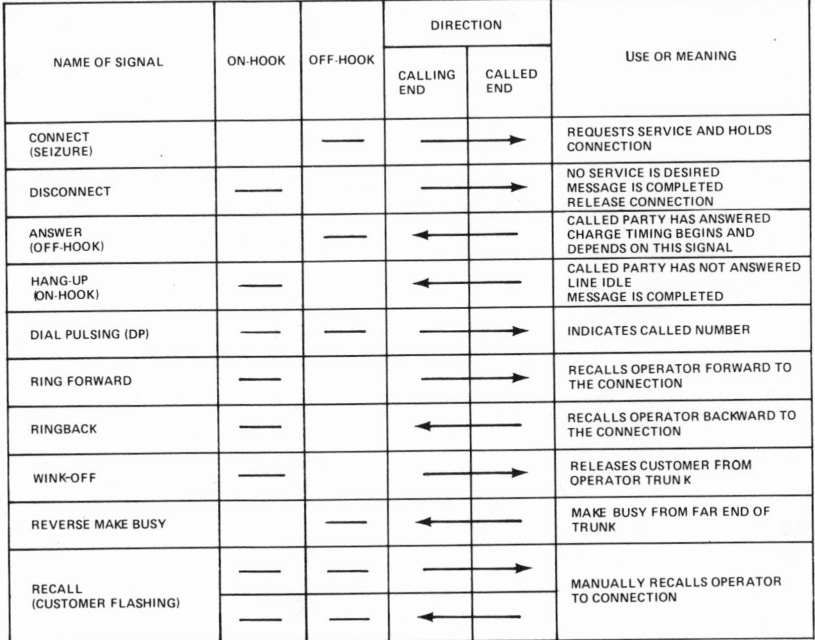

NAME OF SIGNAL	ON-HOOK	OFF-HOOK	DIRECTION		USE OR MEANING
			CALLING END	CALLED END	
CONNECT (SEIZURE)		—	——	——→	REQUESTS SERVICE AND HOLDS CONNECTION
DISCONNECT	—		——	——→	NO SERVICE IS DESIRED MESSAGE IS COMPLETED RELEASE CONNECTION
ANSWER (OFF-HOOK)		—	←——	——	CALLED PARTY HAS ANSWERED CHARGE TIMING BEGINS AND DEPENDS ON THIS SIGNAL
HANG-UP (ON-HOOK)	—		←——	——	CALLED PARTY HAS NOT ANSWERED LINE IDLE MESSAGE IS COMPLETED
DIAL PULSING (DP)	—	—	——	——→	INDICATES CALLED NUMBER
RING FORWARD	—		——	——→	RECALLS OPERATOR FORWARD TO THE CONNECTION
RINGBACK	—		←——	——	RECALLS OPERATOR BACKWARD TO THE CONNECTION
WINK-OFF	—		——	——→	RELEASES CUSTOMER FROM OPERATOR TRUNK
REVERSE MAKE BUSY		—	←——	——	MAKE BUSY FROM FAR END OF TRUNK
RECALL (CUSTOMER FLASHING)	—	—	——	——→	MANUALLY RECALLS OPERATOR TO CONNECTION
	—	—	←——	——	

Figure 4.17. Loop signals required in distance dialing.

This type of signaling uses a ground return which introduces voltage surges during its actuation and release. These surges are of insufficient duration or amplitude to affect the slow-reacting electromechanical equipment but can cause false signals on electronically controlled systems. Hence more recent applications of the signaling system use a pair of leads with loop signaling for each of the functions. Figure 4.18 shows the ATT type-I connections used for electromechanical systems and type-II connections used for electronic systems.

4.4.1.3. Ac Signals between Exchanges

The use of alternating current for line signals is usually limited to frequency division carrier systems. The switching system issues and responds to these signals by the use of E&M signals. In the simplest example a direct current signal on the M lead causes a signaling frequency to be applied to the derived channel. This frequency may be in the speech band of the channel or outside it. Its reception at the distant end causes a direct current signal to the E lead serving the distant trunk interface.

Signals with direct trunk interfaces. Figure 4.19 shows the trunk interface between a direct trunk and a switching system with a metallic transmission matrix. There are three leads from the switching matrix. The transmission path has a

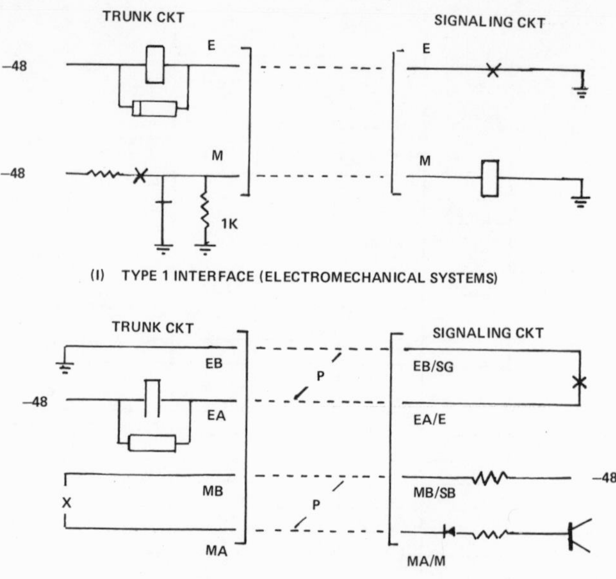

Figure 4.18. E & M signaling interfaces.

and b leads and there is a single control lead. The interface controls the holding of the transmission path through the switching matrix. Line signals are of the loop-disconnect type. The transmission path is two-wire.

Signals with derived trunk interfaces. Figure 4.20 shows the trunk interface between a derived trunk and a switching system with a metallic transmission matrix. The interface with the switching system performs the same functions as the direct trunk interface. The interface with the trunk transmission equipment converts to four-wire transmission and extends E&M signals to the carrier termi-

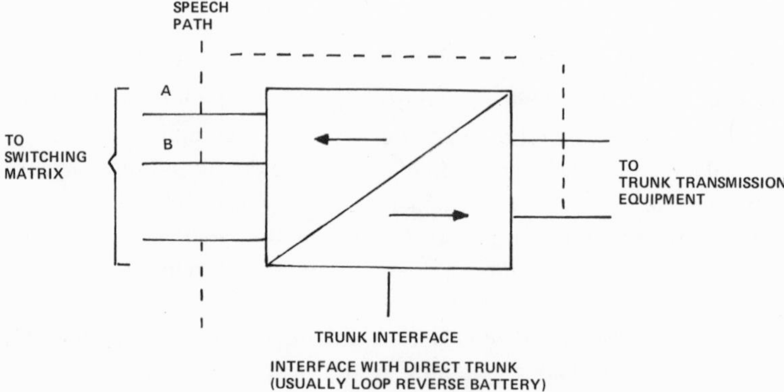

Figure 4.19. Line signaling interface with direct trunk.

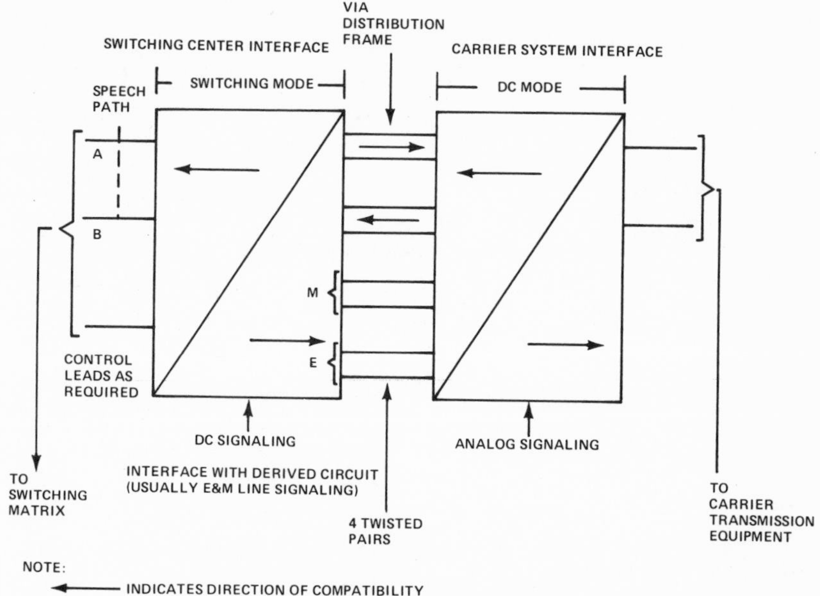

Figure 4.20. Line signaling interface with derived trunk.

nal. The carrier terminal converts the signals to its internal signaling format and converts the metallic transmission path to a derived channel.

4.4.1.4. Digital Signals between Analog Exchanges

In the case of channels derived from a digital carrier system the M lead controls dedicated information positions relating to the channel, if the exchange has an analog transmission path.

4.4.1.5. Digital Signals between Digital Exchanges

No interfaces are necessary in this case since the transmission matrix and its transmission mode are compatible with the trunk transmission equipment.

4.4.2. Line Signaling Modes

Figure 4.21 summarizes the various possibilities for line signaling.

4.5. Control Signals

In addition to the signals used to establish and maintain connections there are signals which are used to effect call processing and to operate and administer the network.

TYPES OF TRUNK TRANSMISSION			SIGNALING MODE	
MEANS			EXCHANGE	TRUNK
WIRE			LOOP DC	LOOP DC
DERIVED	ANALOG SWITCHING SYSTEM	ANALOG	E&M	AC
	ANALOG SWITCHING SYSTEM	PCM	E&M	PCM
———	DIGITAL SWITCHING	PCM	PCM	PCM

Figure 4.21. Line signaling modes.

4.5.1. Call Processing Signals

Call processing signals are used between the various portions of the switching system and are usually digital. They provide the means of conveying information and instructions in the format required by the system which they control.

4.5.2. Operating and Administration Signals

Operation and administration signals include those that produce information relating to call handling, system operation, and call charging.

4.5.2.1. Call-Handling Signals

Call-handling signals are used to inform operators that a call requires special treatment in routing or charging for it. Some of the most common uses are identifying calls from coin boxes or calls that have been intercepted because a call cannot proceed as dialed. These were indicated by special supervisory tone signals in the past. Modern telephone consoles have extensive visual display facilities and special signals in the system mode and format will be required.

Some signaling systems such as CCITT R2, which is described in a later chapter, include these signals in the register signaling repertoire. They must be converted to the relevant signals to communicate with the operator at the terminating exchange. Such signals could be audible or visual signals. The following supervisory signals are in extensive use with the many existing types of toll boards.

Coin box (pay station) tone. Pay station tone may be used to identify calls from pay stations when the calls are routed over a trunk group (which also serves non-pay-station calls) to a manual switchboard. This alerts the operator that the call is from a pay station. The tone is momentarily applied when the operator answers. Its frequency and duration will depend on the local administration since the signal does not cross national boundaries.

Intercept tone to operator. Intercept tone is used to identify intercepted calls routed to an operator over a trunk group which serves both regular and intercept traffic. It is normally a single pulse of tone, applied in both directions immediately when the intercepted call is answered.

4.5.2.2. System-Operating Signals

System-operating signals are signals which are used to indicate or respond to malfunctions in operation or to control traffic routing. At the present times many separate signaling systems, which have been often designed as a result of operating experience, perform these services. Hence they vary from administration to administration. Signaling system R2 has possibilities for including some of the signals, but only the digital signaling systems offer the possibility of a complete integration of register, line, and control signaling.

Alarm signals. These signals are used to alert maintenance personnel that the system is malfunctioning. Arrangements have to be made to provide local and remote indications. Modern systems with stored program control provide detailed fault-indicating information in visual, usually printed form. The information must also be available in a form suitable for cathode ray tube, or other transient visual form.

SIGNALING PURPOSE / SIGNALING CATEGORY	SUPERVISION	ADDRESS	PROGRESS
STATION	LOOP	DIAL PULSE – DTMF	TONES
LINE	LOOP-REVERSAL E&M ANALOG PCM	———	LOOP REVERSAL ANALOG PCM
REGISTER	———	DIAL PULSE IN/OUT BAND	IN/OUT BAND

Figure 4.22. Comparison of signaling categories.

Permanent loop signals. These provide an indication to the customer and to the switching center of a receiver which has not been replaced (permanent loop). After an interval of from 30 sec to 3 min a terminal which is not engaged in a call is returned to a lock-out condition. The system must apply a receiver-off-hook tone to the telephone according to the local practices.

4.5.2.3. Call-Charging Signals

Call-charging signals are used to relay call-charging information to and from the local exchange. They may take the form of a series of reversals of loop current or alternating current signals. The Bell System uses a series of reversals which they term *multiple winks*. These consist of a series of one to five winks each having a duration of 70 to 125 msec at its point of origin and spaced 95 to 150 msec apart. If alternating current signals are used, the frequencies must be within the speech transmission bandwidth.

4.6. Comparison of Signaling Modes

Figure 4.22 summarizes the various signaling functions in relation to the progress of the call and their modes of operation.

5

Basic Techniques

5.1. Introduction

System design can utilize many components and use many techniques. The rapid advance of microelectronic technology has resulted in an abundance of low-cost, reliable circuitry. But the techniques in use in a switching system reflect the state of the art in componentry when the design was conceived, and the relatively long period necessary to bring a switching system into production means a design based on components that may have been surpassed in performance in the period that has elapsed since the design of the system was frozen (which brings to mind the saying "If it works it's obsolete"). Still, all techniques used can be categorized into the application of the fundamental physical laws. The applications of the laws in telephone switching design relate largely to electromechanical techniques utilizing magnetics and to electronic techniques utilizing solid state physics and electro-optical techniques.

Other techniques are those relating to system architecture which results from applications of componentry, transmission, and traffic theory. These also affect switching matrices, packing of equipment, and provision of equipment.

5.2. Techniques

5.2.1. Electromagnetic Techniques

Electromagnetic techniques are based on the production of magnetic forces by the application of electrical currents and the use of inductive coupling. The basis for its use relates to the needs to match impedances and generate mechanical action to close electrical contacts. Its application takes the following forms.

5.2.1.1. *Impedance Matching*

The optimum transfer of power in an alternating current circuit can be obtained as follows.

- The current through a load impedance Z_L, from an electromotive force E having a source impedance $Z_S = E/(Z_S + Z_L)$.
- The power delivered to the load $= P_L = E^2 Z_L/(Z_S + Z_L)^2$.
- The rate of change of the power with respect to $Z_L = dP/dZ_L = E^2(Z_S - Z_L) / (Z_S + Z_L)^3$.

Now the maximum power occurs when the rate of change is zero, i.e., when $Z_S - Z_L$ is equal to zero. The condition obtains when the impedance of the source is equal to the impedance of the load. It can be shown that if the source and load impedances are divided into resistive and reactive portions then maximum power occurs when $jX_S = -jX_L$, i.e., when the reactive portions are equal. This condition is satisfied when the source and load impedances are equal in magnitude but opposite in angle.

5.2.1.2. *Inductive Coupling*

Matching source and load impedance, or at least achieving the best possible balance, is an important function in telecommunications transmission. The relation of the input (primary) voltage (E_p) of a transformer and the output (secondary) voltage (E_s) of a transformer is directly related to that of the number of turns on the primary and secondary transformer windings; i.e., $E_s/E_p = N_s/N_p$, where N_p is the number of turns on the primary winding and N_s is the number of turns on the secondary winding. The actual power transferred will be reduced by the transformer losses. For the sake of illustration it is assumed that there is no transformer loss. The power transfer is

$$E_p{}^2/Z_p = E_s{}^2/Z_s$$

hence $Z_p/Z_s = (E_p/E_s)^2$. Now voltages are proportional to turns in the above equation which can therefore be rewritten as

$$Z_p/Z_s = (N_p/N_s)^2$$

hence $Z_p = Z_s$ multiplied by $(N_p/N_s)^2$ and the ratio of the primary and secondary turns is equal to the square root of the ratio of the primary and secondary impedances.

5.2.1.3. *Electromechanical Coupling*

The electromagnetic coupling principle is based on the application of magnetomotive force, which is $4\pi NI/10$, where N is the number of turns and I is the current through a coil. It may be augmented by the permeability of the magnetic circuit or reduced by the reluctance of the magnetic circuit. The tractive force generated is $B^2A/8\pi$, where B equals the flux density in gauss in the air gap which separates the electromagnet from the attracted item and A equals the area of the air gap in cm². This can also be expressed as

$$6.4N^2I^2/10^5\ S^2A\ \text{g}$$

where S is the reluctance of the magnetic circuit, which includes the return path.

Now, S is the sum of the series of reluctances making up the magnetic circuit for each segment, hence it is necessary to allow for the permeability of each portion.

S equals length in centimeters divided by permeability multiplied by sectional area. Since the flux spreads out at the ends of the electromagnet, practical applications provide return paths of high permeability to minimize the reluctance and loss of flux density.

5.2.2. Applications of Electromagnetic Techniques

Figure 5.1 shows some typical applications of electromagnetic techniques.

5.2.2.1. Transformers

A transformer consists of primary and secondary windings on a high-permeability core. There can be more than one primary or secondary winding depending on the application. It couples a transmission or alternating current input with a transmission output or an ac output. They may be used for interfacing transmission equipment with the switching system or for control functions such as connecting together traffic-carrying units or functional units.

Hybrid transformers. A hybrid transformer connects a two-wire transmission path to a four-wire transmission path. It consists of a number of windings as shown in Figure 5.2. The two-wire transmission path is coupled to the send and receive paths. Impedance matching is provided by the balancing network Z, which is also coupled to the send and receive paths. Alternating current signals, such as those produced by voice or tone signals, produce current in both send and receive paths. An almost identical current is induced into the balancing network circuit. This results in an assisting current in the send circuit. It also produces an opposing current in the receive path. This current cancels out the current induced in the receive path by the alternating currents from the two-wire transmission equipment. Incoming signals on the receive path are prevented from being returned on the send path by a similar process. The degree to which this process works depends on how well the input and balancing impedances match. This is measured in terms of the relative powers of the input signal and the recirculated signal; −27 dB is usually an acceptable level for this trans-hybrid loss.

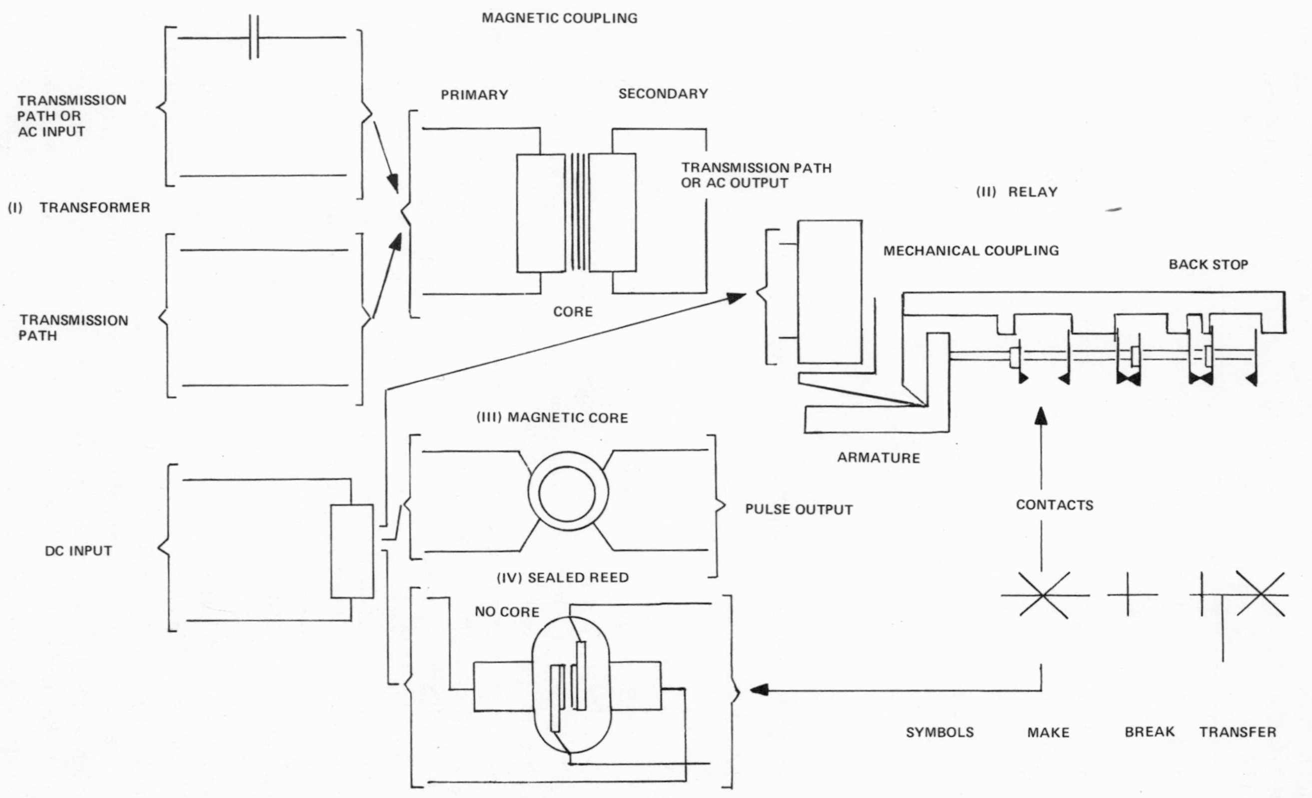

Figure 5.1. Electromagnetic techniques.

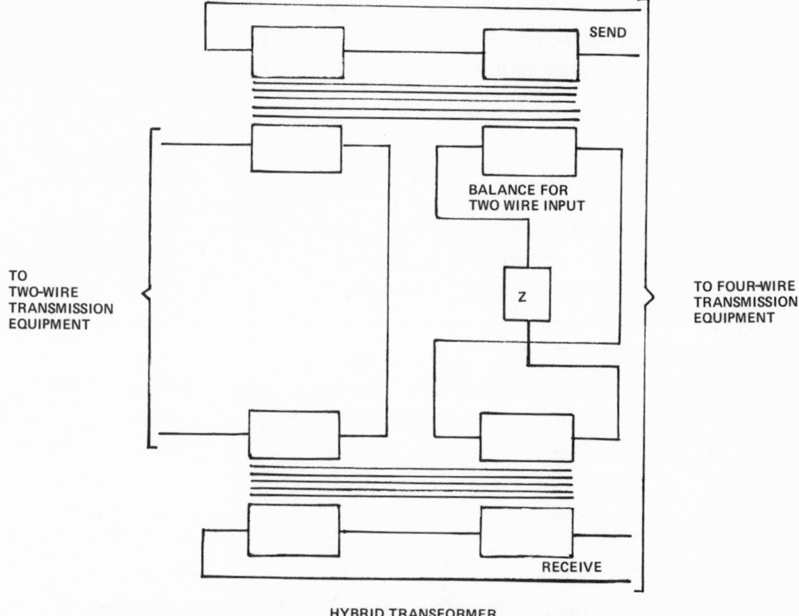

Figure 5.2. Principle of hybrid transformer.

5.2.2.2. Relays

Relays are used to apply, or respond to, signals. They may be used to apply voltages, which are not compatible with the switching matrices, to telephone lines, or to isolate faulty equipment. While relays were used extensively for logic circuits in earlier switching systems, their use is restricted in electronic systems to responding to control or other signals. Relays are being made smaller and require less power to better relate to electronic systems. A typical relay design of an early type operates as shown in Figure 5.1. There are many other differing mechanical designs.

The turns are wound on a core which has a return path of low reluctance which includes an armature which pivots on the edge of the core. The current through the coil causes the armature to be attracted to the core and its movement is transmitted to the contacts.

Three types of contacts are shown: a ''make'' contact, which is normally open and which is closed when the armature is attracted; a ''break'' contact, which is normally closed and which is opened when the armature is attracted; and a ''transfer'' contact, which diverts a path from a normally closed path to one which is closed when the armature is actuated. The figure shows the symbols for the three types of contacts. The contacts are held to a limited amount of motion by mechanical back stops.

There are many ways of closing contacts using electromechanical techniques, and this technique was the basis for the majority of the existing automatic telephone systems. It was applied to switching matrices as well as control circuitry.

5.2.2.3. Sealed Reeds

Sealed reeds are a variation of electromechanical control in which two strips of metal are arranged in a straight line separated by a small space. The free ends of the two strips overlap. The fixed ends are sealed at either end of a glass tube. The tube is filled with an inert gas. The assembly is placed in the middle of a coil winding, replacing the core. When current flows through the winding the two free ends of the metal strips are magnetized and the strips make contact. Conductivity is improved by plating or affixing gold at the point of contact. Normally this application is restricted to make contacts in switching matrices or interface circuits.

5.2.2.4. Magnetic Cores

Magnetic cores are cores made of material having a high degree of magnetic retention. A nonmagnetized core is used for the storage of logic state 0 and a magnetized core is used to store logic state 1. The state of the core is set by energizing a winding which surrounds the core. The core is read by reading the output from the core by changing the state to 0, if the state was 1 a pulse is produced. Cores were used to produce nonvolatile memories but are being replaced by integrated circuits.

5.2.3. Electro-Optical Techniques

Electro-optical techniques are techniques which are mostly being used for trunk transmission equipment. They can be also used for interfacing two dissimilar techniques such as wire lines and electronic switching matrices. They are based on the optical properties of certain solid state devices. Some devices emit light when a suitable current flows through them. These are lasers and light-emitting diodes. Other devices generate an electrical current when they are illuminated. These are photodetectors.

The principal advantages of using fiber optics are broad bandwidth, low-cost material, small size, low weight, excellent electrical isolation of the input and output paths, freedom from cross talk, security of information, widely spaced repeaters. These are of sufficient importance to telecommunications administrations to ensure that they will be used increasingly for new transmission systems.

One application which will undoubtedly be considered is the replacement of existing cables with fiber optic cables. Existing cables generally contain large amounts of copper, which has a high salvage value. In addition to this, the substitution of fiber optics provides much greater bandwidth (20 mbit or higher) and requires less than half the number of repeater access points.

5.2.3.1. Principle of Operation (Figure 5.3)

There are three separate techniques: light emission, light conduction, and light detection.

Light emission. These are either light-emitting diodes or injection lasers, which use the same basic semiconductor material. This has the property of emitting light at the p–n junction when an external voltage drives both electrons and holes into the junction. The recombination of the electrons and holes produces light at a wavelength which is proportional to the energy drop. An injection laser is designed to orient the active area with a number of layers which have a lower refractive index so as to form a resonant cavity. This cavity is perpendicular to the plane of the p–n junction.

Both the devices are modulated by the variation of the driving current. Injection lasers have efficiencies of 10% to 50% as compared with 3% for light-emitting diodes.

Light conduction. The light from the light-emitting device is applied to an optical fiber. This fiber consists of a transparent dielectic material. This material may be silica which has been specially doped to produce a high refractive index. The index is preserved when the fiber is drawn. Maximum single-mode propagation is obtained when the fiber diameter is 2–4 mm. The core is surrounded by a cladding material, usually plastic, which has a lower refractive index. Light is projected through the fiber by being reflected from the cladding. Total reflection occurs if the angle of incidence is greater than a critical angle which is defined by the refractive indices of the cladding and core. Light will only be transmitted if it impinges with the fiber end within the defined critical acceptance angle. Rays entering at other angles than this will be lost.

Single-mode propagation requires extremely small light emission and detection devices, hence multimode transmission, which can be used with larger-diameter devices, is the more usual method at the time of going to press. Multimode propagation involves different velocities and propagation time, hence it restricts the transmitted bandwidth as compared with single mode.

When multimode operation is used the bandwidth becomes reduced as the length of the fiber increases. This is expressed as MHz/km. Signals traveling along the core are attenuated to an extent determined by the wavelength being transmitted. This is expressed as dB/km at a specific wavelength.

Multimode fibers propagate light with a step index or with a graded index. A step index occurs when the core has a uniform index of refraction with an abrupt change in refraction level at the interface between the core and the cladding. A graded index occurs when the core index decreases parabolically from the center outwards. In general step index fibers can handle data rates up to 50 mbit/sec and graded index fibers up to 500 mbit/sec depending upon the application.

5.2.3.2. Fiber Optics in Telephone Transmission

The technique of fiber optics in telephone transmission uses glass fibers to conduct a beam of light. The input device can be a LED or a laser; this emits light into the cable. A sensor at the distant end converts the light into electrical energy. It offers a wide bandwidth and hence can be used in carrier systems. The switching system interface with the transmission system depends upon the carrier techniques and the signaling modes employed. Typically a digital multiplexer would be used to combine the carrier bit streams into the optical fiber bitstream. One early application combined 28 1.544-Mbit/sec streams each comprising one North American primary multiplexing digital carrier system into a 44.736-Mbit/sec stream. (This corresponds to third-level digital multiplexing in the AT & T system.)

5.2.3.3. Circuit Isolation

The latter case, which provides an effective means of isolating, operates on the principle shown in Figure 5.3. A current flows through a light-emitting diode. This produces a light. The light is detected by a phototransistor. The transitor produces an electrical output. The relevant parameters of the devices are the subject or manufacturers' specifications. A typical application is a line circuit interface serving an automatic telephone system with an electronic switching matrix.

5.2.4. Integrated Circuits

Integrated circuits which can provide most logical operations including arithmetical functions, code conversion, etc. are now readily available at low cost. This has freed system designers from much of the calculation and evaluation associated with the design of basic circuits. Designers are now able to devote the majority of their efforts to designing the system architecture. The embodiment of most of the system functions can be realized by commercially available, or specially designed, integrated circuits. This has made feasible new types of interchanging information and opens the door for radically differing system structures. This has been discussed by the author[(18)] and others.

At the time of going to press 200 gates could be formed per mm^2. The gates could operate with delays of 1 nsec with a power dissipation of 0.05 mW per gate. The integrated circuit chips could be built with nearly 40,000 devices on a single chip. 1000 memory cells could occupy only 1000 μm^2.

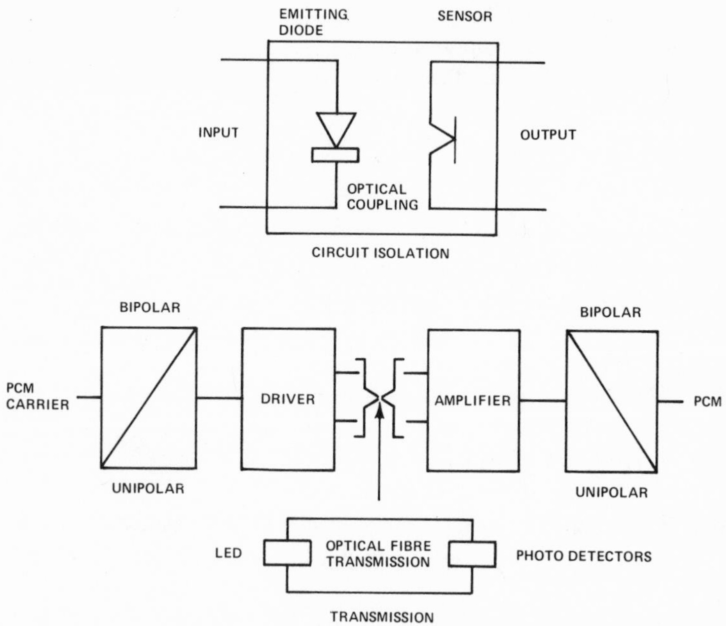

Figure 5.3. Electro-optical techniques.

5.2.4.1. Logic

Logic can be wired or stored. Stored logic is the performance of logical processing sequences by the use of programs stored in memory. Wired logic is the embodiment of a function or functions in the interconnection of components. A switching system can be entirely based on wired logic, as are most existing systems, or can use a combination of stored and wired logic. The application of the rules of logic has been treated extensively in earlier publications.[(19,20)] Its embodiment requires components, and the interconnection of the components must relate to their characteristics, hence new components will bring new relationships and new rules.

The above references show a number of applications of logic to differing components including relays and discrete electronic components. Logic design should be based on flow charts which show the relationship of functions. A number of levels of flow charts should break the functions down in increased detail. This will identify the functions to be performed at each part of the system, e.g., ringing a called line after it has been found to be free.

The sequence is as follows: Ring called line and mark called line busy in data bank.

Then:

- ○ Address signal matrix.
- ○ Command interface circuit to ring.
- ○ Store command at interface or level-1 control.
- ○ Remove address signal.
- ○ Check for called line answer.

Followed by:

- ○ Isolate interface from matrix
- ○ Isolate line sensors, etc.
- ○ Apply ringing to line via ringing trip sensor.

This circuit can be embodied by using an arrangement similar to that shown in Figure 5.3.

5.3. Space Division Matrices

Switching Matrices are made up of stages on interconnected crosspoints. These may be electronic or electromechanical.

5.3.1. Crosspoints

Electronic crosspoints are semiconductor gates which have a sufficient difference between their off (essentially nonconducting) and on (conducting) states. Electromechanical crosspoints are contacts of the types described in Sections 5.1.2.2 and 5.1.2.3.

The term "crosspoint" includes all the controlled paths. For example an electromechanical crosspoint could have two contacts to open and close the transmission path and one or more contacts to control other switched paths. Crosspoints vary from low speed, medium voltage, and medium current, which are typical electromagnetic characteristics, to high speed, low voltage, and low current, which are typical electronic characteristics.

Cost of crosspoints. Modern switching systems use low-cost crosspoints. This is the case in digital switching where crosspoints are fabricated by integrated circuit techniques. This allows the economical provision of networks with very high probability of connecting any inlet to any outlet. This frees the system designer from most of the complex calculations used when crosspoints were expensive. These were designed to minimize the number of crosspoints needed to provide the necessary grade of service. The resultant network had critical dimensions for grouping crosspoints in stages.

5.3.2. Construction of Switching Matrices

The dimensioning and construction of switching matrices is based on the switching characteristics of the crosspoints. Electromechanical crosspoints have many differing mechanical shapes. They also may involve relatively complex operating mechanisms. However, modern electronic stored program switching systems use only simple mechanical crosspoints of the sealed reed type or electronic gates. These are made into arrays and connected together to make switching stages. The matrices may be space divided using either type of crosspoint or time divided, which is only practical with electronic crosspoints.

5.3.3. Single Stage of Switching Matrix

Figure 5.4a shows a single stage of switching which connects any of n inlets to any of m outlets. The number of outlets provided depends upon the number of telephone calls occurring at the same time. This is a function of telephone traffic and is determined by the calling habits of the telephone customer and traffic theory. This is discussed later in this chapter.

The figure shows only a single switched lead, but since most systems operate in a balanced mode, there will be a separate lead for each of the transmission paths. Hence a two-wire matrix will have $2 \times m \times n$ crosspoints for each such stage, plus any gates necessary for control needs.

A typical symbol for such a stage is shown in Figure 5.4b.

5.3.4. Switching Network

A switching network consists of switching stages and their interconnections. The principle of link trunking in a three-stage switching network is shown in

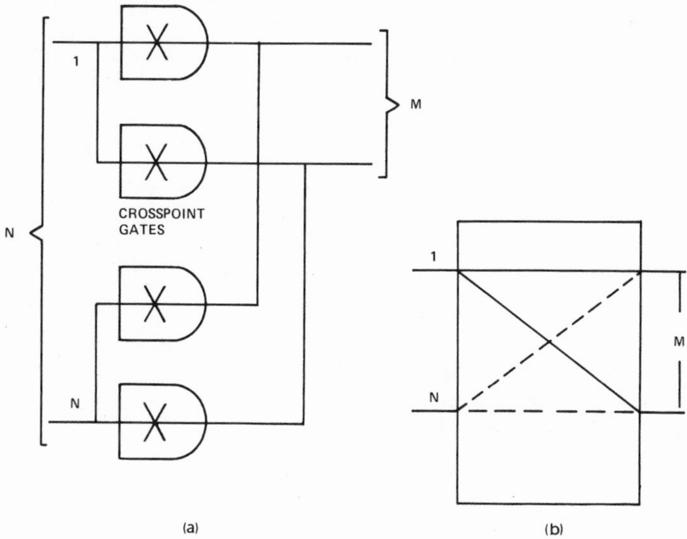

Figure 5.4. Single stage of space division switching. (a) Connection; (b) symbol.

Figure 5.5. The inlets are divided into groups. The size of the group depends upon the mechanical packaging of the crosspoints. It represents a compromise between the optimum size for the traffic offered and the need for providing modular growth.

Modular growth is necessary to dimension systems of varying initial and ultimate sizes. Smaller packaging modules will have a greater utilization of inlets and outlets than larger modules. They will also leave less space vacant on the bays on which they mount. This reduces the number of bays which have to be provided initally. Hence, if the system modules are small the traffic-carrying units will increase proportionately to the expansion of the system. This art of packaging has an enormous impact upon the ability of a system to be cost effective over a range of system sizes.

The primary stage interconnects to a number of secondary stages. Each outlet of a primary group (of which there are *m* for each of the *r* groups) connects to a different secondary group. Hence there are *m* secondary groups each with *r* inlets.

In the example there are *s* such primary/secondary linked groups. These are connected to a tertiary stage. Each of the outlets of a primary secondary group, of which there are *mn,* connects to a separate tertiary group. In the example each tertiary stage has *r* outlets. Hence the network has *snr* inlets and *smr* outlets.

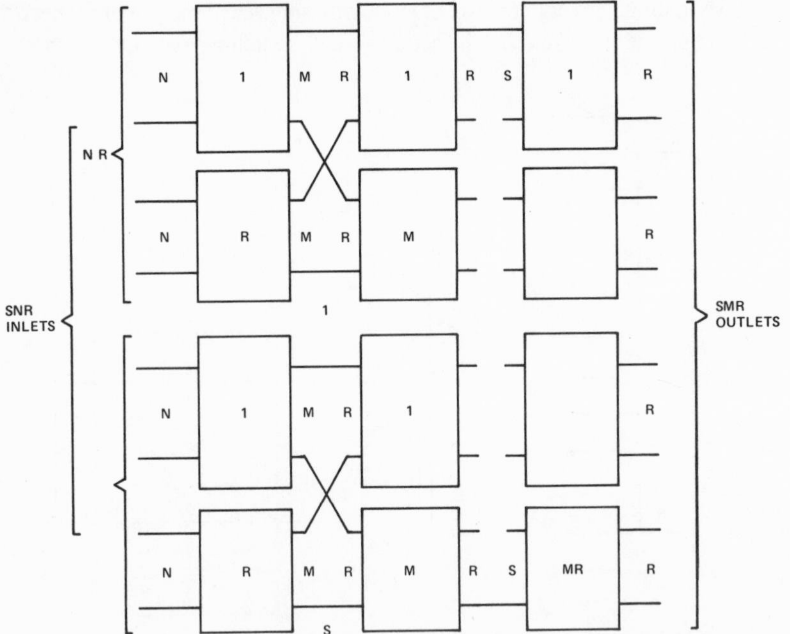

Figure 5.5. Principle of link trunking, space division switching matrix.

5.3.4.1. Number of Primary Outlets

The number of outlets provided on the primary stage provides the number of possible paths (the availability) which can lead to any outlet. Hence the number of outlets is a function of the traffic to be carried. The number can be lower than the number of inlets (concentration) or higher (expansion).

Primary concentration. Concentration is used when the inlets serve sources with a low calling rate (low traffic). This applies to inlets serving customer lines which generate few calls per hour. A typical concentration ratio is ten inlets and five outlets. This is based on a total traffic of 1.5 Erlangs with a loss of one call in 100 due to no path being available to a free outlet.

Traffic unit. The international unit of traffic is the Erlang.[21] This expresses the intensity of traffic as the sum of the product of the number of calls per hour and their duration. It corresponds to the continuous occupancy of a traffic-carrying unit of a switching system for 1 hr. The number of calls per hour is measured during the busiest hour of the day.

Traffic per primary group. The telephone calling habits of customers vary as a result of many factors, some social, some economic. Hence history applicable to a given switching system location is used to estimate the anticipated traffic for new switching system applications. It has been found for some suburban area, operating on a flat-rate call-charging basis, that 0.15 Erlangs per line is a good basis for estimating traffic. This is based on assuming that the average holding time of all calls is 180 sec. Now 180/3600 hr multiplied by the number of calls was used to estimate the traffic of 0.15 Erlangs per line. This is three calls per line per busy hour on the average. These calls comprise both calls from the line, *originating traffic,* and to the line, *terminating traffic.* The ratio of originating and terminating calls also varies according to the habits of a particular group of customers in a particular geographical location.

Application of traffic theory. Telecommunications switching systems are engineered to allow some calls to fail during the busy hour because of a lack of traffic-carrying units. This is done for economical reasons and is called the grade of service. The calculations necessary to meet a grade of service are made according to traffic theory. This calculates the probability of the occurrence of calls requiring traffic-carrying devices at the same time. For example in the case of ten customers served by the same primary group there is the possibility that 30 calls will be made during the busy hours. These calls occur randomly. To be sure that no calls were ever lost at least ten traffic-carrying devices (outlets) would have to be provided. However, traffic theory shows that five, not ten, outlets are sufficient. This is treated in more detail later in this chapter.

Primary expansion. Expansion is used when the traffic per inlet is higher than the traffic that a path through the network can carry without substantial loss. It can be shown that the possibility of reaching a given outlet from a given inlet is reduced as the traffic per path through the network increases. This results from a lower number of possible paths available in the system because of the larger number of calls occupying more of the paths at the same time. The ratio of the number of outlets to the number of inlets is adjusted to allow the desired loading of the transmission path links. When the network is used in this way a similar network is connected in tandem with it, outlet to outlet. This results in *snr* inlets having access to *snr* outlets via *smr* paths at the center of the network.

Fan-type network. A fan-type network provides a single path from a given inlet to a given outlet. Figure 5.6 shows the arrangement. This shows a two-stage network which has four inlets and nine outlets. The primary stage has four inlets each having a path to a different secondary group. Each secondary stage group has three outlets. The three primary stage crosspoints serving an inlet are connected to crosspoints serving the other inlets. The outlets may be multiplied (connected) to other primary groups or to each other depending on the use of the network. Multiplying to other groups increases the traffic offered. Connecting together outlets reduces the number of paths from the secondary stage but improves their availability. That is, it offers alternate paths to an outlet from any of the inlets. A fan-type network can have one, two, three or more stages.

Mesh-type network. A mesh-type network adds a stage to a fan network which increases the number of paths from a given inlet to a given outlet. Figure 5.7 shows the arrangement. The fan-type network shown in Figure 5.6 has a

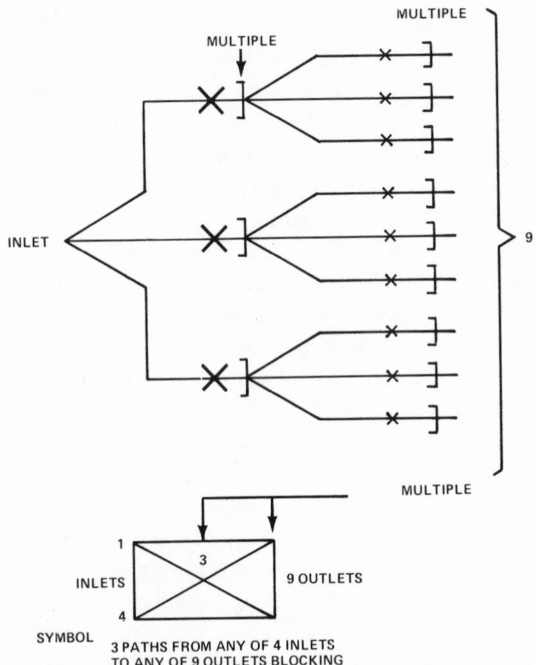

Figure 5.6. Fan-type network.

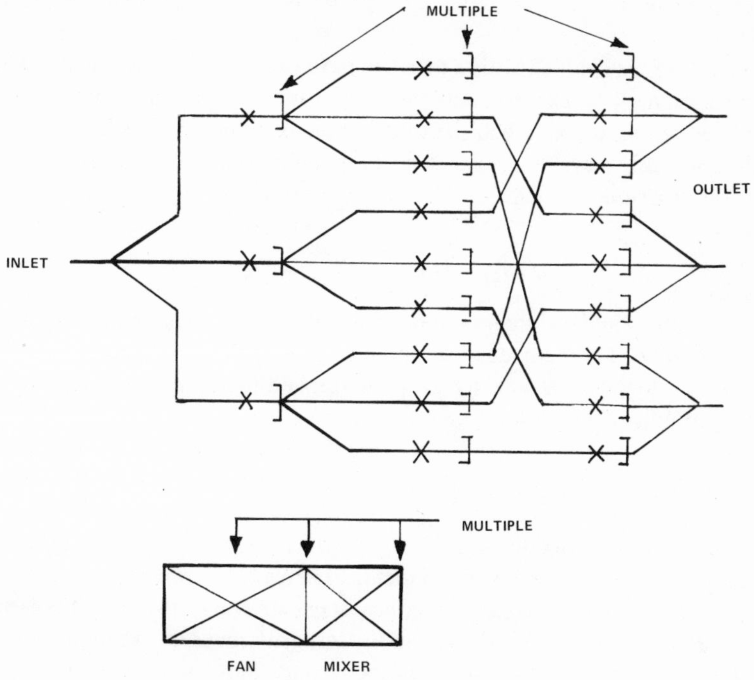

Figure 5.7. Mesh-type network.

tertiary stage added to it. This stage has three inlets each of which connect to a different secondary group. Thus there are three paths from any inlet to any outlet. A mesh-type network can have two or more stages.

Application of fan and mesh networks. In general fan-type networks are used for the concentration function. This is because concentrators connect lines to a small number of groups of traffic-carrying devices. (Usually two groups are provided, one for originating traffic and one for terminating traffic devices.) The possibility of losing calls due to there being no path from an inlet to a free outlet can be compensated for by increasing the number of traffic devices or reducing the traffic offered.

Mesh-type networks are used for selection stages where the outlet traffic is spread over many groups of traffic-carrying devices. (Usually there are a number of groups each serving different traffic routes.)

Availability. This is the number of outlets in a group that can be reached from a given inlet in a switching stage or network.

Full availability. This is the availability that is equal to the number of outlets in the desired group.

Limited availability. This occurs when the availability is less than the number of outlets in the desired group.

Accessibility. This is the ability of a given inlet to reach an available outlet.

Grade of service. This is a measure of the quality of service supplied by an administration. It is expressed as the proportion of total calls that cannot be completed within a prescribed time during the busy hour. (It is usually uneconomical to design systems so that no busy hour calls are lost because of the unavailability of equipment.)

5.3.4.2. *Number of Secondary and Tertiary Outlets*

The outlets in these stages are provided in order to increase the accessibility of the network. The number of accessible paths is directly proportional to the number of outlets per secondary group multiplied by the number of outlets per tertiary group.

5.3.4.3. *Allocation of Tertiary Outlets*

As described in Section 5.3.4.1, when the network is of the expansion type it is connected back to back with an identical network. However, when it is used as a concentrator the outlets in the tertiary stage are connected to traffic-carrying devices. These are distributed over all the tertiary groups as evenly as possible. Hence the ratio of inlets to outlets on the tertiary stage, which should never be less than unity, depends on the traffic. Some outlets will be allocated to originating and some to terminating traffic-carrying devices. The number of outlets provided must be sufficient to allow for the connection of the maximum number of devices that the network can serve. This will result in spare outlets in many applications.

5.3.5. Switching Matrices

Switching matrices can consist of two basic arrangements. They may be of the type shown in Figure 5.8, where terminals connect via a concentrator network to the selector network which provides access to outgoing trunks. Incoming trunks use inputs to the selector and terminating traffic-carrying devices use outputs to the selector.

The other type of matrix is shown in Figure 5.9. Both trunks and terminals connect to the input of the concentrator. Concentrators used in this way have differing possible ratios for the primary stage and a primary group may serve less than the maximum number of inputs. This depends on the amount of traffic which such a primary group can be offered and the allocation of lines and trunks to it. The outlets of the selector network are connected to other outlets.

Variants. There are other ways of forming switching matrices but they are variants of the above. For example, a two-sided network may provide connections between concentrators so that local-originated local-terminated traffic does

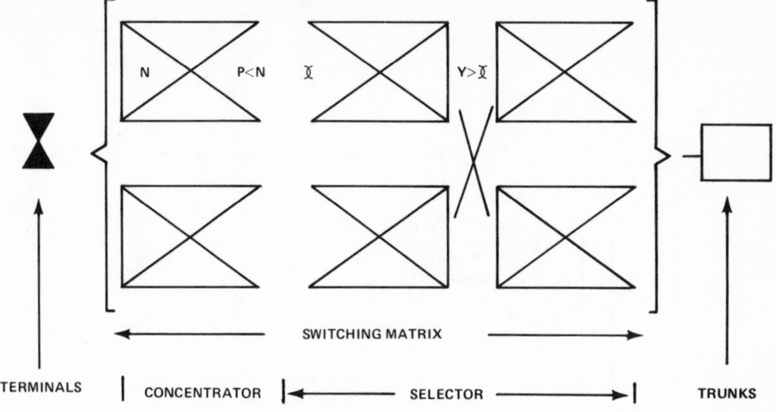

Figure 5.8. Two-sided switching matrix.

not traverse the selector networks. There may be more or fewer stages in the networks than the examples. These arise from supposed or real cost effectiveness or from the crosspoint technique.

5.3.6. Switching System Architecture

Switching systems have three major control categories: direct control, indirect control, and revertive control. In the past the type of control used was based on electromechanical switches and relay circuitry. Present systems based on reed relay or electronic crosspoints use indirect control. New systems use stored-program indirect control. However, in categorizing switching systems the means of embodiment of the system is irrelevant.

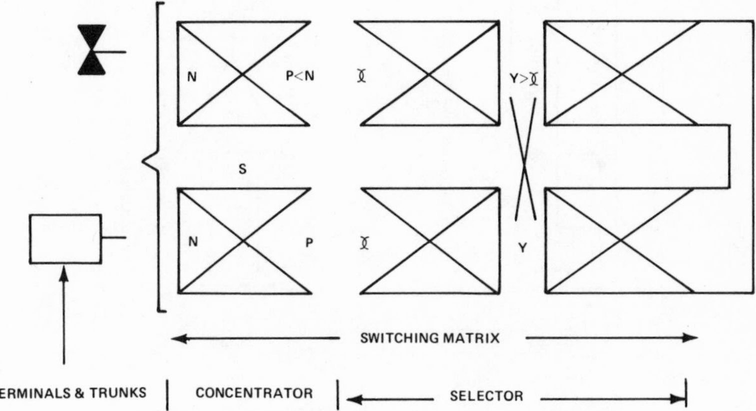

Figure 5.9. One-sided switching matrix.

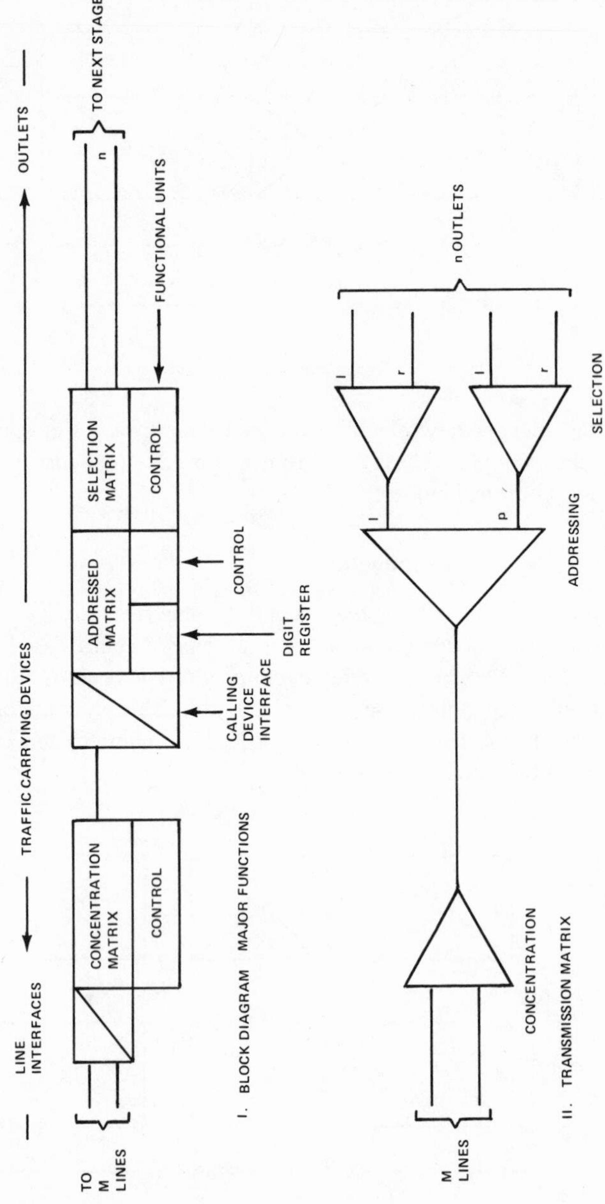

Figure 5.10. Principle of direct control switching system.

5.3.6.1. Direct Control

Direct control systems use selectors which are set by rotary dials. The selectors employ ratchet and pawl setting devices and the selectors are coordinate in action. Their principle of operation is described in a number of publications.[22]

Figure 5.10 shows the principle of operation. The line interfaces connect to a concentration stage. This connects the calling device directly to the selection stage; usually there is one selection stage per decimal digit in the address apart from the final stage, which handles the last two digits. The first selector returns dial tone.

The calling party then commences to dial. The first digit dialed moves the selector to a group of outlets which connect to second selectors. The first selector hunts automatically over the group of outlets to the second selectors serving it. The hunting action takes place in the interval between dialed digits. The succeeding digits set other selectors or are sent via a trunk interface to other exchanges depending on the destination of the call. If the call terminates locally then the last

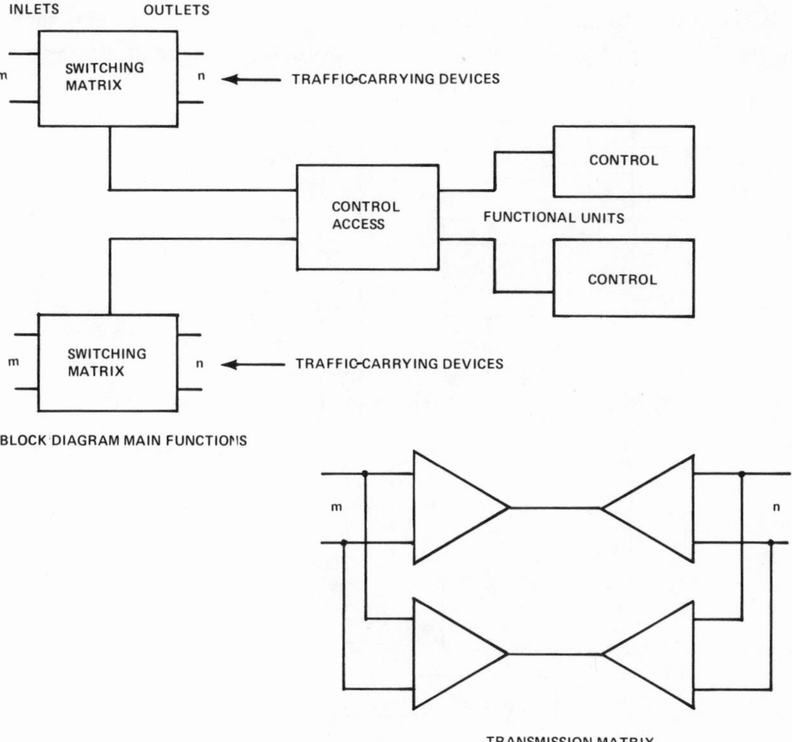

Figure 5.11. Principle of indirect control switching system.

two digits set the selector on the line interface of the called line. The final selector tests this line and either applies ringing and returns ring tone or returns busy or number unobtainable tone. Special types of final selectors are used for multiparty lines and groups of PBX lines.

Systems of this type occur all over the world; many have been converted to indirect control by the addition of registers and translators.

5.3.6.2. Indirect Control

In the indirect control operating mode, which is shown in Figure 5.11, the calling device is extended to a register and the switching matrix set under the control of the register. This is the mode of operation described in the switching systems described in the text. It has been applied to all types of crosspoints including electromechanical. In addition to the control method described in the text, which is a distributed control (Figure 5.12), it can use traffic divided controls.

Traffic divided controls. Systems organized on the basis of traffic divided controls have controls, any of which can accomplish the same functions, provided according to the size of the system. The principle of operation is shown in Figure 5.13. An executive, duplicated, control allocates a control to a call.

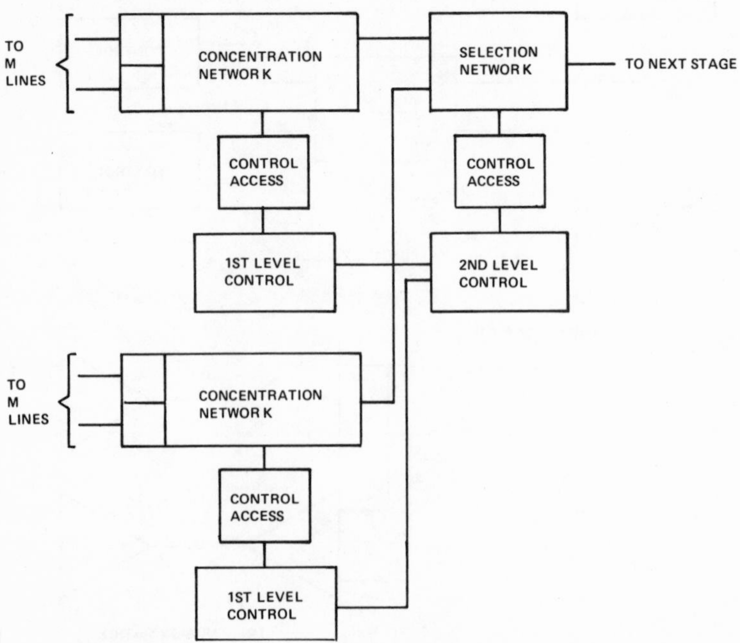

Figure 5.12. Principle of distributed control switching system.

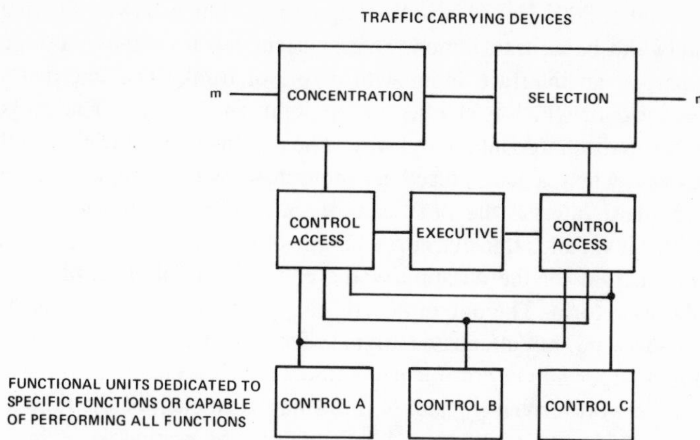

Figure 5.13. Principle of traffic divided control.

5.3.6.3. Revertive Control

The principle of operation of revertive control is shown in Figure 5.14. Revertive control was used to control certain types of electromechanical selectors, which were not usually capable of direct control.

The selectors were designed to move as a result of a signal extended from the register/sender. The selector moved in a series of steps sending a signal back to the register/sender after each step. The register/sender removed the signal when the selector had taken the relevant number of steps to correspond to the address stored in the register. Selectors designed to operate in this mode could be, and often were, based on nondecimal coordinates.

5.4. Time Division Multiplexing

Time division multiplexing is a transmission technique in which intervals of time are allocated on a recurring basis to telephone calls. Since the path is allocated to a number of calls in an interleaved pattern, only samples of the information can be sent. This information can be sent by pulse amplitude modulation or pulse code modulation.

5.4.1. Pulse Amplitude Modulation

Pulse amplitude modulation is a method of coding analog signals which are continuously-varying-level information signals. This method uses samples of the waveform being transmitted. The principle of operation is shown in Figure 5.15.

In the example a cyclic channel selector cycles through 32 steps continuously at a speed of 8000 Hz. This is the Nyquist rate, which is twice the frequency of the bandwidth being transmitted. During each step it enables the input gate which is part of an interface input with a line or trunk. The interface circuit includes a capacitor which is charged by the input voice signal. The gate causes this capacitor to discharge into a highway. The resultant signal travels to the end of the highway, where a gate, pulsed in synchronism with the input gate, extracts it. After a guard interval the next pair of gates is actuated and information relating to the next call is transferred. A sample of the waveform is transferred, at a recurring interval for the duration of the call. The amplitudes of the signals indicate the waveform. The output signals can be applied to a low-pass filter to recover an adequate, but attenuated, signal. The 32 intervals, *time slots,* can be allocated to a larger number of inlets and outlets. The arrangement consists of a distribution network driven by address information in each of the time slots.

A time slot is associated with an inlet and outlet by writing the address of the relevant input and output interfaces into the appropriate positions of the counter. This was the basis for the transmission matrices of early electronic switching

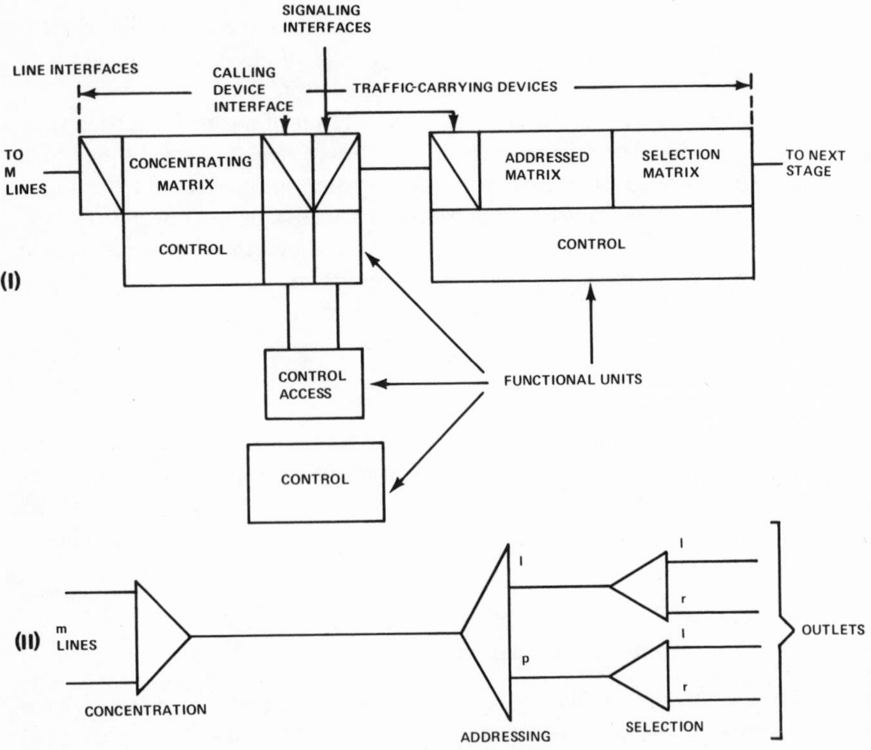

Figure 5.14. Principle of revertive control. (I) Block diagram; (II) transmission matrix.

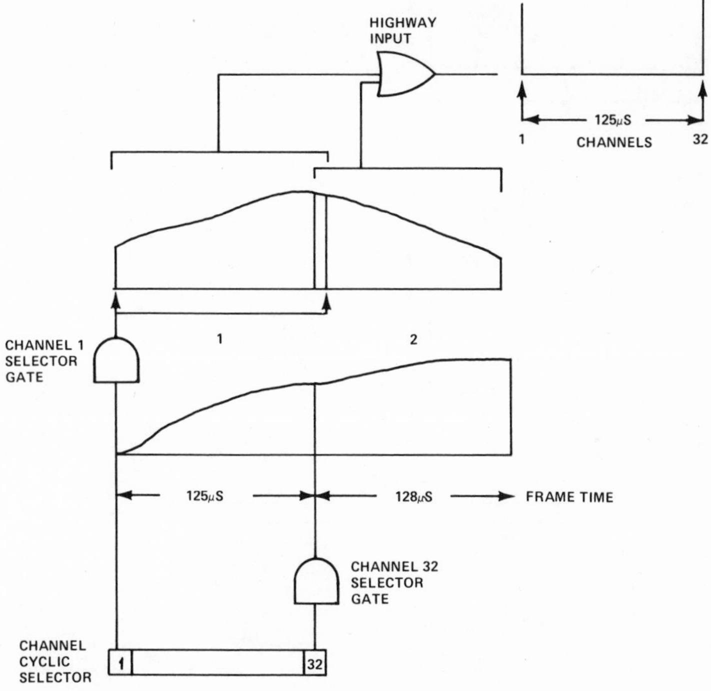

Figure 5.15. Principle of pulse amplitude modulation.

systems but is now only used for private branch exchanges. However, the principle of pulse amplitude multiplex coding is a necessary step in pulse code modulation. Practical switching systems generally used higher sampling speeds than 8 kHz. This is to simplify the filter design and hence reduce its cost. A filter has a characteristic impedance curve which is lowest over part of its bandwidth but rapidly increases as the cutoff frequency is reached.

The objective for cost effectiveness is to use only a portion of the flat part of the curve for the transmission bandwidth. There will be a continuation of the flat portion of the curve which is not used. This allows wider tolerances on the design of the filter, which makes it easier to manufacture.

5.4.2. Pulse Code Modulation

Figure 5.16 shows the principle of pulse code modulation coding. The pulse amplitude samples are compared with a series of reference values; eight values is typical. Each sample will either be of greater or lower amplitude than the reference value. Reference values are arranged according to coding laws. Hence an amplitude sample will be coded as a binary series, each sample defining in

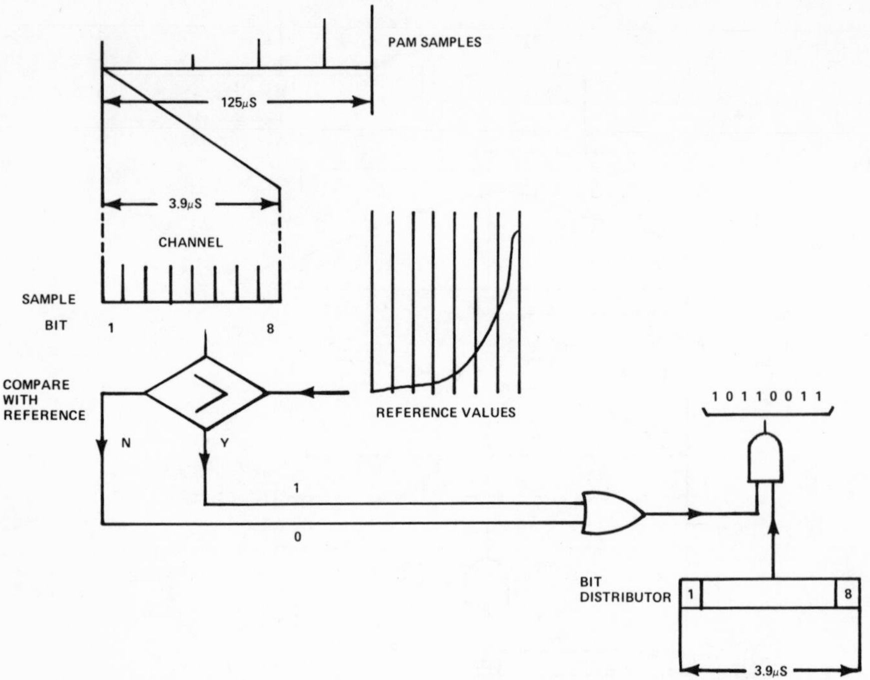

Figure 5.16. Principle of PCM encoding.

greater detail the amplitude of the sample. The sampling has to occur at a fast enough speed for the coding to be completed during the pulse code amplitude sampling time. (With a sampling rate of 8 kHz and 8 quantizing levels the speed will be 64 kHz.)

5.4.2.1. Advantage of Pulse Code Modulation

A bit signal used in pulse code modulation has only two states. One of the states is used to indicate binary 1 and one binary 0. These represent a signal having an amplitude of greater than a reference and less than a reference level. Hence the state can easily be detected and regenerated. The need for precise amplitude preservation, which is necessary in pulse amplitude modulation, is avoided. However, this is at the expense of much greater bandwidth and more elaborate logic functions and control circuitry.

5.4.2.2. Bit Stream Format

The highway carrying pulse code signals carries a series of time slots each corresponding to a transmission channel. The eight bits of information used for coding signals can be in one of two basic formats.

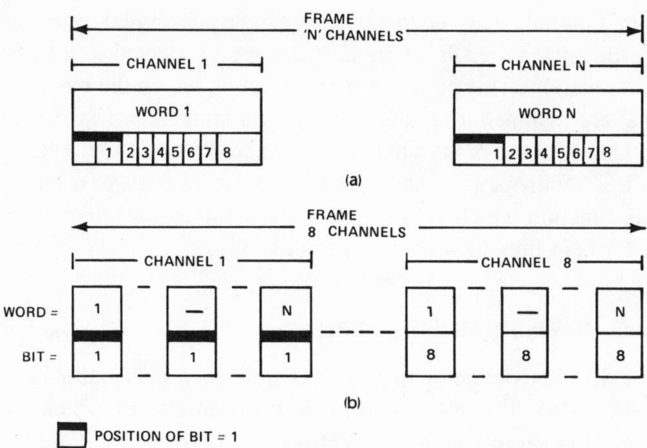

Figure 5.17. PCM frame format. (a) Word switching, (b) bit switching.

The first is word switching, in which all eight bits occur in a time slot. Figure 5.17a shows the principle: *n* channels occur in sequence in a frame; each channel corresponds to a separate telephone transmission. The second method is bit switching. In this method all bits having the same value occur in the same time slot. In this case there are eight time slots each divided into *n* bits. Figure 5.17b shows the principle.

5.4.2.3. *Principle of Pulse Code Modulation Transmission*

The sequence of functions is shown in Figure 5.18.

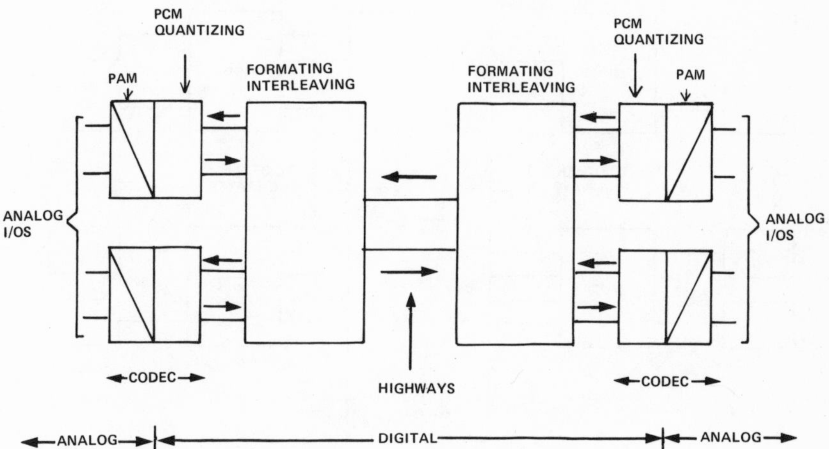

Figure 5.18. Principle of PCM switching system.

Analog input devices connect to an analog-to-digital interface called a *codec*. The quantized signals in digital form are connected to a formatting and interleaving function. This produces a bit stream in which the bits corresponding to a channel are assigned to a time slot. The bit streams are carried by a pair of highways. One of these is the send highway and the other the receive highway, hence this is a "four-wire" system. The highway connects to a formulating and interleaving function which extracts the signals and applies them to the relevant codec stage where they are converted to analog.

5.4.3. Time Division Switching Matrix

A time division switching matrix interconnects inputs and outputs via highways (Figure 5.19). It consists of line or trunk interfaces which connect to a multiplexer. This serves the two directions of transmission: The sending side encodes signals into time slots and applies them to highways. The receiving side extracts time slots from the highway, decodes them, and applies the resultant analog signal to the interfaces. It can operate in either the pulse amplitude modulation or pulse code modulation modes.

5.4.3.1. Pulse Amplitude Modulation Switching

The pulse amplitude modulation switching arrangement needs individual inlets and outlets for each interface. Regeneration of signals is not practical and a space division matrix is used to connect the highways together. Time slot inter-

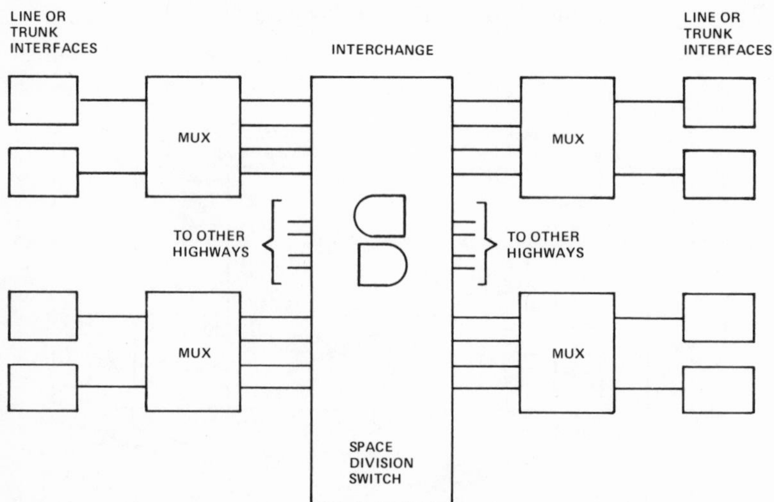

Figure 5.19. Principle of TDM switching matrix.

change is not practical and the inlet and outlet gates have to pulse in a fixed time relationship which considers the propagation time of the signal through the matrix. The number of time slots per highway is limited to a maximum of 32. This is necessary to provide a long enough interval (guard time) between signals to discharge the highway to minimize cross talk (overhearing). A low-impedance path to ground is connected to the highway during the guard time to minimize the residual charge. Such residual charges will distort the next signal.

5.4.3.2. Pulse Code Modulation Switching

The pulse code modulation switching arrangement may be used either to work with individual analog line and trunk interfaces or to connect directly to digital carrier systems. The highways are connected together by a time slot interchange. This connects channels having the same or differing time slots in the two highways. This is accomplished by temporarily storing the information in one frame and retransmitting it in another frame. The interval between storing and retransmitting is used to change the time slots used in the first highway to those used in the second. This is necessary in cases where the time slot position allocated in the first highway is in use on another call in the second highway.

5.4.3.3. Digital Transmission Matrix

The digital transmission matrix is a "four-wire" matrix. Figure 5.20 shows the principle of a digital switching matrix. The first stage concentrates m inputs to p time slots provided by the highway. The second stage concentrates r highways into a superhighway with pr time slots at a proportionately higher bit rate. The selection stage converts back into individual highways which in turn connect to trunk interfaces or the line concentration stage. Two time slots are used per call, one in the send highway and one in the receive highway.

5.4.4. Digital Switching System Architecture

Figure 5.21 shows the basic trunking arrangement for a digital switching system. Line interfaces connect to a concentration stage. This can be time division matrix or a space division matrix. The matrix connects to the first-level control via a control access such as a scanner. First-level controls can associate with the second-level controls to set up the connection and provide other functions. The highways are multiplexed onto superhighways. The superhighways are interconnected via a time slot interchange. This controls a space division network with electronic crosspoints pulsed in synchronism with the superhighways.

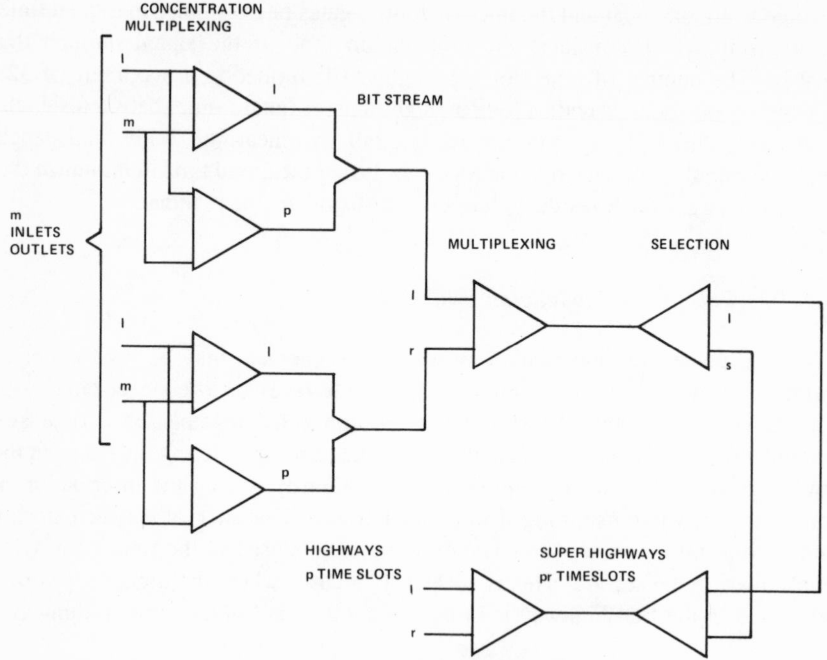

Figure 5.20. Digital transmission matrix.

5.5. Transmission through Switching Networks

Switching matrices are used to carry voice and data information and hence have to switch the information with minimal degradation and reduction in volume. This means that they have to conform to rigid transmission specifications. The interfaces to the switching centers need to provide transmission functions. These range from supplying power to telephones to providing impedance matching and in the case of a combined local–trunk exchange a means of equating losses on trunk-to-trunk and trunk-to-line calls.

This is implemented by the provision of switchable transmission pads which insert a loss to replace the normal trunk-connecting trunk loss. The transmission loss is obtained either by inserting an attenuator network in series with the analog speech path or it may be introduced digitally in a pulse code modulation system. In the latter case the time slot encoding is modified to indicate an appropriate reduction in amplitude for each of the samples.

Transmission losses in a system are described by relating to reference values.

5.5.1. Allocation of Switching Losses

The majority of switching systems in present-day operation use a metallic speech path. The networks of which they are part include both two-wire and

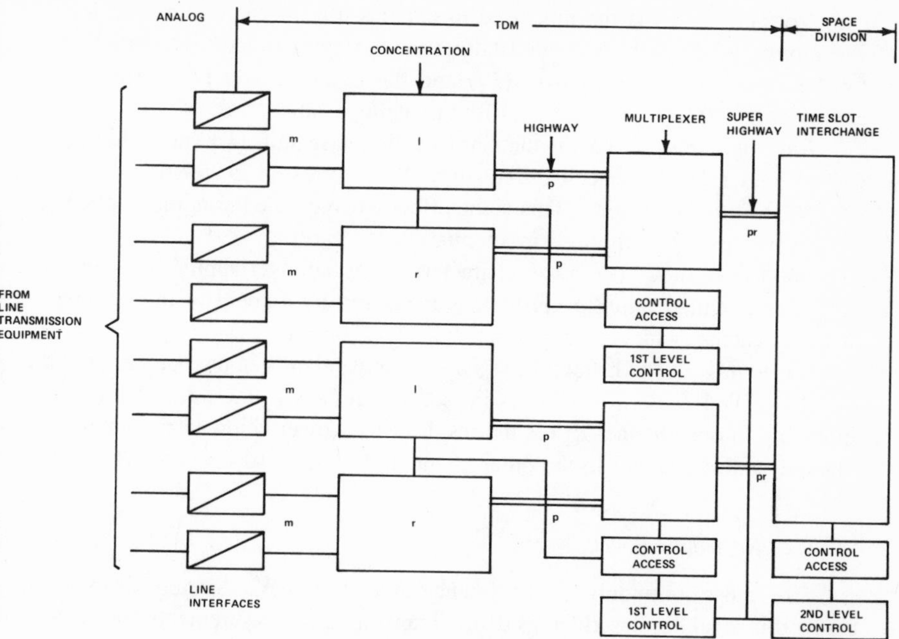

Figure 5.21. Basic trunking arrangement digital switching system.

four-wire telecommunications exchanges. Transmission losses have been allocated to each part of the network by the telephone administration. The CCITT transmission recommendations[23] reflect this situation and hence the parameters which relate to all types of switching systems are based on an analog network.

The possibility of treating the allocation of losses on a system basis exists with an all-digital network. CCITT study group D has been studying this since 1968 but has not issued any recommendations at the time of going to press.

Local and trunk exchanges have to conform to recommendations for transmission when these are used in international calls. This means that when digital exchanges are used in analog environment, analog transmission parameters will prevail unless the telephone administration rules otherwise. In the case of integrated networks the losses can be reassigned by the local administration. They will decide what portion is assigned to each class of telephone exchange. The overall CCITT requirements will set the limits for the total losses in the connection.

5.5.1.1. Transmission Terms

The terms used in Section 5.5 are defined below.

Attenuation. This denotes a decrease in the magnitude of the signals. It is expressed as a ratio of input power and output power.

Decibel. This is the unit used to express transmission gains and losses. The number of decibels indicates 10 times the logarithm to the base 10 multiplied by the ratio of the input power (P_1) and the output power (P_2). The loss is expressed as 20 $\log_{10}$ times the current or voltage ratio.

Envelope delay. This is the slope of the phase shift (B) versus frequency (W) characteristic at a specific frequency. It is expressed as dB/dW.

Harmonic distortion. This is the effect whereby the harmonic frequencies are not transmitted with uniform attenuation.

Impulse noise. The noise characterized by nonoverlapping transient disturbances is called *impulse noise*. It is measured by a count of the number of times transients (which exceed a threshold level) occur.

Cross talk. The leaking of signals to paths other than the one carrying the signal is called *cross talk*. It is expressed in decibels as the ratio of the power appearing in one channel P_1 (as the result of the power being carried in another channel) and the power in the other channel, P_2, i.e., $10 \log_{10} P_1/P_2$.

5.5.1.2. Reference Levels

The most commonly used reference level is 1 mW. References above or below this level are described as dBm. Transmission measurements are made at 1000 Hz in North America or 800 Hz for CCITT specifications of switching system insertion loss.

A common expression for measuring noise expresses the result in *dBrn,* which is decibels referred to the noise reference (rn). Hence 20 dBrn is equal to −70 dBm when rn = −90 dBm at 1000 Hz.

C message weighting which covers a bandwidth of 300 to 3400 Hz is used in North America. This is expressed as dBrnC.

The CCITT standard is similar using the same frequency weighting.

5.5.2. Requirements for a Local Telephone Exchange

The following assumptions are made:

- The telephone exchange interfaces with two-wire line and trunk transmission equipment.
- It uses time division multiplexed four-wire internal switching in the PCM mode.

The degree to which the objectives are met can be determined by tests from main frame to main frame with the transmission equipment disconnected. The tests are made at the impedance levels indicated, over all ranges of loop current to be encountered. Appropriate test methods are given below.

The impedance level in 600 Ω.

The maximum subscriber loop resistance is 1900 Ω. This includes the telephone set and the transmission bridge.

The transmission bridge has two coils each having a resistance of 200 Ω, connected to a 48-V power source of low impedance. This completes the telephone loop.

The following parameters should obtain.

Insertion loss.

Line to line: 1.0 (maximum)
Line to trunk: 0.5 (average) Measured in decibels at 800 Hz
Trunk to trunk: 0.5 (maximum)

Variation of insertion loss.

Line to line: −0.5 dB
Line to trunk and trunk to trunk: −0.3 dB

Frequency response. Measured in decibels relative to loss at 1000 Hz.) Relative to 1000 Hz, "+" is more loss, "−" is less loss. There must be no insertion gain at any frequency. The test level for insertion loss and frequency response is −10 dBm.

Frequency (Hz):	60	180	300–3000	3400
Response:	20	0–3.0	−0.5–1.0	0–3.0

Envelope delay distortion. There is a need to establish envelope delay distortion parameters due to its effect on data transmission. No values are available at the time of going to press.

Harmonic distortion. (In decibels relative to level of fundamental at output with input at 200 Hz and 1000 Hz at a test level of −10 dBm.) Requirement: second and third harmonics greater than 43 dB below fundamental.

Linearity (at 1000 Hz).

Input level (dBm)	
+3 to −37	± 0.5
−37 to −50	± 1.0
Less than −50	± 3.0

Idle channel noise. Less than 23 dBrnO (C message).

Impulse noise. Less than five counts in 5 min above 59 dBrn (C message weighted) on no more than 50% of the test calls with busy hour traffic loading.

Longitudinal balance. In talking state using AT&T test method.[24]

Balance (dB)			
Frequency (Hz):	180	1000	3000
Balance (min):	61	55	53
Balance average:	66	60	58

Cross talk (coupling loss in dB). Should be greater than 75 dB for either line to line, line to trunk, or trunk to trunk call, within the frequency range of 200 to 3200 Hz.

Hybrid impedance matching (against 900 or 600 Ω + 2.16 uf termination). Trunk to line (on trunk side)—measured in terms of return loss in decibels:

- Echo return loss: greater than 27 dB.
- Singing return loss: greater than 20 dB.

Trunk to trunk—terminal balance application:

- Echo return loss: 27 dB.
- Singing return loss: 20 dB.

Through balance application:

- Echo return loss: 27 dB.
- Singing return loss: 20 dB.

The exchange may have to switch data. If it does then it must handle data rates up to 4800 bits per second with an error rate not to exceed 1×10^{-7} measured at the busiest hour. The input levels of the data can be expected to vary from −12 to −26 dBm.

5.5.3. Analog-to-Digital Conversion

In addition to the above requirements the analog-to-digital conversion process at the interface between the digital switching matrix and the outside analog plant requires that the signal-to-noise ratio be as follows. This is measured at 1020 Hz, C message weighted.

Input level (dBm):	0 to −30	−40	−45
Output signal/noise (dBm0 min):	33	27	22
Overload Point:	+3 dBm		

5.6. Packaging

The components, traffic-carrying devices, switching matrices, and functional units are assembled and interconnected together to form the parts of switching systems. The provision of a telephone exchange requires the supply of sufficient parts to meet the user's requirements. The methods by which the quantities are calculated depend on the way in which the equipment is packaged and the system architecture. The dimensioning of systems is best made by computer programs. The basis for packaging is as follows.

5.6.1. Components

Components include discrete components and integrated circuitry which are used to make up the functional units and other parts of the system. These represent the means of embodying wired logic circuits and of accessing and storing information. The rapid exploitation of integrated circuit techniques has resulted in low-cost circuits which provide, on a single chip, many functions.

These include scanning, storing, coding, decoding, etc. The bulk of the control circuitry can be accomplished by using commercially available integrated circuits. Interface circuits use a mix of integrated and discrete components. The latter consist of transformers, miniature relays, and special semiconductor devices as necessary.

5.6.2. Printed Circuit Boards

Miniature relays and integrated circuits and other discrete components are mounted on circuit boards. These usually consist of one or more layers of printed circuits. A single-sided printed circuit board is made of a conducting and a nonconducting side.

The nonconducting side is used to mount the components. The conducting side is etched so as to leave a series of conducting paths. The pattern of the paths interconnects the components to form a circuit. The circuit is formed by applying an acid resist to the conducting side according to the circuit pattern desired. When such a circuit has been formed the relevant components are inserted into the locations determined by the circuit. This is mostly done by machines. One end of the board is used as a connector. A number of connector points provide for external connections to the board. These points are usually plated with a precious metal, such as gold, to reduce the electrical resistance and reduce atmospheric contamination.

Printed circuit modules can use more than one printed circuit layer and double-sided boards connected by plated through holes or other means. Some of the layers can be used to provide grounding or power.

5.6.3. Cells

Printed circuit modules plug into cells; these contain the sockets for the printed wiring board connectors. The sockets are wired to provide connections between the printed circuit boards according to the function of the cell. External connections to the cell are made either by direct wiring or by sockets provided at the end of the cell for this purpose.

5.6.4. Bays, Frames, or Racks

The synonymous terms *bays, frames,* and *racks* refer to the next level of packaging. The cells are mounted on a bay usually being bolted on. The bay may contain such equipment as power, fuses, and tone distribution. It may also have local maintenance devices such as lamps, keys, access, jacks, etc. This is located in the middle of the bay in Figure 5.22, which shows a family of packaging modules. It can include provisions for directly connecting power, tones, etc. and sockets for some cables. The general basis for bay interconnection is to use plug-in cables between cells.

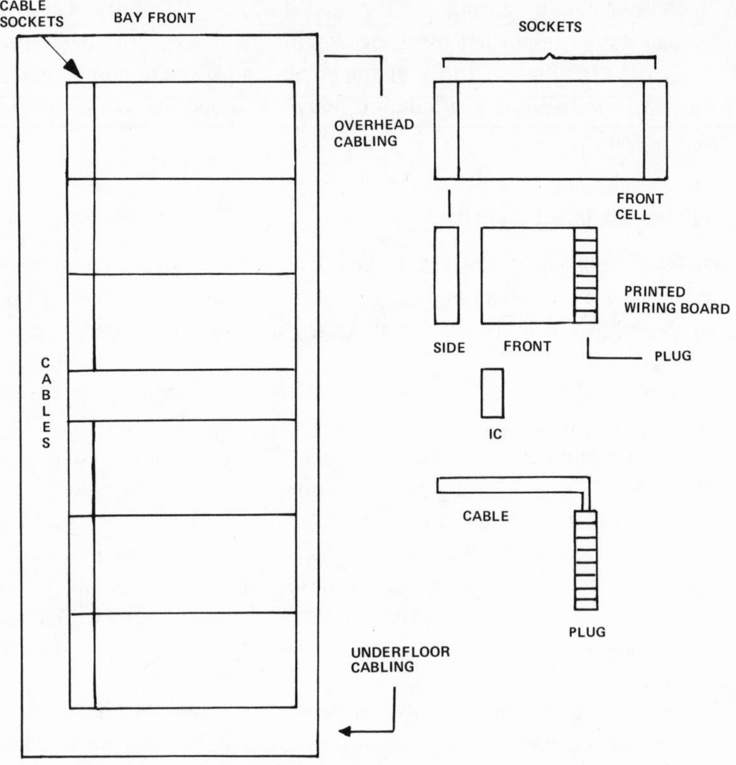

Figure 5.22. Family of packaging modules.

Dimensions of bays. Earlier systems used bays as high as 3.5 m with widths of 1.4 m. This required ceiling heights of 4.5 m in the equipment room. The miniaturization of components permits much higher density of packaging than was previously possible. This has resulted in bay heights of the order of 2.4 m. The bays should be designed either for under floor or overhead cabling.

5.6.5. Packaging Philosophy

Care should be taken in packaging equipment. Packaging must be considered when the earliest design studies are made. This is important for the following reasons.

5.6.5.1. Compatibility with System Techniques

Problems of distribution and association of components increase with the use of high speeds in the system operation. These can introduce noise or system malfunction and place severe limitations on lengths of cabling and packaging methods.

5.6.5.2. Heat Emission

Unless stringent limits are placed on the maximum heat dissipation per square foot of bay face, it will be necessary to provide forced-air cooling. This introduces an additional element in reliability since failure of the cooling system can result in failure and possible damage to the componentry.

5.6.5.3. Modularity

Switching systems use traffic-carrying devices and functional units. A switching system for a particular application will require a quantity of traffic-carrying devices depending on the number of lines and trunks needed and the traffic that it must carry. A switching system is planned to meet the long-term requirements of the user. This means that a switching system is only partially equipped at its installation. These factors lead to a need to so package the system that the initial installation does not contain too much unneeded equipment.

This is a difficult compromise; if it is not carried out properly the system will not be cost competitive. In this respect costs based on a so-called baseline system may fit only the average case, which of course never occurs.

5.7. Traffic Distribution in Switching Systems

Switching systems are designed to handle traffic. Traffic originated by customers and the administration is concentrated into suitable densities and distributed through the telecommunications network to various terminations. The traffic distribution in a local exchange can be shown in many ways. One method is shown in Figure 5.23.

This shows the source of customers originating traffic entering the switching system and the sinks of customers terminating traffic serving as traffic terminations. The originating traffic either terminates in the local exchange or is extended to other switching systems via trunks. Terminating traffic arrives either from local lines or trunks. Some traffic may be routed from incoming trunks to outgoing trunks. This is called *transit traffic*.

5.7.1. Originating Traffic

Originating traffic includes all call attempts which originate from customers' terminals. Some of these calls will be abandoned prior to a completion, some will be blocked at congested points, some will reach the destination and be

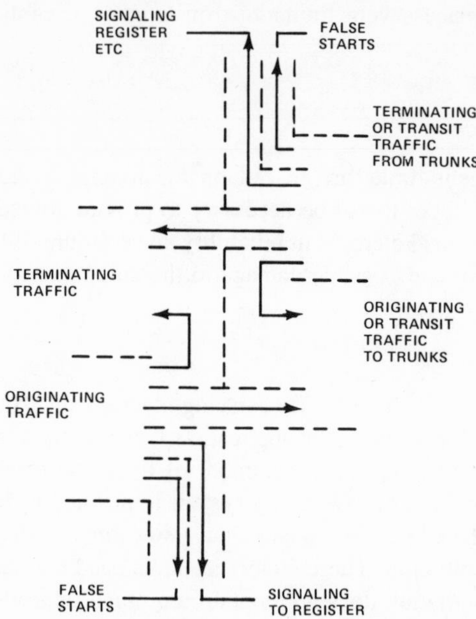

Figure 5.23. Traffic distribution, local exchange.

unanswered, but some will result in completed calls. Part of the time for which a call exists will be used to signal the called destination address by using a calling device. This is register traffic and is so shown in Figure 5.23. Some of the calls will not progress beyond the register stage. These are called *false starts*. Calls which are identified by their addresses are either routed to local terminations or to outgoing trunks.

5.7.2. Terminating Traffic

Terminating traffic reaches the terminals either as part of the local originated traffic or as traffic received from other exchanges via incoming trunks. Trunk traffic consists of traffic due to signaling and to a much lesser degree false starts.

5.7.3. Transit Traffic

Transit traffic does not affect local terminals since it is traffic intended for outgoing trunks to other exchanges from incoming trunks.

5.7.4. Traffic Measurement

Figure 5.24 shows traffic distribution without reference to the system organization.

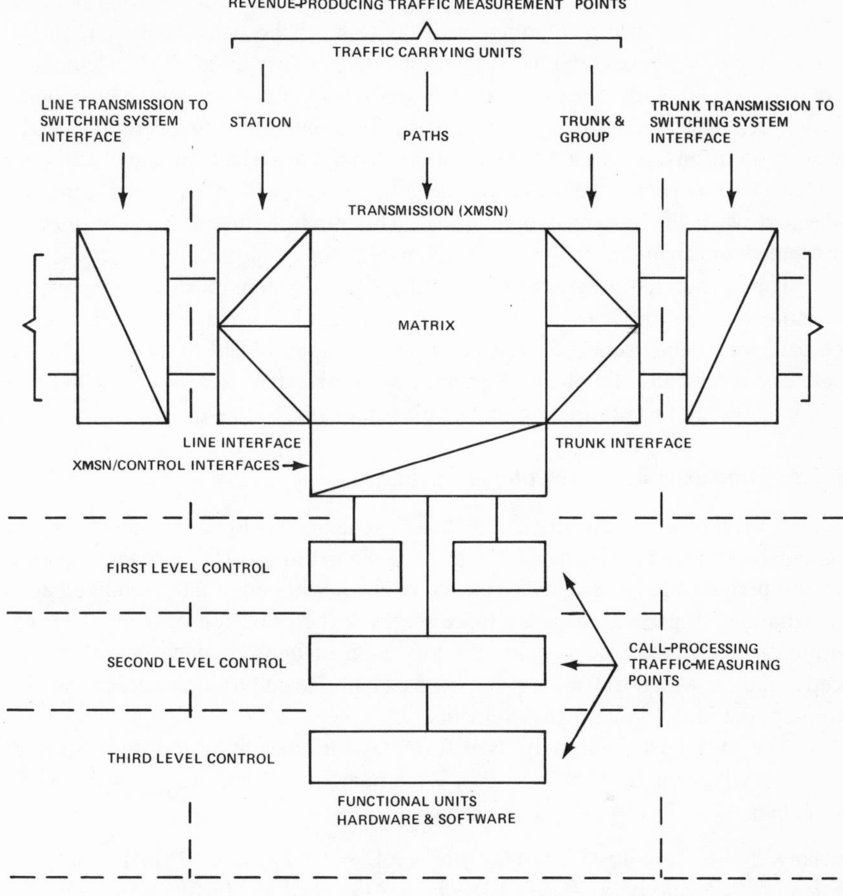

Figure 5.24. Traffic measurement areas.

The revenue to support a switching system is obtained by charging for telephone calls. Hence there must be a means of measuring the traffic through the system. Typical measuring points are at line interfaces (station traffic), switching matrix paths, and trunk matrices. In addition to this, measurements of traffic must be made in the control for checking how well the system is performing. Information obtained from the controls is used to check the efficiency of routing, etc.

Measurement of Traffic on Controls

Modern electronic systems have transferred many of their functions to common control units. This has resulted in a critical area of traffic loading on the control. This is especially true of stored-program control. This technique uses stored logic which involves the processing of many instructions to control a call. The effect of the traffic loading on a control can be calculated in terms of processing time per call and the number of calls per busy hour. Such calculations are based on allowing a permissible average delay per call during the busy hour. However, the calculation is best verified by simulating the operation of the system by means of running a program on a computer. The results will still only indicate the expected behavior of the system and will need to be carefully checked when the system is operational. This kind of support activity must be estimated and included in system development cost.

The operational program for a stored-program system must include a means of indicating the traffic in the first- and second-level controls. This traffic has to be paid for by the users and must be priced and apportioned to the cost of local calls and trunk calls. Trunk call costs may be recovered by agreements with other administrations for terminating and originating calls.

5.7.5. Dimensioning a Telephone Exchange

A local or trunk automatic telephone exchange has to be engineered to suit its location. This means that it has to be designed to meet its projected growth over a period of ten or twenty years and hence it will be partially equipped at its installation. In general there are three stages in the provision of equipment: the ultimate capacity of the system, the provision of unused space in frames and cells, and the actual traffic-carrying devices and line and trunk interfaces needed to meet the initial system requirements.

The two most commonly used formulas for calculating the probability of loss are (where the probability is P, n is the number of trunks, and A is the traffic in Erlangs), for n trunks and A Erlangs,

$$\text{Erlang } B \text{ (lost calls lost)} = (A^n/n!)/(1 + A + A^2/2! \cdots + A^n/n!)$$
$$\text{Erlang } C \text{ (lost calls held)} = (A^n/n!)(n/n - A)/[1 + A - A^2/2! + \cdots A^n/n!]$$

The B loss formula is used for such things as trunk quantities; the C formula is used for delay calculations as applied to originating registers, etc.

5.7.5.1. Ultimate Capacity

Modern telephone switching systems have two limits on the system size: the amount of traffic that the switching matrix can carry and the amount of traffic that the controls can carry without unacceptable losses. (This supposes no packaging constraints.)

Ultimate capacity due to traffic. The ultimate capacity of the system is limited by the amount of traffic that it can carry and the maximum number of line and trunk interfaces that it can provide. In the case of line connections a line group will consist of a number of line circuits. The line circuit traffic per unit will have parameters determined by the system design. The maximum number of line circuits allowed in the group is a matter of choice. Some systems offer a number of different packages to cater to different calling rates. The calling rates per line vary according to many factors, community of interest, calling rates, social habits, etc., and must be considered carefully when a new exchange is being designed, as well as when it is being engineered.

The traffic has two parts, the number of calls and their duration. Although systems are dimensioned using 100-sec intervals (CCS or UCS) in North America, they are dimensioned using erlangs elsewhere. Systems are always designed using erlangs. The cost per switched erlang is a commonly used means of comparing design efficiencies. One erlang equals 36 CCS.

There will be physical limits on the number of trunk interfaces that the system can accommodate and also on the traffic that can be carried by the serving trunk transmission channel.

The ultimate traffic capacity of the system limits the number of calls and their duration.

Ultimate capacity due to control. Controls operate on a delay basis with requests for processing queued or scanned at regular intervals. Usually level-1 controls operate on a scanned basis and interface with telephone customers. They serve to gather address information which is signaled at variable intervals. The central, level-2, processors are only involved when all the data necessary for the request have been received.

The delay on a processor can be calculated from the following formula, which gives the average delay on calls delayed:

$$\text{delay in seconds} = h/2(1 - a)$$

where a is the traffic on the processor and h is the holding time in seconds, that is the total time that the processor takes to handle a call (i.e., the sum of the total number of instructions required for all required functions divided by the number of instructions processed per second). This figure will vary for different types of calls and different services. Hence in applying dimensioning formulas to a system it is necessary to calculate and verify the processor holding time for each type of call. The application of the system to a particular location requires a breakdown of the calls into various types. This permits a realistic, as opposed to a worst-case, average for the installation.

The ultimate capacity of the control limits the number of calls that the system can handle. The duration of the call is irrelevant.

5.7.5.2. Assigning a Value to the Processor Traffic

Experience has shown that it is extremely difficult to estimate accurately how a system will behave with real traffic. Simulation of the entire switching system by making a mathematical model of the system is essential but it is a guide rather than a bench mark.

Calls and holding times can be obtained from random number table lists and the distribution and traffic can be adjusted to provide different behavior patterns.

The success of the simulation and its verification by calculation must be weighed against the fact that apart from the physical dimensions of the system networks and its logic organization all traffic data are assumed.

Hence the best approach for assigning values to the traffic that the processor will carry is to take a conservative approach. The initial loading should be that adopted for the initial applications of No. 2 ESS by Bell System engineers.[25] This allowed 5% for eventualities, then 30% for false traffic, calls abandoned before dialing, during dialing, etc., then 30% for peak traffic. This left 0.465 erlangs for call processing. However, this system did not use level-1 controls. Hence the value applies for a system using a level-2 control on a demand basis.

Level-1 controls normally operate on a scanning basis. The size of the control group depends on the maximum tolerable duration of the scan cycle and also on the way in which the equipment is packed (i.e., the number of interfaces that share the control).

The use of level-1 controls in a system reduces the load on the level-2 control by the resultant absorption of most of the false traffic in the first control level. Hence a level-2 control can be initially loaded to 0.95 minus 30% for peak traffic or 0.665 erlangs. However, this assumes a total processor holding time of less than 100 msec for the average call at the exchange being designed. This would result in a delay of $100/2(1 - 0.665) \times 10^3 = 150$ msec.

The average delay on all calls during the busy hour equals (150×0.665), which would be approximately 100 msec. This value is acceptable.

The traffic loading can be adjusted as the result of field experience on the systems.

5.7.5.3. Application to System

The switching network must be able to provide terminations for the lines and trunks needed for the ultimate size.

The switching network must provide sufficient traffic-carrying devices between lines and trunks to provide the grade of service listed in Table 10 when specified by the administration.

Table 10. Typical Grades of Service in a Switching Network

Type of connection	Grade of service required
Terminating calls (trunk to subscriber or trunk to trunk)	$P = 0.02$ (2 lost calls in 100)
Originating calls (subscriber to trunk)	$P = 0.01$ after dial tone is received
Intraoffice calls (subscriber to subscriber)	$P = 0.02$ after dial tone is received
Dial tone connections (subscriber to register)	Not more than 1.5% of busy hour calls delayed over 3 sec

The information for the system is obtained from a questionnaire which is completed by the administration. This defines all the requirements for the system. It lists the numbers and types of lines and the distribution of traffic. The first step is to break down the system into groups of lines and appropriate traffic blocks, according to the system packaging rules. This will establish the initial and ultimate number of frames. The second step is to determine the trunk groups and calculate the number of trunks in each group. Both the above calculations can be made by the use of full availability, traffic tables (using pure chance for lines and smooth traffic tables for other applications except overflow trunks). However, if the line group is ten lines or less special tables will be necessary for lines because the requirement for traffic equilibrium (which postulates infinite sources) on which the Erlang tables is based does not apply. When there are very small groups, the probability of a second call occuring in the group is lessened. This means that more traffic can be offered to and received from the lines. This fact is used to advantage in No. 1 ESS of the Bell System(26).

Sometimes a group of ports can serve either line or trunk interfaces. If this is the case then the total traffic that can be carried, assuming a traffic limited matrix of the blocking type, is a balance between lines and trunk traffic. Usually such arrangments are inefficient since ports are either wasted because of traffic limitations when used for trunks only or underoccupied because of the low traffic when occupied by lines. Hence line traffic should always be concentrated.

Careful consideration should be given in applying traffic data to the fact that busy hours on trunk groups may not coincide, as a result of traffic being too and from different time zones. The traffic basis used must consider only the busiest hour for switching network and control dimensioning.

Number of line bays. The system will be packaged so that there will be a line group, or groups, per bay. There will also be a level-1 control per line group, or groups, and a portion of the switching network. The equipment lists for the system will indicate the relationship between the number of line interfaces and the number of cells and bays. This information can be obtained manually or by means of a computer program. It will also indicate the power requirements. The ultimate quantity of bays will be used to derive the floor plan for line equipment.

Number of trunk bays. The number of trunk bays is determined by the number of trunk groups, the number of trunks in the group, the type of trunk and associated trunk transmission equipment, and its signaling mode. This information is used to derive the number of trunk bays and the associated switching and level-1 control equipment from traffic tables and equipment lists.

Categorizing trunk groups. The trunk groups may be divided into high usage and overflow groups. The high-usage trunk groups are dimensioned to permit a loss of traffic under busy hour loading conditions. This traffic is routed over alternate direct routes which are dimensioned for a low loss such as five calls in 1000. A theory was developed by Wilkinson of Bell Laboratories to apply to direct routes carrying random or rough (i.e., peaked traffic) traffic. This is called the "equivalent random theory" and can be used to compensate for the roughness of the traffic by indicating additional trunk quantities than those that would be provided for smooth traffic. This theory applies when the ratio of the variance of the overflow traffic to the mean of the overflow traffic (VMRo) is greater than unity.

The mean of the overflow traffic and the variance of the overflow traffic are obtained from traffic tables.

The mean of the overflow traffic (Mo) is obtained as follows: A_1 Erlangs of traffic is offered to the a trunk group having N trunks with a mean carried traffic per trunk of A_2 Erlangs equal to A_1 multiplied by the loss obtained from Erlang B tables for NA_2 Erlangs of traffic.

The variance of the overflow traffic (Vo) can be obtained from the following equation:

$$\text{Vo} = \text{Mo}\,[1 - \text{Mo} + A_1/(N + 1) \times (\text{Mo} - A_1)]$$

The equivalent random theory states that the A Erlangs of rough traffic offered to a group of N circuits can be considered to be the overflow traffic from an imaginary group of C trunks which was offered a Erlangs of random traffic. An equation to derive a value for a has been derived by the LM Ericsson engineer Y. Rapp.(27) This states that

$$a = \text{Vo} + 3 \times \text{VMRo}\,(\text{VMRo} - 1)$$

The value for C,

$$C = a(\text{Mo} + \text{VMRo})/[(\text{Mo} + \text{VMRo} - 1) - (\text{Mo} + 1)]$$

is derived from Erlang's theory.

The above applies to trunk groups with overflow or alternate routing trunk groups. If there is only a single trunk group to a destination then the group is dimensioned using random traffic Erlang B tables.

Number of registers. The data on which the number of registers is calculated must be based on locally obtained data. An application based on North American data is given in Section 5.10. The basic approach is to use the anticipated number of origination calls per line group served by a level-1 control. This number is multiplied by the average dialing time for calls; this will be a weighted average depending on the calling devices and call destinations of the group.

The resultant figure gives the traffic in erlangs for the register function in that group. It can be referred to full availability tables (Erlang C) based on lost calls held, using a grade of service of 1 in 1000 to determine the number of register functions, basically memory areas. Similar calculations are made for the register-signaling registers and senders.

The performance of the system should be monitored during its early period of use. This will usually result in the provision of registers at a greater loss than the 1 in 1000 provided at the initial installation.

Provision of common equipment. In addition to the provision of traffic-carrying units, which are proportional to the number of line and trunk interfaces and the traffic that they carry, common equipment has to be provided. This includes level-2 and level-3 controls with related memories and power and ringing and tone equipment. It also includes distribution frames. The number of frames depends on the method of packaging. The manual or computer lists for the system will indicate the number required. The main distribution frame is made up or a number of verticals providing 300 or so pairs per vertical. Hence the MDF can easily be dimensioned once the number of line and trunk interfaces have been calculated.

Provision of power system. Automatic telephone systems operate on a full floating battery mode. That is, a battery which is at least capable of providing the full busy hour load of the system for four hours is driven by battery chargers powered by the main's power. Hence it is necessary to calculate the power required for each of the bays. This requires calculating the transmission power which is provided at a nominal voltage of 48 V. This provides the necessary current to drive line and trunk transmission interfaces and station and line signaling. Additional voltage levels are normally necessary to drive the electronic circuitry. These voltages are best provided by direct current convertors provided on a modular per bay or cell basis. The direct current convertors (dc/dc) are driven from the transmission battery. Filtering is applied at the input to the battery from the chargers. It may also be necessary to filter power at low voltage distribution points, and at interfaces with transmission equipment. The power drain per circuit for traffic-carrying and functional units must be tabulated as part of the design process. The requirements for a particular application are obtained from manual or computer reference tables.

The power system should employ a single-point ground window.[28] Any ac power equipment such as test equipment must use isolating transformers.

Provision of progress tones and ringing. Digital switching systems can produce progress tones by using appropriate pulse-code-modulation (PCM) codes. Earlier switching systems needed separate tone generators. However, in either case it is necessary to generate and distribute ringing and, when required, metering pulses. The method of calculating the requirements is treated in Section 5.10.

Floor plan. When the calculations are complete the number of equipment frames, the number of distribution bays, and the power requirements, both initial and ultimate, will be known. Hence a floor plan can be drawn. Modern equipment usually employs bays of 2.4 m or less in height. This requires a minimum ceiling height of 3.2 m. The floor loading should be no greater than 732 kg per square meter, including the power plant. Aisles between rows of equipment frams should be of the order of 0.75 m wide. Large systems may use separate rooms for power equipment and main distribution frames.

The initial requirements for the system specify the number of completely and partially equipped bays needed. These are located in appropriate positions on the floor plan depending on the specific requirements and constraints of the system.

Forced air cooling. The system should be designed so that its ambient temperature can be maintained by commercial heating and cooling units provided for human comfort. However, if special means are necessary these must be included with the normal calculations for sizing the equipment. In addition the relevant auxilliary equipment must be provided, as needed, to meet the needs of the system.

5.7.6. Associating Traffic-Carrying Devices with Functional Units

It is necessary to provide a means of connecting line and trunk interfaces, and other traffic devices, to the switching matrix. This requires the use of distribution frames. These frames provide a means of connecting outside plant facilities, street cables, or carrier equipment, to the switching equipment. These also provide facilities for cross-connecting traffic devices with other portions of the switching equipment. The interconnection means have to be easily changeable in order to simplify rearrangements due to changes and additions in the terminals and trunks served.

In some older systems a delicate traffic balance is necessary, to achieve cost-effective traffic loading, to suit the switching matrix. Modern systems have reduced the amount of traffic balancing necessary by providing lower loss probabilities in lower-cost switching matrices. However, there is no way to avoid work due to new or discontinued customer lines and additions to trunk transmission facilities as the switching system grows.

5.8. Distribution Frames

Distribution frames interconnect line and trunk transmission equipment to the interfaces with the switching system main distribution frames and interconnect traffic-carrying devices and functional units with intermediate distribution frames. The first type is called a MDF and is still in extensive use in modern systems.

This frame has been miniaturized but still performs the basic function of cross-connecting street-order cable appearances with equipment-numbered order appearances. It also serves as an interception point to test the station equipment and provide protection for the switching system from lightning and other extraneous potentials. The use of frames (IDFs) for associating traffic-carrying devices with switching matrices is diminishing with modern systems. This is because time slot interchanges using miniature electronic devices replace the mass of wire interconnections in analog systems. Direct digital interfaces will replace the amount of interconnection with trunk transmission equipment.

5.8.1. Main Distribution Frames (MDF)

Main distribution frames connect lines and trunks to the switching matrix. Lines or trunks appear on one side of the frame and inputs to the interfaces on the other side. The design of the MDF must simplify work at the frame. Provision must be made for 2-, 4-, and 6-wire terminations from lines and trunks.

5.8.2. Intermediate Distribution Frames (IDF)

The intermediate distribution frames serves to distribute trunk and other traffic-carrying devices to the inputs of the switching matrix. It should use plug-ended cabling from the frame to the cells on the equipment bays.

It may be combined with the MDF or be a separate entity.

5.8.3. Combined Distribution Frames (CDF)

The combined distribution frame is a basic frame assembly which is arranged for MDF/IDF facilities, which is used for small local exchanges generally of 1000 to 1500 protector pairs size.

5.8.4. Protection

The MDF provides a means of protecting the switching system from external power and lightning as well as a means of access for testing the external plant. It includes on the line side, on a per line basis, a test jack which isolates the line and the equipment and the protection equipment such as carbon block or gas tubes which provide low-impedance discharge paths for lightning strikes, line fuses to protect against high currents, and sometimes heat coils which protect against permanent low, but dangerous, current in the loop. Electronic systems are more susceptible to damage than earlier electromechanical systems and care must be taken to match the protection to the system.

5.8.5. Grounding

MDF grounding requirements are critical to electronic systems. Grounding connections must be made to the ground window and between protector blocks they should be as short and direct as feasible. There can be no loops and differences in potentials must be avoided.

5.8.6. Mechanical Design of Frames

The outside dimensions of the frame are usually no greater than $2.4 \times 0.568 \times 1.022$ m, in order to conform with present miniature packing techniques.

The frame is designed so that it can be fastened to the floor so that it can be electrically isolated from the surface on which it stands. Frames are usually designed to be fastened together in rows. Cross connection between the two sides of the frames is by means of jumper wires and jumper rings to facilitate this.

5.9. Redundancy

Modern electronic systems allocate level-1 and level-2 to a telephone call as necessary to process a call. A control may not be available at a given time, either because it is in use, or because it is faulty, and hence has been taken out of service. Nonavailability of a control because of its temporary use on another call is allowed on a basis limited by the grade of service offered. Nonavailability of a control because it is faulty is compensated for by the provision of redundant or multiple equipment.

5.9.1. Provision of Redundant Equipment

Figure 5.25 shows the principle of provision of redundent equipment. The controlled equipment can be connected to one of two control functions via a changeover device "c/o." Both of the control functions are supplied with all the

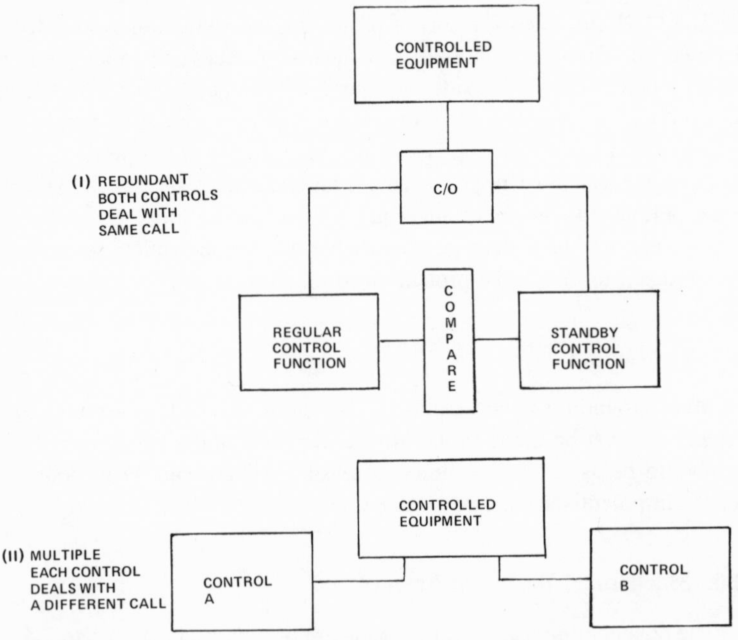

Figure 5.25. Switching system redundancy principles.

information relating to the processing of a call. Only one of the controls is controlling the equipment. The output of the regular and standby controls are compared in a monitoring function, "compare." If this function indicates discrepancies in the output the faulty control is removed from service and the other substituted.

5.9.2. Multiple Controls

With multiple controls the traffic load is divided among a group of controls. Each of these controls can perform the same function. If one of the controls becomes faulty the traffic-handling capacity of the system is correspondingly reduced. This reduction can be compensated for by providing more controls than necessary to handle the traffic.

5.10. Equipment Quantity Calculation

This section will describe briefly the methods used in calculating the amount of traffic-dependent equipment required in a local telephone exchange. This information is given as a general guide and is not intended to be used for a specific installation. The data can be compiled only after an extensive traffic survey is taken of the originating and terminating traffic of the proposed location. The data used in the following example are applicable only in some North American locations.

5.10.1. Quantity of Originating Registers

The originating register traffic in erlangs is determined by multiplying the number of subscriber-originated calls (obtained from the traffic study) times the average originating register holding time given in Table 11. Holding time, if expressed in seconds, should be divided by 3600. This will determine the total originating register usage.

The originating register functional units are arranged in groups; each functional unit includes a means of storage. (The grouping depending on the system architecture.)

The required quantity of originating registers at a grade of service of 0.001 is obtained from Table 11. This is based on Poisson theory[29] using lost calls held, based on full availability and random traffic.

5.10.2. Quantity of Incoming Register

The incoming register traffic (expressed in erlangs) is the number of incoming calls for all trunk group served by the level-1 control multiplied by the

Table 11. Average Holding Time of Originating Register Function Units

Item	Holding time (sec)	
	Rotary dial	DTMF
Sum of intervals:		
Dial tone to dialing start and dialing end to originating register release	2.7	2.1
Dialing time per digit	1.5	0.8
The following intervals should be added when required:		
Digit timing (station delay)	3.6	3.6
Call from pay station using 0 and three-digit operator codes with coin return	1.5	1.5
Total time for typical calls:		
0 operator, noncoin	4.2	2.9
Seven-digit local number	13.2	7.7
DDD foreign NPA 1 + 10 digits	19.2	10.9

register holding time, in seconds divided by 3600, taken from Table 13, for each type of register signaling specified. The required number of incoming registers is obtained from Table 12.

5.10.3. Quantity of Senders

The quantity of sender functional units is determined from the busy hour calls of the outgoing trunk groups served by a level-1 control and the sender holding times shown in Table 14. The number required is obtained by using Table 12.

5.10.4. Quantity of Analog Tone Generation

The following formula and calculations for a precise tone plant are based on U.S.A. data which shows that dial tone requirements are the largest factor in tone generation. A uniform modular design of tone generators is desired. Hence all modules have this capacity even though the requirements for other tones may be less than for dial tone. The following assumptions apply.

Delay before dialing. The average subscriber reaction to dial tone in the U.S.A. is 2.7 sec for rotary dials and 2.1 sec for DTMF dials. Calls using dials are used in these calculation. 20% is added to allow for variation in dialing habits.

Register holding time. The average register holding time is 15 sec assum-

Table 12. Loading Capacity Grade of Service 0.001, Full Availability

Calling rate		
Erlangs	UC	Number of Traffic-carrying units
0.003	0.1	1
0.05	1.6	2
0.20	6.9	3
0.40	15	4
0.75	27	5
1.10	40	6
1.55	55	7
1.95	71	8
2.45	88	9
2.95	107	10
3.50	126	11
4.05	145	12
4.60	166	13
5.20	187	14
5.80	208	15
6.40	231	16
7.05	253	17
7.65	276	18
8.30	299	19
8.95	323	20

Table 13. Average Holding Time (Incoming Registers)

Source		Time (sec)	
		4 Digits	Per additional digit
Dial pulse automatic exchange	10 pps[a]	6.4	1.5
Dial pulse operator	10 pps	7.0	1.3
Dial pulse operator	20 pps	5.5	0.9
Dial pulse automatic exchange	20 pps	3.3	0.6
Multifrequency automatic local exchange		1.8	0.14
Multifrequency automatic trunk exchange		3.5	—
Key pulsing operator		5.6	0.6

[a] pps = pulses per second.

Table 14. Average Holding Time of Sender Functional Units

Type of call	Time (sec)			
	Non-LAMA[a] 4 digits	Per additional digit	LAMA 4 digits	Per additional digit
Dial pulse automatic exchange 10 pps	5.1	1.2	5.1	1.2
Dial pulse automatic exchange 20 pps	3.1	0.6	3.1	0.6
R1 automatic local exchange	1.9	0.14	2.4	0.14
R1 (seven-digit address) automatic exchange	3.0	0.14	3.0	0.14
CAMA senders ANI	4.6	—	—	—

[a] LAMA = local automatic message accounting.

ing rotary dials only. This is a weighted average of a number of calls including local 7-digit dialing (13.2 sec), 11-digit trunk calls (19.2 sec), and other traffic.

Output at line interface. A dial tone power of 14*MVA* is provided.

Busy hour coincidence. All registers have a simultaneous busy hour.

Use of dial tone. Dial tone is not used for purposes other than to indicate that station address signaling may start.

Required dial tone power. The dial tone traffic per register group equals

$$\text{register traffic} \times 1.2/\text{register holding time} = A \text{ Erlangs}$$

The total dial tone traffic is A multiplied by the number of register groups. Reference to Table 11 will give the number of trunks that is the corresponding number of simultaneous calls. This number of simultaneous calls multiplied by 0.014*VA* gives the total power requirements for dial tone.

5.10.5. Provision of Digital Generators

A digital system generates supervisory tones at the CODEC (digital to analog conversion) from the relevant bit pattern and the tone source need only drive a single port with an output of 0.014 *MVA*, if the tone generator is analog. The signal can be transferred to its point of application by using the highway bit stream of the line interface.

5.10.6. Provision of Positive Potential

As far as possible individual subscribers' meter registers should not be provided with automatic switching systems since modern message-recording systems offer a better way of accounting for local call charges. However, the need for working with an existing call-charging system may dictate their provision. The following are guides for the sizing of the positive potential supply. It should be noted that (while the calculations which lead to the number of simultaneous connections apply to any system) the current values for operating the meters are specific to North America. The application to any other system must be based on the relevant technique and parameters of the system. The method can be applied to other portions of the system using shared functions.

5.10.6.1. Possibility of Simultaneous Operation of Customers' Meter Register Based on the Poisson Distribution

If a set of originating calls is distributed individually and collectively, at random in an interval t, then the probability of n calls occurring within any subinterval of t of length equal to the operating time of the meter is $P\,(m,x) = [(kx^n/m!]e^{-(kx)}$, the Poisson distribution, where k equals the expected number of calls per unit interval and kx equals the expected number of calls within the subinterval. Therefore the expectation of operation of more than one meter is as follows:

$1 \times P_2$	One meter actuated simultaneously
$2 \times P_3$	Two meters actuated simultaneously
$3 \times P_4$	Three meters actuated simultaneously
$4 \times P_5$	Four meters actuated simultaneously, etc.

where

$$P_2 = e^{-(kx)} \times (kx)^2/2! \quad \text{and} \quad P_3 = e^{-(kx)} \times (kx)^3/3!, \text{ etc.}$$

The maximum number of meters that will be pulsed simultaneously is needed to calculate the capacity of the positive battery source.

The following calculations are based on message registers which operate on 0.3 W (typically a 500-Ω coil which operates on 25 mA).

Example of application (10,000 busy hour calls). Let the number of originated calls equal 10,000. Then the number of calls per unit length is k = 10,000/3600 = 2.78 calls per second.

Let the length of the meter pulse be 150 msec; then the number of calls per 150 msec = kx = 2.78 × 0.150 = 0.417. Now $e^{-0.417}$ = 0.6590. Hence the probability of five simultaneous actuations of the meter is $0.6590 \times 0.417^5/L^5$ = 6.92×10^{-5}. The values for the possibility of multiple simultaneous actuations are listed in Table 15.

On the basis of those values the system should be capable of simultaneously pulsing five meters. The need to pulse five meters simultaneously with 10,000 busy hour calls originated and assuming 8 busy hours per day and 300 days per year will arise on the average every 6 yr. The need to pulse more than five meters simultaneously can be neglected.

Application to 20,000 and 30,000 calls per busy hour. Applying the same reasoning to systems generating 20,000 originated busy hour calls, we obtain a

Table 15. Probability of Simultaneous Actuations with 10,000 Busy Hour Originated Calls

Number of simultaneous actuations	Probability	Once per
2	0.06	2 days
3	8×10^{-3}	15.6 days
4	8.3×10^{-4}	6 months
5	6.9×10^{-5}	6 years
6	4.8×10^{-6}	86 years
7	2.86×10^{-7}	
8	1.48×10^{-8}	

possibility of once every 2×10^{-4} busy hours for the simultaneous operation of 6 meters or once per 10 years. In the case of 30,000 originated busy hour calls the interval for six meters is 1.5×10^{3} busy hours or once per 0.625 yr. The results for other numbers of meters are tabulated in Table 16.

On this basis the maximum number of simultaneous actuations to be provided for would be 7 and 8, respectively.

5.10.6.2. Multimetering

The application of multiple meter pulses per call does not affect the maximum load but only the increased probability of occurrence, since there are more meter pulses per call. Up to six impulses can be applied per call.

Table 16. Probability of Simultaneous Actuations with 20,000 or 30,000 Busy Hour Originated Calls

	Probability	
Number of simultaneous actuations	20,000 busy hour originated calls	30,000 busy hour originated calls
5	1.45×10^{-3}	7.29×10^{-3}
6	2×10^{-4}	1.5×10^{-3}
7	2.4×10^{-5}	2.7×10^{-4}
8	2.4×10^{-6}	4.24×10^{-5}
9	2.3×10^{-7}	5.89×10^{-6}
10	4×10^{-8}	7.36×10^{-7}
11	1.45×10^{-9}	8.38×10^{-8}
12	1.01×10^{-10}	8.7×10^{-9}
13	6.48×10^{-12}	8.4×10^{-10}

Now the synchronized metering cycle picks up all calls that require metering within the 3-sec period of the metering cycle.

Then for 10,000 originated calls there are 2.78 calls per second but the interval is now 3 sec. Hence calls per subinterval = 2.78 × 3 sec = 8.34:

$$e^{-8.34} = 2.39 \times 10^{-4}$$

Probability of 5 = $2.39 \times 10^{-4} \times 8.34^5/15 = 0.080$ or (8×10^{-2}), i.e., once per 100 busy hours. If we assume 8 busy hours per day this will probably occur once per 12.5 days.

5.10.6.3. Time-Variable Metering Pulses (Bulk Metering)

In the case of single-pulse variable-time metering the metering pulses occur at random and the time intervals start at random. The effect of this is to decrease the interval of time between the occurrence of simultaneous metering pulses according to the number of timed intervals per call. The average number of pulses per call should be obtained from the country of application. The interval for the occurrence of the simultaneous operation of m meters established above will be decreased by the average number of timed intervals per call. An interval of two years should be used for the probability of simultaneous operation and the number of meters adjusted upwards accordingly.

5.10.6.4. Type of Supply

The supply may be obtained from duplicated dc to dc convertors with alarm and automatic changeover in the event of fault. If a battery is specified by the country of application it should have the same reserve period as the main −48 battery for that application.

5.10.7. Provision of Ringing (General)

Interrupted ringing should be distributed in at least three phases to reduce the power of the ringing generator. The parameters used in the calculation of ringing are based on U.S. experience; they can be used in the absence of specific country requirements.

At least one of the duplicated ringing generators should be battery powered.

The initial power should be sufficient to provide for 10–15 years of system growth.

The calculations do not include ringing drain requirements for any circuits which use continuous ringing. These circuits may be provided for by allowing an additional 1.7 W for each simultaneously rung telephone in the relevant application.

5.10.7.1. Basic Formula

The formula for the calculation of the power per ringing phase is based on the following assumptions:

- The answering delay is time which is equal to three ringing cycles, which equals 3×6 sec for a 1.5-sec ring and 4.5 sec of silence.
- An answering time of 18 sec per call.
- A ringing drain per call of 1.7 W which is the composite average of 20 Hz; ringers located from 0 to 32.18 km from the local exchange.
- F equals factor of 1.2 or 20% of the ringing load varies from the average.

The calculation proceeds as follows: The average number of busy hour calls is equal to the terminating traffic, in erlangs, per group of ringing circuits (A) multiplied by 3600 divided by the holding time (t) in seconds: $(A \times 3600)/t$. The average holding time on intraoffice calls is assumed to be 120 sec. Hence this becomes $A \times 3600/120 = 30A$ multiplied by the proportion of the holding time used by the ringing cycle. This equals $30A$ multiplied by 18/3600 or $0.15A$.

Effect of ringing phase. Ringing is equally distributed in phases; if there are n phases then the power is reduced to $0.15A/n$.

5.10.7.2. Quantity of Ringing Power

Ringing is distributed as a single-frequency continuous ringing in many modern systems with the three phases of ringing generated at the individual ringing application points. As an example, if a group of 60 traffic-carrying devices is provided at a grade of service of 0.001, then it will carry 38.9 Erlangs using Table 11.

And the amount of ringing required is equal to

$$\frac{0.15\,A}{3} \times 1.2 \times 1.7\ \text{W} = 0.15 \times \frac{38.9}{3} \times 1.2 \times 1.7 = 3.96\ \text{W per line unit.}$$

6

Transmission Applications

6.1. Introduction

Transmission equipment used in conjunction with telephone exchanges utilizes a number of transmission modes. These include analog, frequency division, and time division channels. The means include wire, coaxial, or optical cable; radio; microwave; and satellites. This book is only concerned with interfacing with the modes and means. The details of the modes and means are covered in other texts.

The interface with the switching system is in analog form in the case of wire equipment and frequency divided carrier systems. The interface between a digital switching system and a pulse code modulated system can be on a direct digital basis. This technique obviates the need for conversion to analog.

Frequency range of transmission media. Figure 6.1 shows the frequency ranges with various transmission media. The broadband characteristics of some of the media promise low-cost audio band channels. Broadband characteristics are also of importance for carrying data. Switching systems capable of carrying higher-than-audio bandwidth signals are simpler to realize with digital switching. This is because digital systems use bit stream switching, and interfacing with such a system can be at higher bit levels than the 64 kbit per second per audio channel.

6.2. Interfacing with a Telephone

Figure 6.2 shows the transmission path from the telephone to the local telephone exchange. This terminates at the telephone in one of two ways, depending on the state of the line. If the telephone is not in use, a telephone bell coil in series with a condenser provides the termination. If the telephone is in use the telephone transmitter with its associated network is connected in series with

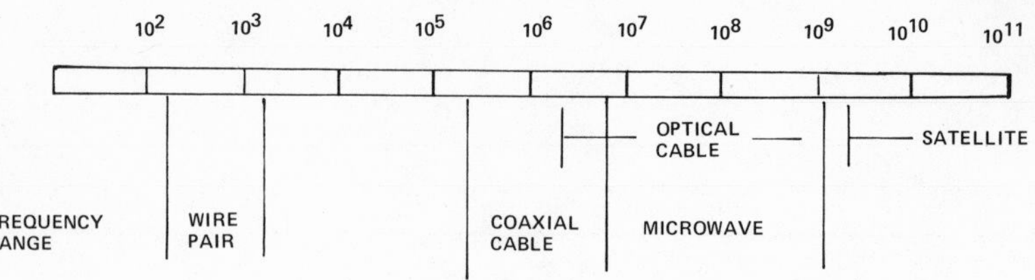

Figure 6.1. Frequency ranges of transmission media.

the line in lieu of the bell. When the telephone is in use a direct current circuit replaces the alternating current circuit. The change of state produces the station signals discussed earlier.

6.2.1. Characteristics of a Wire Line

The connection from the telephone to the local exchange usually consists of a small gauge wire pair. As small a gauge as possible is used to reduce the cost of the cable in which the pairs are included. This is a compromise which reduces the cost by reducing the amount of expensive copper but increases the line resistance. The characteristics are as follows.

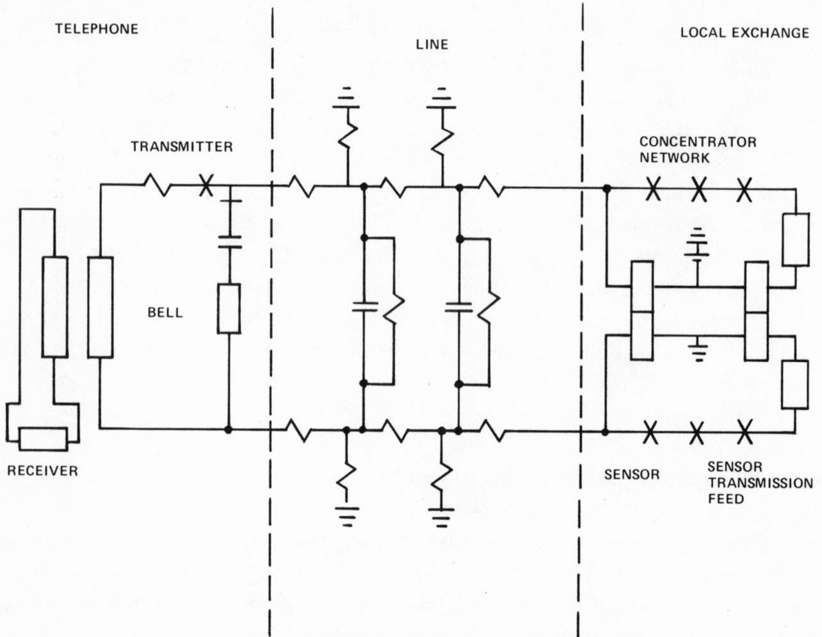

Figure 6.2. Wire line transmission path.

Conductor–loop resistance. This is the series resistance of the conductors of the line or trunk loop, excluding the telephone or the local exchange line interface.

Station–loop resistance. This is the sum of the conductor–loop resistance and the resistance of an off-hook station terminal.

Line insulation resistance. This is the resistance between the loop conductor and ground or between the two conductors in the pair.

Line capacitance. This is the capacitance between the two conductors in the pair. This can build up to the point where it is necessary to add inductance to correct distortion and minimize speech attenuation.

Characteristic impedance. This is the impedance of the telephone loop which must be balanced in a hybrid. It is nominally 900 Ω + 2.14 mF for two-wire lines and 600 Ω + 2 mF for four-wire lines.

6.2.2. Line Loading

Generally speaking when the line lengths are 5.4 km or greater, 88-mH coils should be provided at 1.6 km spacing. This, however, represents a compromise between better quality speech and a lower bandwidth. Such a spacing introduces a filter effect with a cutoff frequency of approximately 3.5 kHz. This can affect the upper end of the nominal bandwidth requirements for satisfactory speech quality (300 to 3400 Hz) depending on how sharply tuned the resultant loop circuit is. (Sharper tuning moves the upper frequency of the usable bandwidth nearer to the cutoff frequency.)

6.2.3. Local Line Interface

The interface shown in Figure 6.2 relates to a switching matrix concentrator network which uses crosspoints which are compatible with the wire line. The local line loop includes a sensor which detects the change in current between the two telephone states.

As a result of a response to a request for service another circuit is connected, via the network, in lieu of the line sensor. This includes the circuit which powers the telephone and couples it inductively to the next stage of the switching matrix. This circuit also includes a line sensor to monitor the state of the telephone loop. This circuit represents a shunt to battery for the transmission of speech signals. Hence it must be of high impedance when carrying the loop current.

The loop current can vary from a high value to a low value due to long loops or short loops, respectively. The minimum loop resistance will always be greater than the resistance of the telephone in its off-hook state plus the resistance of the interface circuit. These are of the order of 200 and 400 or 800 Ω, respectively. The maximum loop resistance must permit a current of approximately 22 mA to flow through the telephone circuit. This is necessary for the transmitter and, if used, the DTMF calling device to operate satisfactorily.

Use of loop compensation. Studies in Bell Telephone Laboratories[30] have shown that improvements in the transmission performance can be obtained by the use of dc isolating capacitors. This improves the transmission by reducing the bridging loss. This allows the use of more effective inductors since they do not have to carry direct current.

6.3. Interfacing with Analog Trunks

Interfaces with analog trunks are treated as individual trunk interfaces providing the relevant functions, whether they are direct or derived from a carrier system. Space division systems with compatible crosspoints can achieve better impedance matching through the use of loop compensating networks. This results in an improvement in the return loss.

6.4. Interfacing with Digital Bit Streams

Pulse code modulation systems can connect to digital switching systems without modulation and conversion to analog at the interface. However, some processing will be necessary at the interface. This will provide for synchronization of the carrier bit stream and the switching system highways bit stream so that the time slots can be identified. It may also require speed buffering if the bit rates of the carrier and highway differ. It may require conversion from the carrier format to the system format, as for example if a word-organized bit stream is connected to a bit-organized bit stream. The actual interfacing depends on the type of carrier system used.

6.5. Digital Carrier System—Analog to Analog

Figure 6.3 shows the general arrangements of a digital carrier system. This is used to connect analog trunk interfaces between switching systems. It includes the following items.

6.5.1. Trunk Interface

The trunk interface connects the switching matrix outlets or inlets to a channel unit. It includes signaling functions and, in the case of a two-wire matrix, a hybrid transformer. This provides the separate send and receive transmission path necessary for carrier operation.

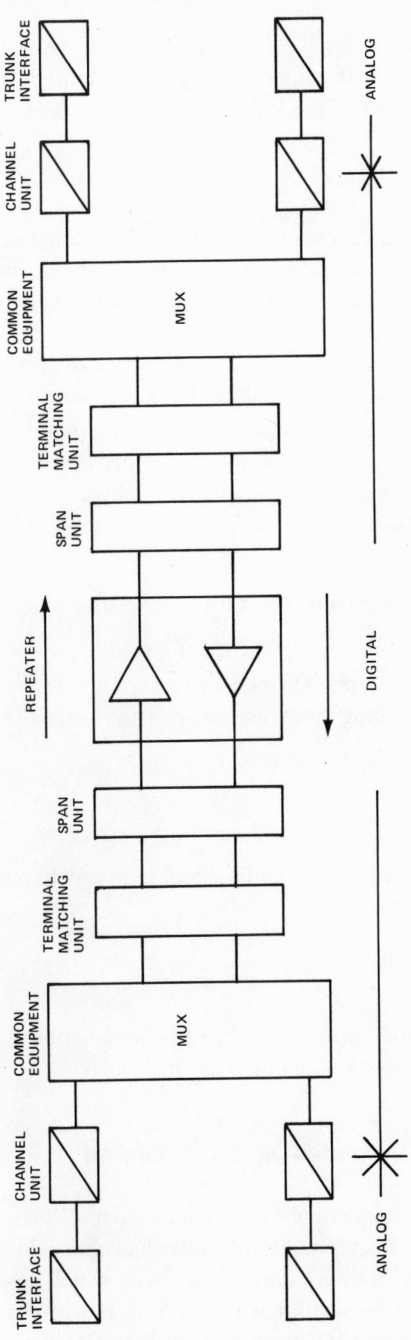

Figure 6.3. Basic digital carrier system (analog to analog).

6.5.2. Channel Unit

The channel unit provides the interface between the trunk circuit and the common multiplexing unit. It provides for pulse amplitude modulation of the analog signal and includes control leads so that it may be sampled cyclically.

6.5.3. Common Equipment

Common equipment provides the means of associating each of the channel units with its time slot in the bit stream and generating the recurring series of time slots. In the case of the sending highway it encodes the pulse amplitude signals from the channel units in an eight-bit binary code. It should be noted that earlier systems provided functions in common since this was cost effective with the existing components. New systems will undoubtedly take advantage of the low cost of logic provided by large-scale integrated circuits to perform more functions in the channel unit. The channel unit performs functions concerned with converting the bit stream from the receive highway to analog (a reversal of the process described above).

6.5.4. Terminal Matching Unit and Office Repeater

The terminal matching unit and office repeater provides the impedance matching and driving functions between the common equipment and the trunk transmission means. Signaling over the transmission means is accomplished by bipolar signals.

6.5.5. Span Unit

The span unit provides a means of switching to an alternative transmission path under fault conditions.

6.5.6. Repeaters

Repeaters are used to replace the losses encountered in the transmission means and are usually spaced at about 2-km intervals.

6.6. Digital Carrier System—Digital to Digital

Figure 6.4 shows the principle of a digital carrier system, digital to digital. The common multiplexing equipment, the individual channel units, and the trunk interfaces are replaced with direct digital interfaces which connect directly into the system highway. The direct digital interface is synchronized with and controlled by the first- or second-level controls in the telephone exchange.

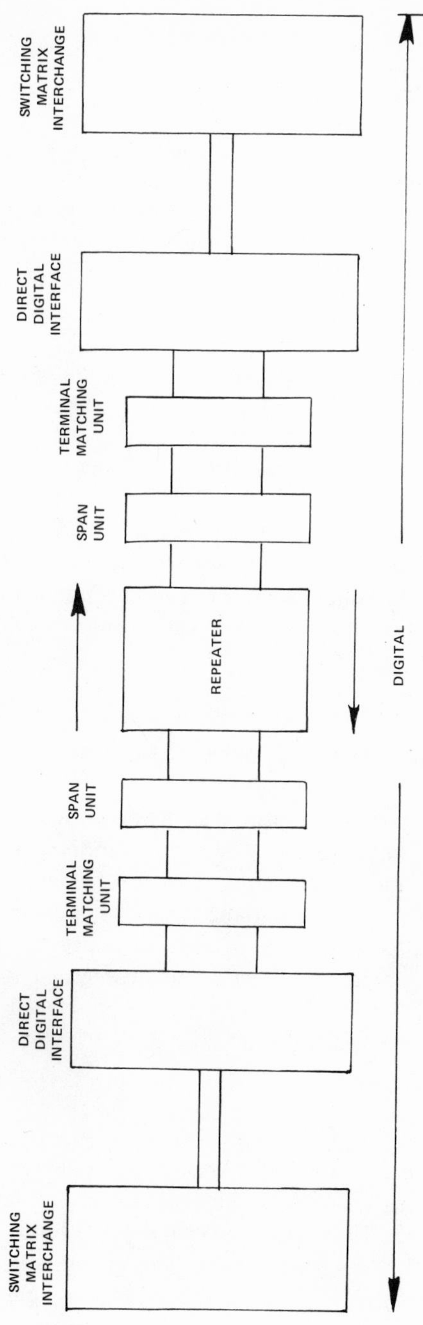

Figure 6.4. Basic digital carrier system (digital to digital).

6.7. Advantages of Applying Digital Carrier to Existing Routes

The installation of a digital system presents an opportunity to review the existing trunk transmission means with a view to replacing analog trunk groups with PCM channels.

In the case of existing networks with cables *in situ* advantage can be taken of the reduced pair requirements for PCM channels as compared with analog channels. For example, 30 PCM channels use only two analog channels for transmission. Now a 56-pair cable will carry 180 four-wire trunk circuits by using six groups of 30 PCM channels each. The remaining pairs can be made spare or used as required for control purposes.

The conversion of the cable transmission mode from analog to PCM would require, in addition to power equipment, a frame of equipment to house the six PCM groups. This is obviated when the cable interfaces with a digital system, since the control circuitry is included in the switching system and is mostly software. If the distant exchange is also digital there are similar savings in equipment at that end.

Cost comparisons must be based on a specific location but must consider the following: costs of interfaces including prorated bay floor space and power costs; cost of PCM channel including prorated common equipment, repeaters, power, span equipment, and mounting facilities; cost or value of cable including ducts, cable installation, and materials.

6.8. Principal Time Division Carrier Systems

There are two primary multiplex systems in worldwide use. They are not directly compatible. The reason for this stems from their separate development times and the state of the art in componentry at the time of their introduction.

The systems which was first introduced (which operates at a speed of 1.544 kbit per second) is the 24-channel T1 system developed at AT&T and used very extensively in North America and Japan. The other system is the 32-channel (30 speech channels) CEPT system (which operates at a speed of 2.054 kbit/sec), which is increasingly being used in other areas.

6.8.1. Comparison of PCM Carrier Systems

Figure 6.5 shows the basic differences between the two systems. There are only three related items which have the same value and these have to do with basic coding principles. These are bits per time slot, frame repetition rate, and the resulting number of bits per channels. The timing signal is obtained in the same way in both systems. The format is different for signaling and alignment, hence the methods of interworking differ, needing different designs for direct digital interfaces in each case.

GENERAL CHARACTERISTIC	OPERATING SPEED	
	2048 KBIT/S	1544 KBIT/S
ENCODING LAW	A LAW	μ LAW
# QUANTIZED LEVELS	256	255
TIMING SIGNAL	INTERNAL, I/C DIGITAL SIGNAL, EXTERNAL	SAME
BITS PER TIME SLOT (TS.)	8 (1 TO 8)	8 (1 TO 8)
TIME SLOTS PER FRAME	32 (0 TO 31)	24 (1 TO 24)
BITS PER FRAME	256	193
FRAME REPETITION RATE	8000 H_3	8000 H_3
MULTIFRAME/SUPERFRAME	16 FRAMES	12 FRAMES
SIGNALING	TS, 16	BIT 8 (1) FRAMES 6 & 12
ALIGNMENT	TS, 0	S BIT (BIT 1)
FRAME ALIGNMENT	TS, 0	S BIT ODD FRAMES
MULTIFRAME ALIGNMENT	TS, 0	S BIT EVEN FRAMES
BIT RATE	2048 KBIT/S	1544 KBIT/S
BITS PER CHANNEL	64,000	64,000
CHANNEL BAND WIDTH D2 = 2 WIRE	(3) 3400 −5 TO +1 300 TO 3000 ± .5	(2) 3300 −1.5 : 3400 −3 300 TO 3000 ± .25
CHANNEL BAND WIDTH D2 = 4 WIRE	———	(2) 3400 −3 300 TO 300 +.5 TO −1

NOTE:
1. D2 FORMAT (D1 ALL FRAMES)
2. DB RELATIVE TO 1000 H_3
3. DB RELATIVE TO 800 H_3

Figure 6.5. Comparison of PCM carrier systems.

6.8.2. T1 Carrier System Format

Figure 6.6 shows the T1 system format. A super frame consists of 12 frames (numbered 1 to 12). Each frame has 24 channels (time slots) 1 through 24 with one additional bit S per frame. Each channel has eight bits. Frames 1 through 5 and 7 through 11 use all eight bits for speech encoding. Frames 6 and 12 use only seven bits for speech encoding and one bit (The A and B bit, respectively) for signaling (speech digit signaling). The S bit in each frame is assigned a value of 1 or 0 to produce a bit pattern of 100011011100 (out-slot signaling). This pattern is used for frame alignment. While the pattern could occur in different bit positions, it is unlikely.

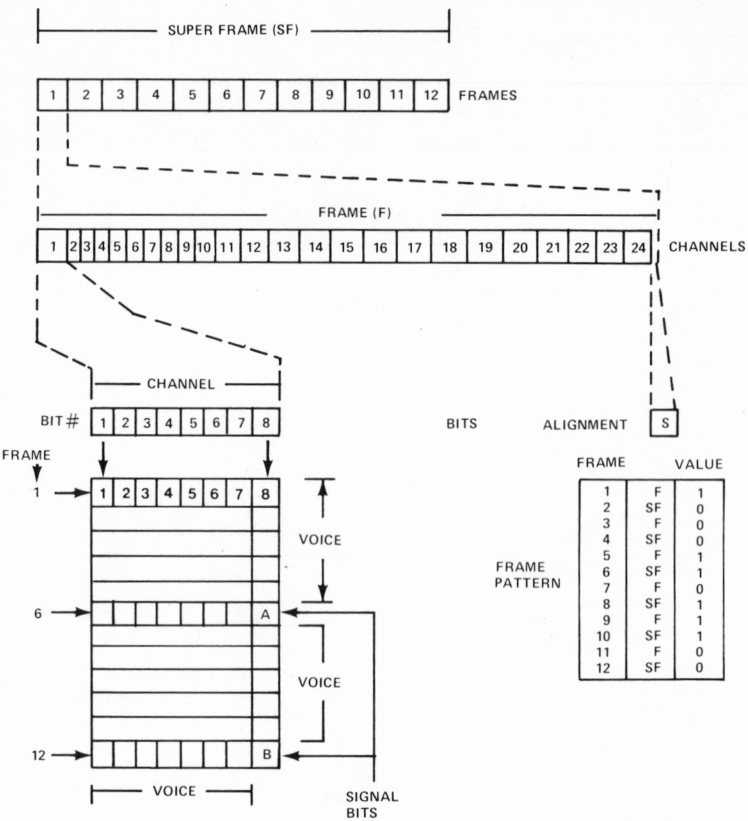

Figure 6.6. Format T1 carrier.

6.8.3. CEPT Carrier System Format

Figure 6.7 shows the CEPT carrier system format. A multiframe consists of 16 frames (numbered 0 to 15). Each frame has 32 channels (numbered 0 through 31). Each channel has eight bits. The first frame (0) is used for alignment and other administrative functions. Frames 1 through 15 and 17 through 31 are used for the transmission of signals and voice. Channel 0 is allocated for national and international use including the transmission of alarm signals. Channel 16 is used for line signaling (in-slot signaling). The time slot in each frame is allocated to a pair of the transmission channels in frames 1 through 15. The remaining channels are used for speech transmission.

6.8.4. Format of Signaling Bits

The carrier systems transmit signal information in separate bit positions to the speech channels. The allocation of the bits and their positions in the frames varies as described below.

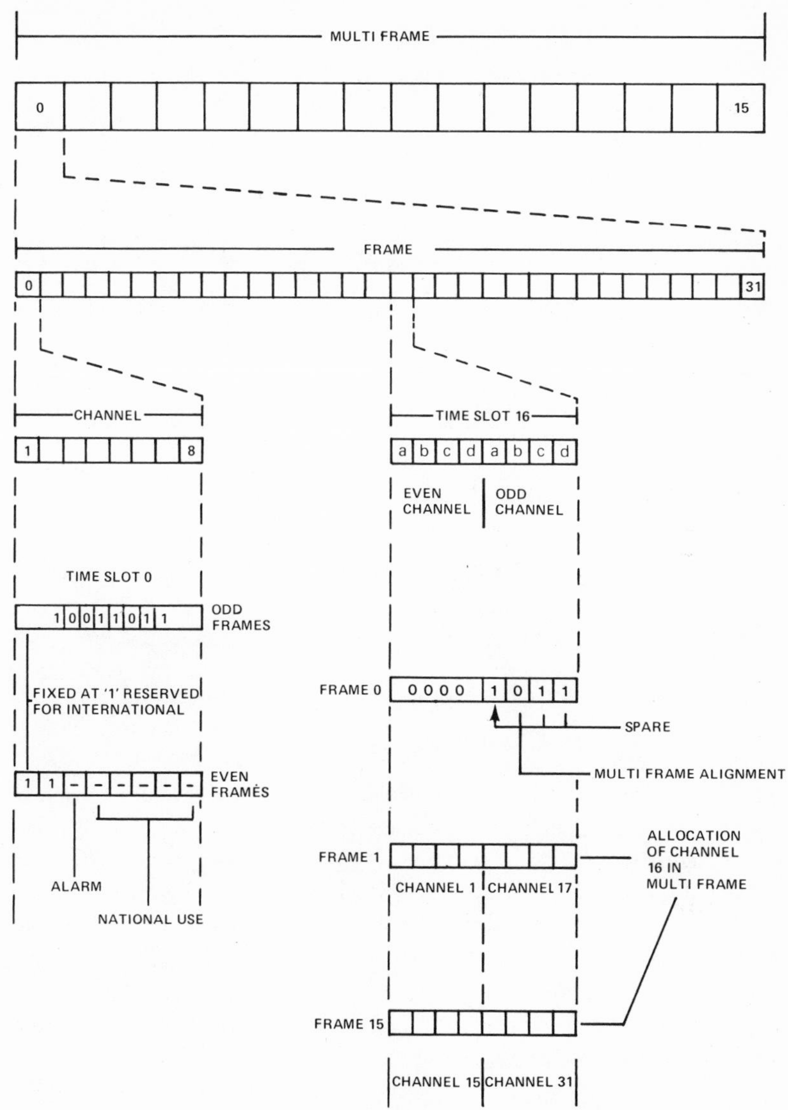

Figure 6.7. Format CEPT carrier.

6.8.4.1. T1 Carrier Signaling Format

Figure 6.8a shows the position of the signaling bits in a T1 carrier. These occur in the eighth bit position of frames 6 and 12 in each of the 24 channels per frame. Signaling bits occur in each of these two channels every 1500 usec.

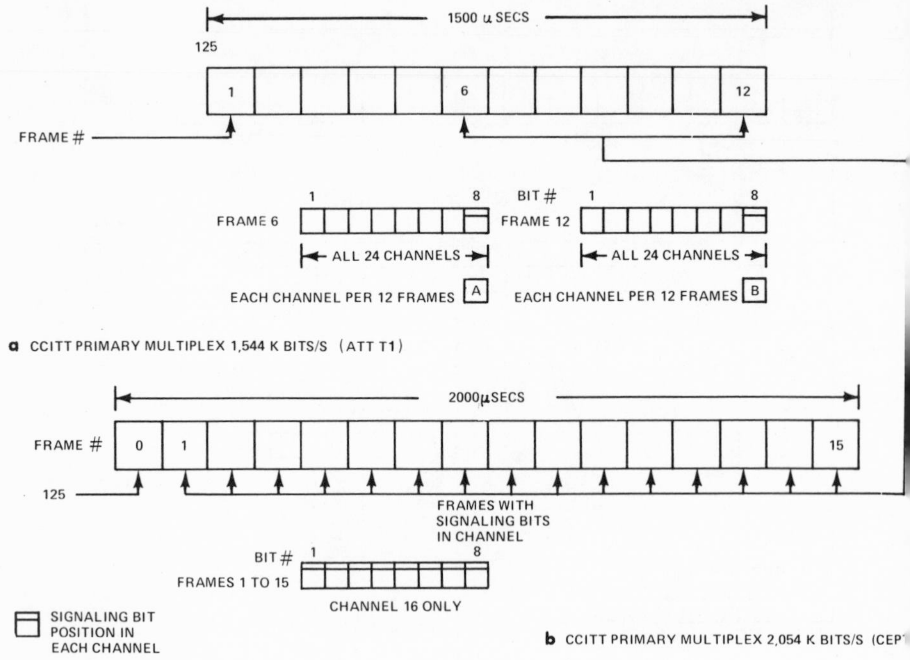

Figure 6.8. Signaling bits PCM carrier systems.

6.8.4.2. CEPT Carrier Signaling Format

Figure 6.8b shows the position of the signaling bits in CEPT carrier. These occur in frames 1 through 15 in channel 16 only. There are eight bits in each channel and these are divided into two sets of four-bit words. Each of the words is allocated to one of the 30 transmission channels in the following way: Time slot 16 in frame 1 is allocated to channels 1 and 17; Time slot 16 in frame 2 is allocated to channels 2 and channel 18, and so on. Signaling bits occur in these positions every 2000 μsec.

6.8.5. Principle of PCM Signaling

Figure 6.9 shows the principle of PCM signaling between analog trunks. The carrier system employs separate send and receive transmission paths. This means that the channel allocated to the connection has a send and receive time slot in each direction. These signals travel separate paths in the carrier system but are combined into a single analog path at two-wire interfaces.

Each trunk interface has the means to separate the signals into send or receive signals. In the application shown in the figure only line signals are sent in the signaling bit positions. The register signals, which are in-band voice signals, are encoded as voice signals and sent in whichever of the 30 channels has been allocated to the connection.

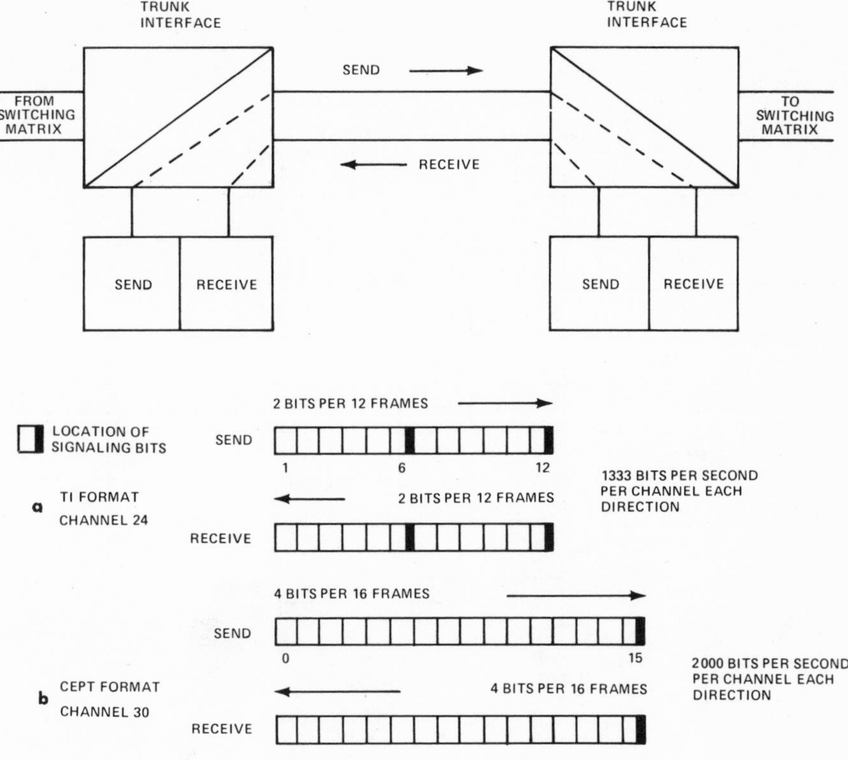

Figure 6.9. Principle of PCM signaling.

6.8.5.1. Signaling Bit-Rate T1 Carrier

Figure 6.9a summarizes the signaling bit allocation for T1 carrier time slot 24, which has two signaling bits per channel per 12 frames. This corresponds to 1333 bits per second in each direction per channel.

6.8.5.2. Signaling Bit-Rate CEPT Carrier

Figure 6.9b summarizes the signaling bit allocation for CEPT Carrier time slot 30 which has 4 bits per channel per 16 frames. This corresponds to 2000 bits per second in each direction per channel.

6.8.5.3. Typical Use of PCM Carrier Signals

Figure 6.10 shows a typical use of a PCM carrier channel for signaling between analog trunks. E&M line signaling is used at either end. The signals are transmitted by activating an M lead. This causes the generation of a bit pattern in the relevant signaling bit positions. In the case of a calling signal this corresponds

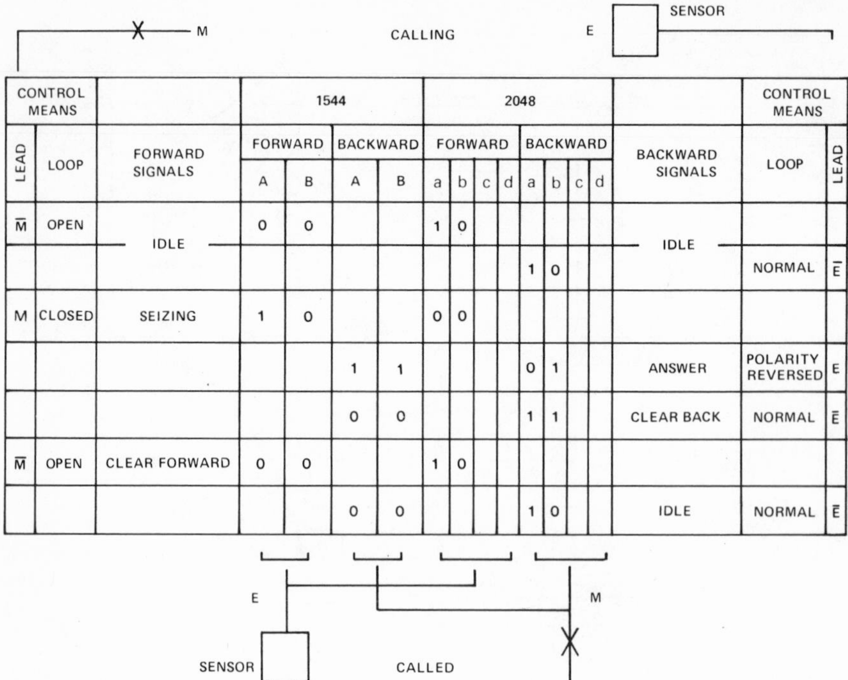

CONTROL MEANS			1544				2048									CONTROL MEANS	
LEAD	LOOP	FORWARD SIGNALS	FORWARD		BACKWARD		FORWARD				BACKWARD				BACKWARD SIGNALS	LOOP	LEAD
			A	B	A	B	a	b	c	d	a	b	c	d			
$\bar{M}$	OPEN	IDLE	0	0			1	0							IDLE		
											1	0				NORMAL	$\bar{E}$
M	CLOSED	SEIZING	1	0			0	0									
					1	1					0	1			ANSWER	POLARITY REVERSED	E
					0	0					1	1			CLEAR BACK	NORMAL	$\bar{E}$
$\bar{M}$	OPEN	CLEAR FORWARD	0	0			1	0									
					0	0					1	0			IDLE	NORMAL	$\bar{E}$

Figure 6.10. Typical use of PCM carrier signals.

to 10 in bit positions A and B for T1 or 00 in bit positions ab for CEPT. At the called end this signal is decoded and detected by the sensor on the E lead at the trunk interface. The figure shows the usage of the possible signals which fully uses the possible codes for T1 but only uses two of the four-bit positions available in the CEPT case. Other uses of the signaling bits are possible though the use of suitable interfaces. The two basic uses of the CEPT system are as follows.

Channel-associated signaling. In this arrangement, the 16 frames of the multiframe are each (with the exception of frame 0 which is used for multiframe alignment) used for two telecommunication channels as described above.

Common channel signaling. Channel time slot 16 may be used for common channel signaling at a rate of 64 kbit/sec. The arrangements for the use of channel time slot 16 had not been specified at the time of going to press.

6.8.6. Multiplex Systems

The carrier systems described above are the primary order multiplexing levels of multiplex systems. Present systems employ four levels.

Figure 6.11 compares the capacity and bit rates of the CEPT and ATT systems. The systems differ, partly because of the existing analog, frequency

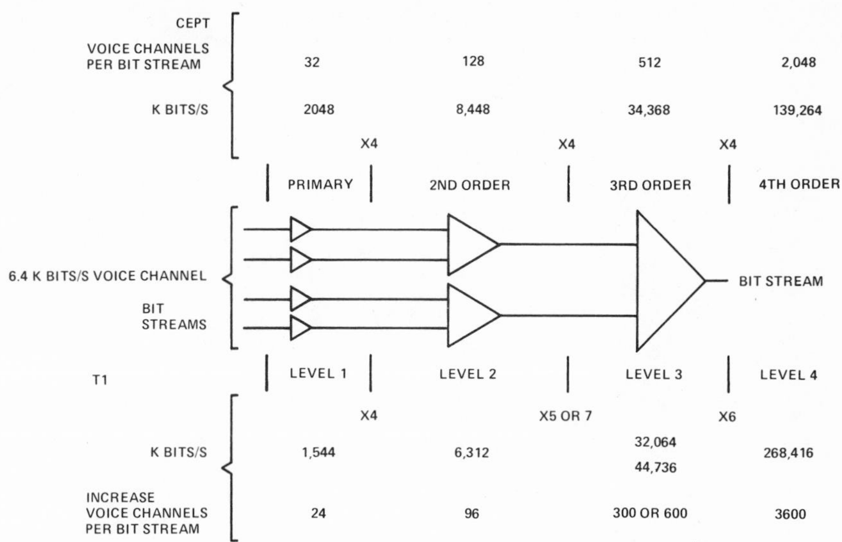

Figure 6.11. Comparison of digital multiplex systems.

division multiplex, system channel capacities. The CCITT-recommended family expands the bit rate on a binary base. The ATT family does not. The CEPT system multiplexes groups of 32 channels (30 speech channels plus signaling and alignment channels) into a fourth-order bit stream of 139,264 kbit/sec. This corresponds to 2048 speech channels. The ATT system multiplexes groups of 24 channels into 3600 speech channels, at a bit stream of 268,416 kbit/sec.

6.9. Direct Digital Interfacing

Direct digital interfacing requires the alignment of the carrier time slots with the carrier channels. In the case of T1 carrier this requires the recognition of the alignment pattern of the S bit. In the case of CEPT carrier the indentification of frame 0 is obtained by examining the bit pattern of time slot 16 in frame 0 and time slot 0 in all frames. A method of alignment for each system is described below.

6.9.1. Aligning with T1 Carrier

Figure 6.12 shows a procedure for alignment. A random check of each bit is made for a 1 state. When this is encountered a count of 193 is made and the corresponding bit in this frame is checked for its state. The result of this state and the previous state are stored. This can result in 1,0, which can indicate the

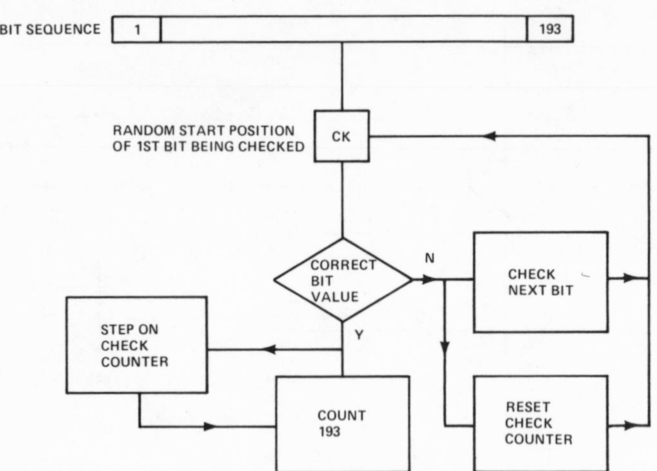

Figure 6.12. Alignment procedure T1 carrier.

following possible positions for the S bit: frames 1 and 2, frames 6 and 7, or frames 10 and 11.

This process continues until either the alignment pattern is stored or the sample does not conform to the expected bit value. If this is the case then the store is cleared and the next bit sampled. The logical operations involve comparing the sampled bit pattern with the alignment bit pattern and instructing the comparison circuit of the value or values which can be accepted as being possible in the S bit alignment pattern. Figure 6.13 shows the functions involved in making such an alignment, and the relative association of bits.

On the average this process will have to be repeated 96 times. The frame time is 125 usec and hence a relatively long interval is necessary. This is the simplest method of alignment but practical systems use patented procedures which involve storing and manipulating groups of bit samples to minimize the matching time.

6.9.2. Aligning with CEPT Carrier

Figure 6.14 shows the frame alignment signals used in this system. Unlike T1 carrier where the alignment bits are bit switched the alignment bits are word switched. This means that they occur in the same time slots and hence alignment is simpler.

6.9.2.1. Alignment Procedure

Figure 6.15 shows that the alignment method involves two phases and shows how the loss of alignment is detected. The system is constantly checked

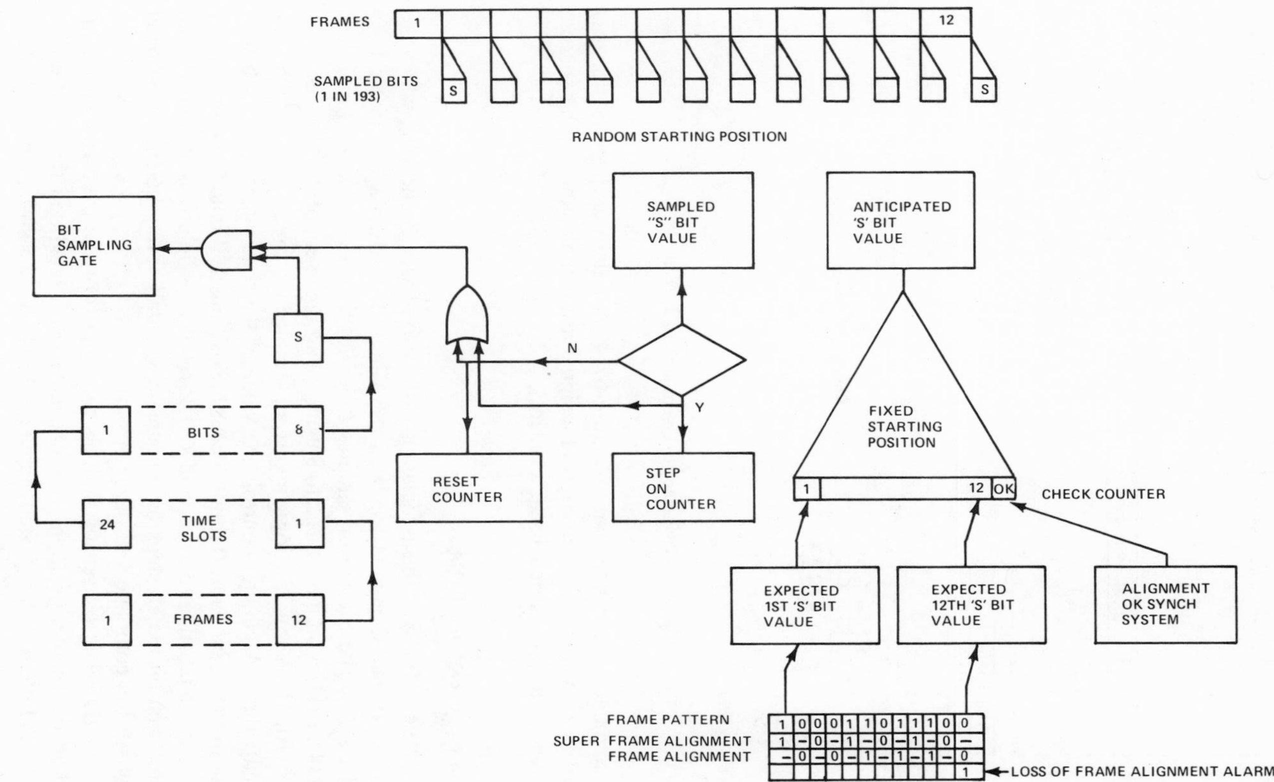

Figure 6.13. Alignment principle T1 carrier.

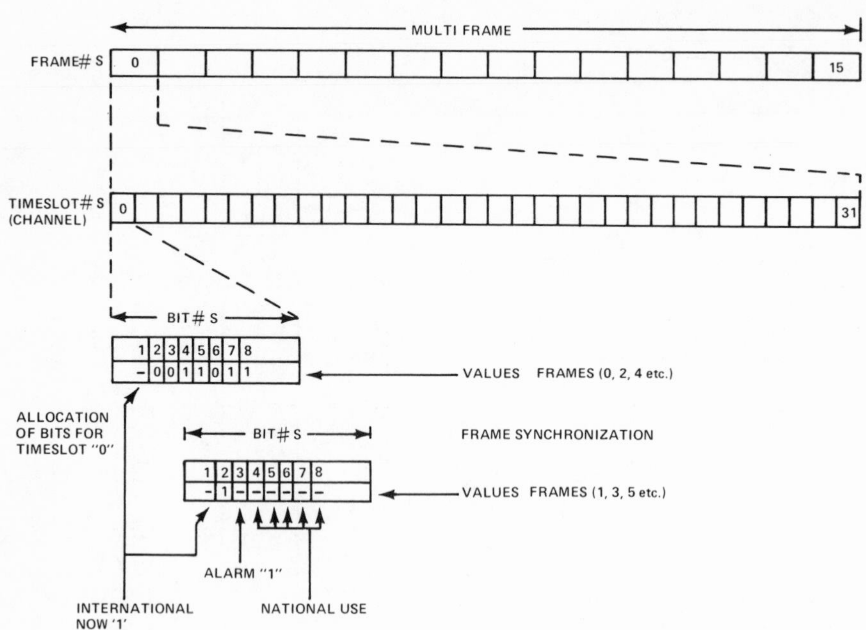

Figure 6.14. Frame alignment signal CEPT carrier.

for the alignment signals shown in Figure 6.14. A failure to detect a signal is recorded in a counter. The detection of a signal resets the counter. Four successive losses of the signal results in an alarm.

6.9.2.2. Recovery of Alignment

Figure 6.15b shows the association of the two phases of checking which are used to detect that alignment has been recovered. The first stage in phase I.

Phase I. The bit sensor samples time slot 0 of even frames for the bit pattern 0011011 in bit positions 2 to 8 using an arrangement of the type shown in Figure 6.16. This shows the logic involved; sampled bits are compared with their expected values. A counter controls the value of the state which is being examined; for instance, bit value 0 should occur in positions 1, 2, and 5 and bit value 1 should occur in positions 3, 4, 6, and 7. The relevant weighing is applied at each position. Each bit match steps on a counter; when all seven matches have been made phase I is complete and phase II starts.

Phase II. The bit sensor samples time slot 0 of the even frames for bit value 1 in the second position. If it encounters this condition then phase II is complete and phase I is repeated. Figure 6.17 shows the arrangement.

Alignment achieved. Alignment is achieved when the second application of phase I is made.

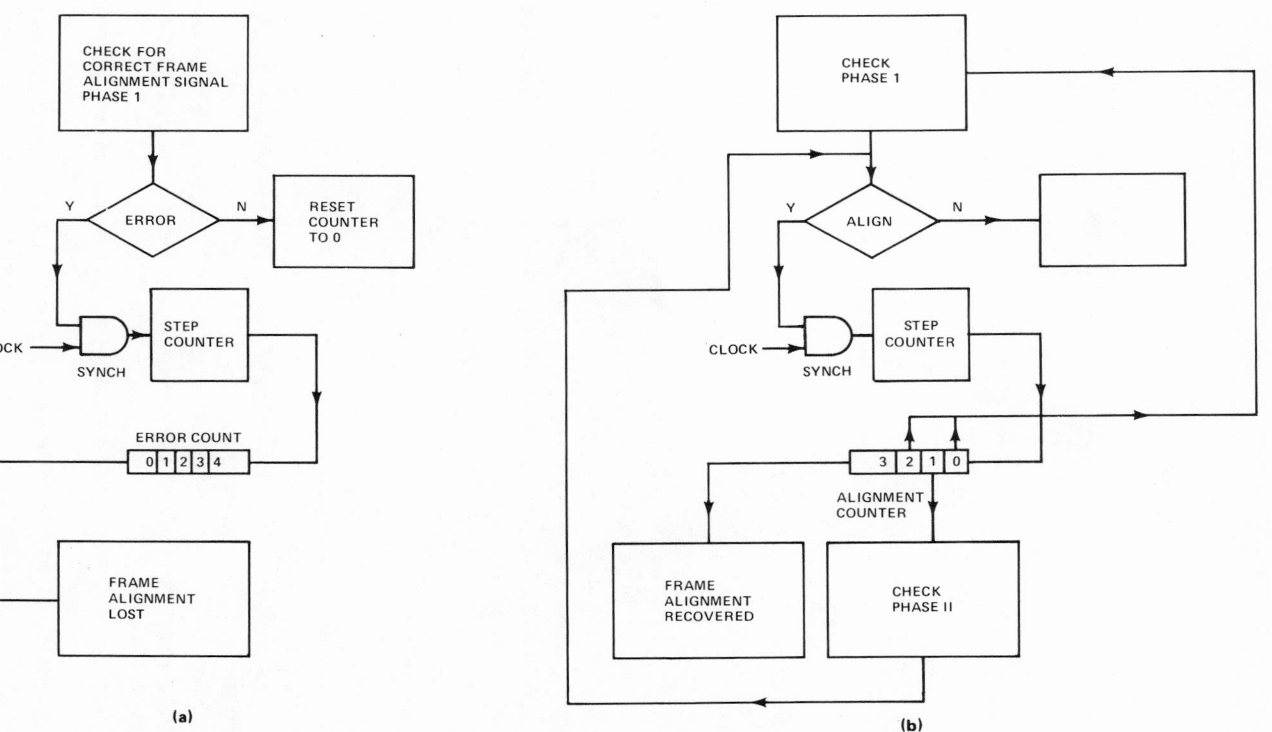

Figure 6.15. Alignment procedure CEPT carrier: (a) loss of alignment, (b) recovery of alignment.

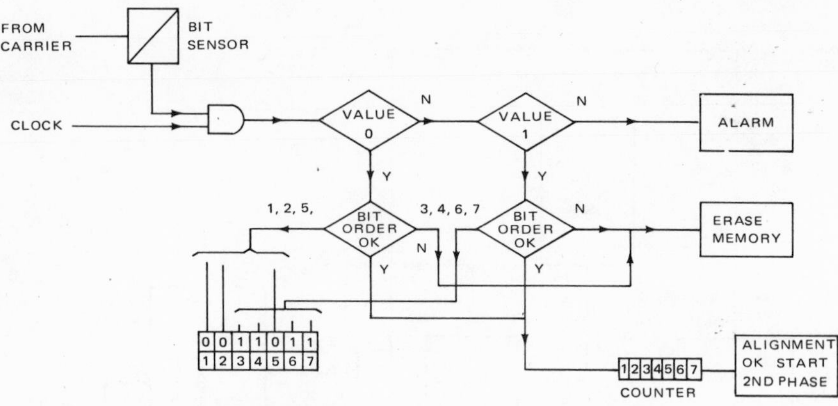

Figure 6.16. Alignment principle CEPT carrier phase 1.

6.9.3. Switching System Interface

The interface with the digital carrier system and the switching system occurs at the bit stream level. The first stage in this process is the identification of the channel time slots in the incoming carrier bit stream. Timing pulses are derived from this bit stream to synchronize the interface bit rate and the carrier system. The resultant information is usually stored in a temporary memory. This facilitates alignment and allows for processing to convert to the switching system format and bit rate. The interface converts bipolar carrier signals to unipolar switching system signals.

6.9.3.1. Basic Functions of Interface

Signaling. The interface extracts the signaling information and sends it to the relevant control function for processing. It also provides the reverse procedure for outgoing signals.

Frame detection. The interface aligns the bit stream in terms of frames, as described above, of the incoming bit rate.

Time slot generator. The interface aligns the time slots in relation to the frame and provides a means of converting from serial to a parallel output for word processing.

Diagnostics. The interface provides for fault indication and alignment alarms.

Other functions. Functions are provided on a cost effective basis according to the availability of low-cost componentry. Hence there will be a tendency to include functions in the interface as this becomes cost effective.

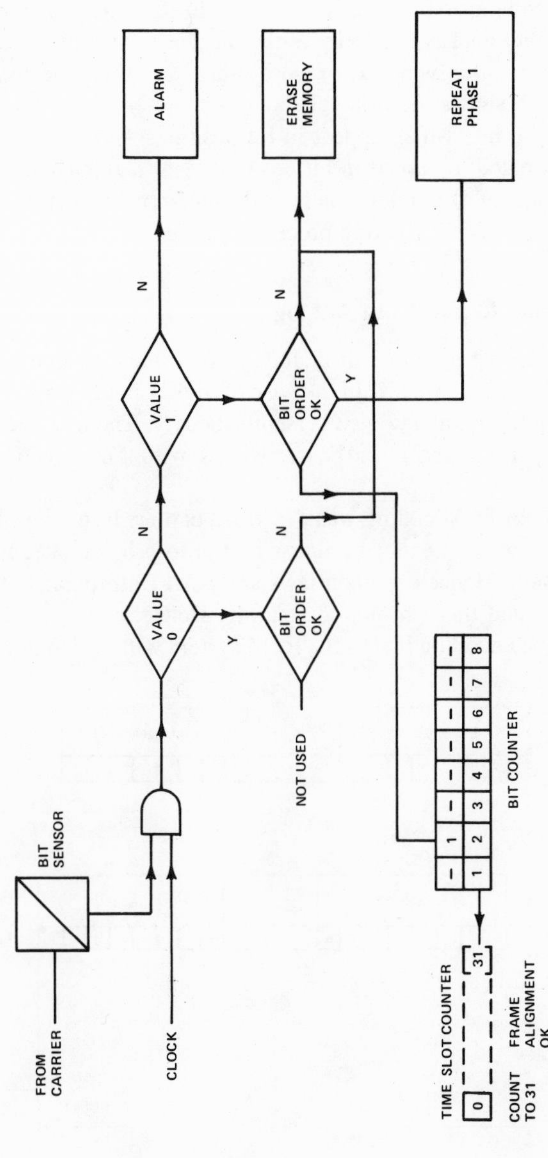

Figure 6.17. Alignment principle CEPT carrier phase 11.

6.9.3.2. Conversion, Word to Bit Order

The PCM carrier systems described in this chapter operate on a word-order format of speech channels. Digital switching systems can operate either in a word-order bit stream or bit-order bit stream. In the latter case the temporary storage of the input data is necessary at the interface. This provides a means of retaining the information which is in word order until it can be converted to bit order. Figure 6.18 shows the principle.

The incoming bits which appear in bit positions 1 to 8 in channels 1 to n have to be reformated to appear in bit positions 1 to n in channels 1 to 8. This requires the storage of a complete frame of the incoming data. The delay consists of the frame time plus reformating processing time.

6.9.4. Conversion to and from Analog

A local switching system will include analog line interfaces in addition to possible analog trunk interfaces. In the case of digital switching systems conversion from analog to digital and vice versa must occur. The line interfaces can be provided on a per line basis or be associated with lines via a concentrator network.

The method of PCM coding was described earlier. It involves the use of an encoding method which cannot be linear if it is to provide adequate low-level signal transmission. Hence encoding laws are used. Unfortunately there are two such laws in use and they are not compatible. Hence, either the digital system must use the same encoding law as the PCM system with which it interfaces or it

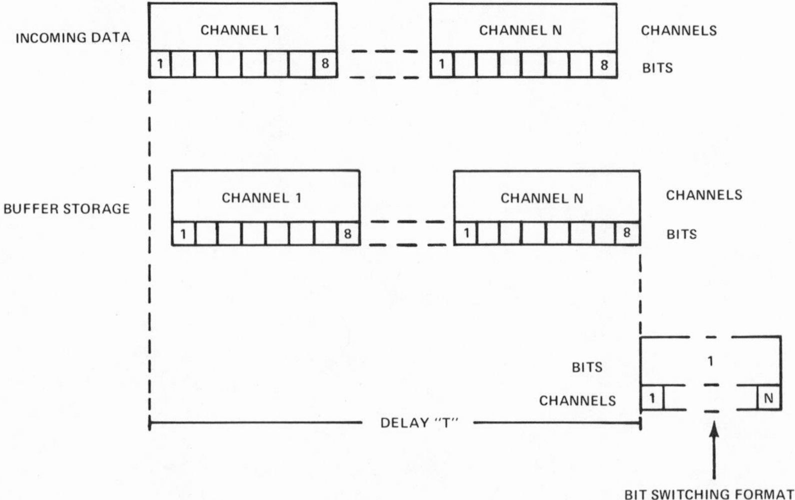

Figure 6.18. Conversion word–bit order bit streams.

must be able to convert to and from it. The obvious way is to use individual analog trunk interfaces by demodulating the carrier channels. However, other methods such as converting the digital words into which speech has been coded in one law into the corresponding code in the other. This can be done by a table look-up function which is well suited to stored-program control working.

There are two encoding, companding, laws: the A law used in the CEPT system and the μ law used in T1 carrier.

Legend for Companding Laws

Let v be the instantaneous value of the input voltage; V be the maximum input voltage level possible without limiting; n, the position of a quantization level related to the mindpoint of the range; and, N, the total number of quantization levels. Then we have the following laws.

A Law.

$$\begin{aligned} n/N &= (Av/V)/\,(1 + \log A) && \text{when } v \leqslant V/A \\ &= (1 + \log Av/V)/\,(1 + \log A) && \text{when } V/A \leqslant v \leqslant V \end{aligned}$$

In the case of the CEPT carrier where this companding system is used $A =$ 87.6. This gives a companding advantage of 24.1 between small ratios of v and V and large ratios.

μ Law. The μ law is expressed as

$$n/N = \log\,(1 + Nv/V)/\log(1 + v)$$

The present preferred value for μ is 255. Earlier PCM systems used a value of 100 for μ. This, however, was not usable for switching purposes.

7

Interfacing

7.1. Introduction

Any telephone exchange has to interface with its existing environment. Hence, there will always be the need for adaptive engineering for specific exchanges. The design approach taken must recognize these needs and allow the system to adapt gracefully to it by modifications of software and/or hardware. In particular, it must be able to adapt to varying requirements for line and trunk signaling and two-, three-, or four-wire trunks. These trunks will interconnect with existing direct and indirect control exchanges both public and private.

7.2. Allocation of Interfaces

A system interface is often termed a *port*. This is defined as an interface to a switching system transmission matrix with access to the system control. The system design must permit interfacing of a port with any of the following (via the existing line or trunk transmission equipment):

- Various types of station equipment
- Various types of interexchange trunks
- Lines and trunks to local or remote automatic or manual test facilities
- Lines and trunks to local or remote test and repair desks
- Test line terminations
- Intercept trunks
- Recorded announcement machines
- Such other circuits as may be required for processing, routing, advancing and terminating any kind of call
- Lines and trunks required for local or remote fault reporting, maintenance and administration of the system

Direction of trunk traffic. Ports may serve either one-way or two-way trunks.

7.3. Types of Line Terminations

There are various kinds of station equipment which connect to line transmission equipment. Figure 7.1 shows the association of the most common types of stations.

These include interfaces for the following:

- Single-party residence or business telephones
- Foreign exchange lines
- Manual lines
- Private branch exchanges
- Coin boxes
- Multiparty lines
- Data modems
- Mobile radio

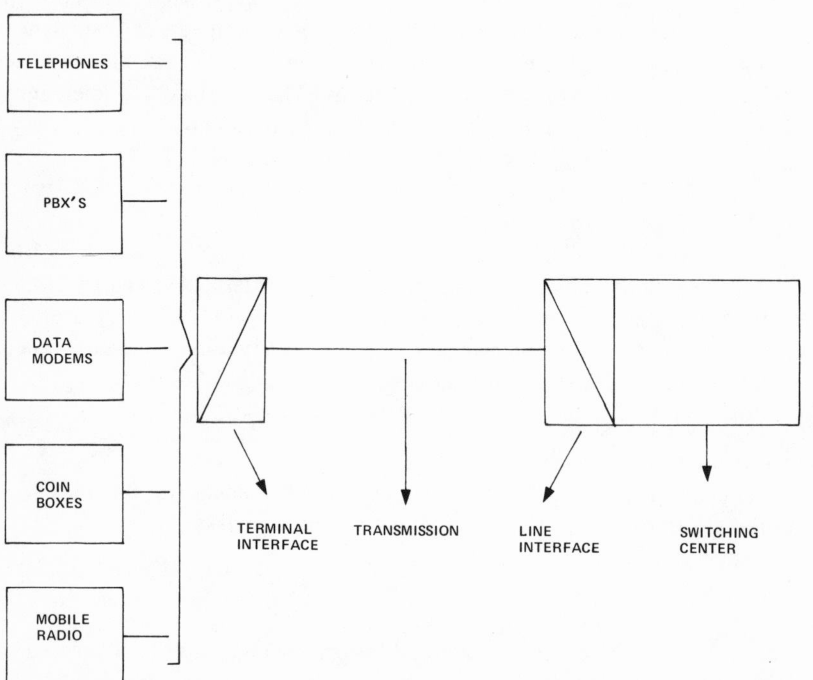

Figure 7.1. Types of stations.

7.3.1. Line Features

The features provided must include the following:

- Automatic number identification
- Rotary dial calling device
- DTMF calling device
- Remote indication of telephone "charges" to customers' premises
- Hotel or motel services for charging guests for calls
- Automatic lockout of faulty lines
- Line load control to reduce nonessential service during emergencies
- Called party answer signals

7.3.1.1. Line Lockout

A lockout line function is arranged to remove service from a line if, after a suitable delay, no calling signals are received. Calls to a line which is locked out receive a busy tone. Removal of the condition automatically restores the line to service.

7.3.1.2. Terminating Only

Lines marked with terminating only class of service are not allowed to originate calls.

7.3.1.3. Originating Only

Lines marked with originating only class of service are not allowed to receive calls. An unlisted directory number should be assigned to enable test access to these lines.

7.3.1.4. Free Terminating

Calls made to lines marked with free terminating class of service are not charged.

7.3.1.5. Line-Load Control

Line-load control is a means of taking certain lines out of service during emergency conditions.

7.3.1.6. Loop Start

Loop start is a method used by station equipment to indicate to the local exchange that service is required.

7.3.1.7. Multiline Hunting Private Brand Exchange (PBX Hunting)

Multiline hunting is a method of routing calls to one of a number of line interfaces when all such interfaces serve the same customer.

The equipment position numbers need not be consecutive but can be located wherever convenient. Directory numbers can be assigned to each interface but hunting only occurs if only one of the numbers is dialed. Dialing other numbers results in the call being treated as though an individual line had been dialed.

It may be necessary (for traffic balancing purposes) to distribute the interfaces over the line equipment groups. Multiline hunt always starts from the same point and proceeds in a predetermined sequence through the allocated equipment numbers.

7.3.1.8. Reverting Call

Reverting call is a method of alerting a called subscriber when both the calling and called subscriber share the same line. It is also a method used by operating company installers to check out the ringing-related functions of the station equipment from the subscriber's premises whether it is a single-party or a two-party installation. In the case of a single-party reverting call feature, the term *inspector's (faults man's) ringback* is normally used.

Interparty call. The calling subscriber dials only the number of the called subscriber. The caller hears a distinctive steady tone which is specified in tone and cadence by the country of use. The calling party disconnects, causing the called party to be rung in the normal manner. When the called party answers, ringing is tripped, a burst of tone or a recorded announcement, according to the country requirements, is applied, and the line is released from other switching equipment to allow the parties to converse. The line is placed into "busy" status but is not placed into lockout status since this could cause false trouble indications.

However, if the calling party fails to disconnect within approximately 28 sec after receipt of the tone, the line is placed into lockout. If ringing is not tripped within approximately 60 to 90 sec, ringing is removed and the line is restored to normal service.

Inspector's ringback. Inspector's ringback, single-party lines, occurs when the calling station dials a special code (usually unlisted). When the caller replaces the handset the calling station is rung. This continues until time-out or the handset is removed as described for reverting call service. When the handset is replaced for the second time the call is disconnected.

7.3.1.9. Loop Range Extension

Station lines which are beyond the system limits require special treatment.

Common mode operation. In this method loop range extenders are included in the switching matrix. The class of service of the line indicates when this is necessary; however, this may not be practical in any given system design. An alternative method is as follows.

Individual loop range extenders. These are connected between the line transmission equipment and the line circuit assigned to that station.

Message rate recording to hotel or motel Private Automatic Branch Exchange (PABXs). This provides remote message rate recording on calls originated from stations of that PABX to chargeable local or EAS points. The control signals are applied when the call is answered by one of the following methods, depending upon the type of PABX equipment at the hotel or motel:

- Reverse battery supervision toward the PABX when a chargeable local call is answered.
- Battery or ground pulse or pulses of 75 ± 25 msec each applied to a third wire associated with each trunk from the PABX when a chargeable local call is answered.
- Pulses of 50 Hz or 12 kHz or 16 kHz applied via an applique circuit at the port in response to metering signals from the interface.

7.3.2. Allocation of Line Features

Line features can be divided into two classes; the first relates to the type of lines being served, the second to the specific services which can be provided such as type of calling device, etc. The available services are indicated by the relevant class of service. This indication is usually data stored in the memory area assigned to the interface. The information is obtained when the call is being set up.

7.3.3. Standardization of Line Interfaces

The administration of an exchange which included both record keeping and load balancing could be greater complicated by providing multiple kinds of line circuits. Systems should therefore utilize one kind of line circuit which can serve the most commonly encountered needs such as single party, PBX, or coin lines, etc. The use of line circuit adapters should be restricted to such functions as remote metering and equipment to extend the operating loop limits for long lines.

7.3.3.1. Basic Calling Line Interface Functions

Figure 7.2 shows the basic calling line interface functions.

Loop sensor. The loop sensor monitors the loop from the station for station signals. It provides an input to a scanner which regularly checks each line for a request-for-service signal. In the example it is assumed that the register is accessed from a concentration matrix. Hence when the request is honored the

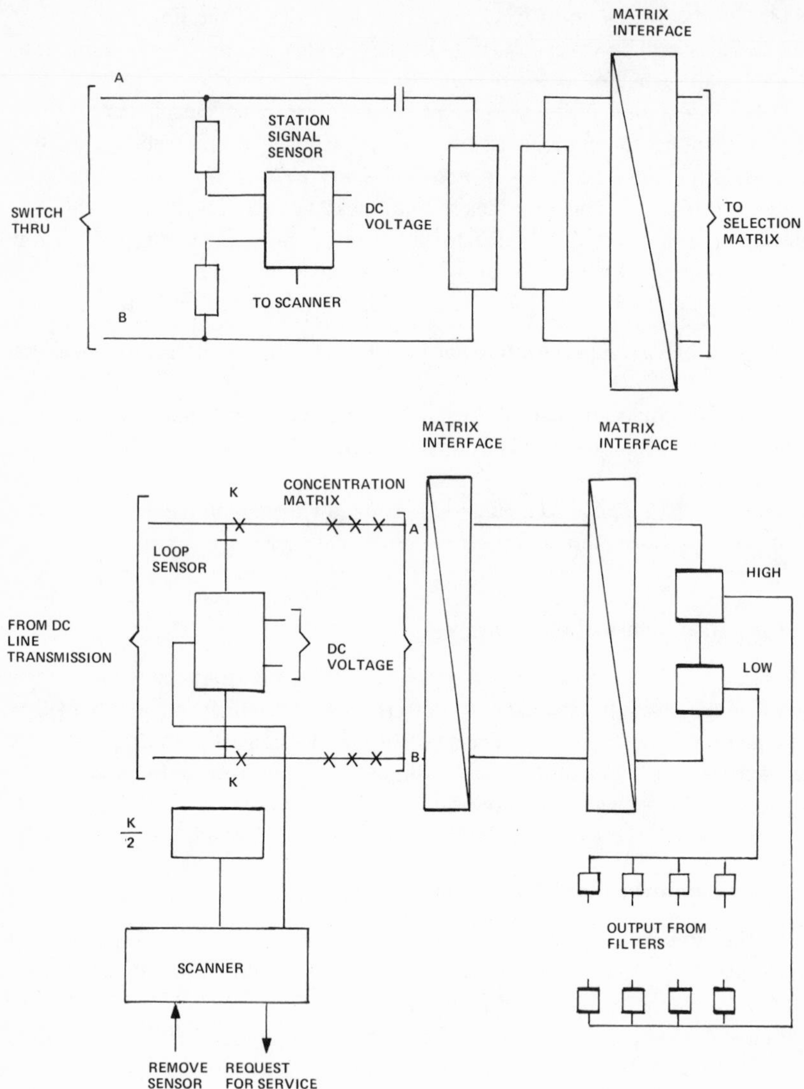

Figure 7.2. Basic line interfaces calling functions.

loop sensor is disconnected and the station loop connected to the calling detector (lower right hand) via the selection matrix.

The figure shows the transmission path from the loop extended to a DTMF detector. However, if the calling line used a dial and the switching transmission matrix was not compatible with loop/disconnect pulses then the loop sensor would detect the dial pulses.

Modern signaling detectors should operate on a station loop resistance of 3000 Ω with an insulation resistance of 30,000 Ω leg to leg or leg to ground with a loop sensing voltage of 44–54 V. It should, however, be remembered that loops of this length may require transmission treatment or higher battery voltages to provide adequate transmitter current for the telephone. The system should recognize any closed circuit interval of 10 msec or more as an off-hook station loop condition.

It should recognize any open circuit interval of 10 msec or longer duration as a station open loop disconnect signal. It should recognize a sustained closed loop condition of 180 msec or longer duration, following a series of disconnection signals, each of which lasts for less than 180 msec, as an end of digit signal. It should recognize a sustained open circuit condition of 180 msec or more as a disconnect signal, unless the system uses hook switch flash signals.

Typical loop lengths for a 1500-Ω loop without the telephone are 540 m of 26 American standard gauge wire or 870 m of 24 American standard gauge (AWG).

The corresponding lengths for station loops which include 200 Ω for the telephone is 468 m of 26 gauge or 750 m of 24 American standard gauge (AWG). The cross-sectional areas of 26 AWG and 24 AWG are 0.129 mm^2 and 0.204 mm^2, respectively.

Tone detection. Tone detection separates the signal into one in the lower band and one in the higher band and decodes the signal into corresponding system signals. The tones are decoded by analog filters or by digital analysis. In either case the detector must be designed to respond to signals having the following characteristics.

The receiver must register a digit if the frequencies are within plus or minus 1.5% of their nominal value, but not register a digit if either frequency deviates from its nominal value by more than plus or minus 3.5%. It must accept signals, with a power per frequency of −25 to 0 dBm when any of the four highest frequencies signal powers are no more than +4 to −8 dB relative to the low-frequency signal of the digit pair. The measurements are made with a 900-Ω termination bridged across the receiver. It must accept signals recurring every 85 msec with a minimum DTMF signal pulse of 40 msec duration and minimum interval between signals of 40 msec.

Calling transmission battery supply and sensor. The calling transmission battery supply and sensor provide the battery power to drive the telephone transmitter and detect station signals in an established call. Older types of switching systems used capacitive coupling with a capacitor, usually of 2 or 4 mF, in series with each leg of the transmission pair and a high impedance loop sensor, generally a telephone-type relay. Newer types of switching systems use inductive coupling between the calling and called transmission pair. Systems employing a transmission matrix which is noncompatible with line transmission equipment (such as digital systems) use split transmission bridges. These are often of the

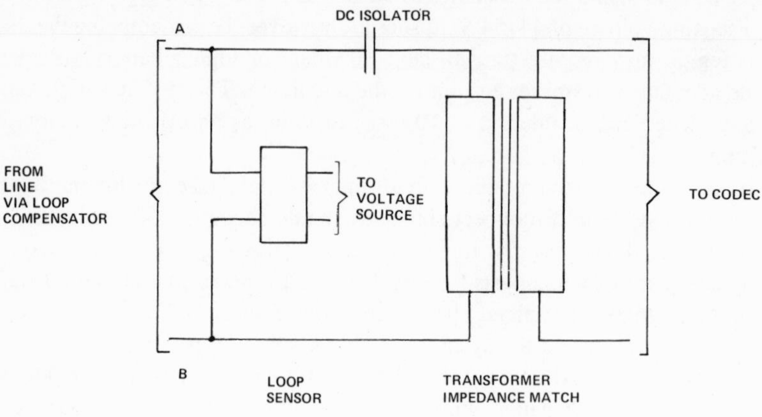

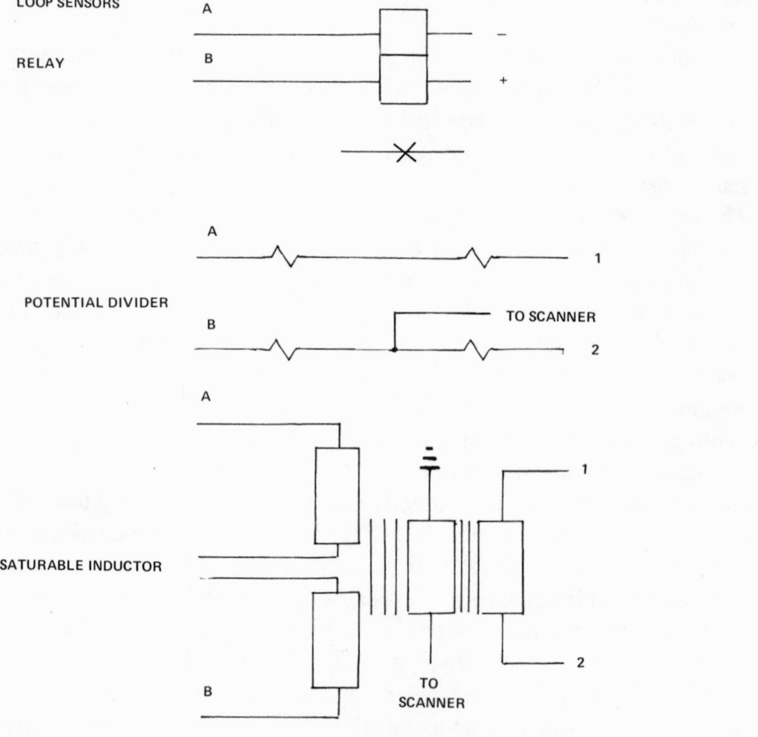

Figure 7.3. Line sensors.

type shown in Figure 7.3. The condenser prevents direct current from flowing through the coupling transformer. This allows the use of smaller transformers, since smaller cross-sectional areas can be used for the cores, due to the use of higher permeability magnetic core material.

Types of line sensors. Figure 7.3 shows some typical types of line sensors.

These include two current-sensitive devices: a relay which was described above and a saturable inductor. The saturable inductor is based on the use of a transformer which is connected in series with a station loop. An alternating current flows through one of the windings and when the circuit is in a quiescent state induces a current into a sensing winding. This winding produces output signals to the control. When a request for service occurs the current which flows through the station loop saturates the transformer, which loses its induction. Thus the output signals stop and the control detects the request.

The other type of sensing principle involves measuring the change in potential across a resistive network connected in parallel with the station loop. This potential varies from that due to the normal on-hook to that due to the greatly reduced off-hook condition. Other types of sensors may be used depending upon available technology but in all cases care must be taken to protect sensors from excessive voltage and currents due to lightning or cable faults.

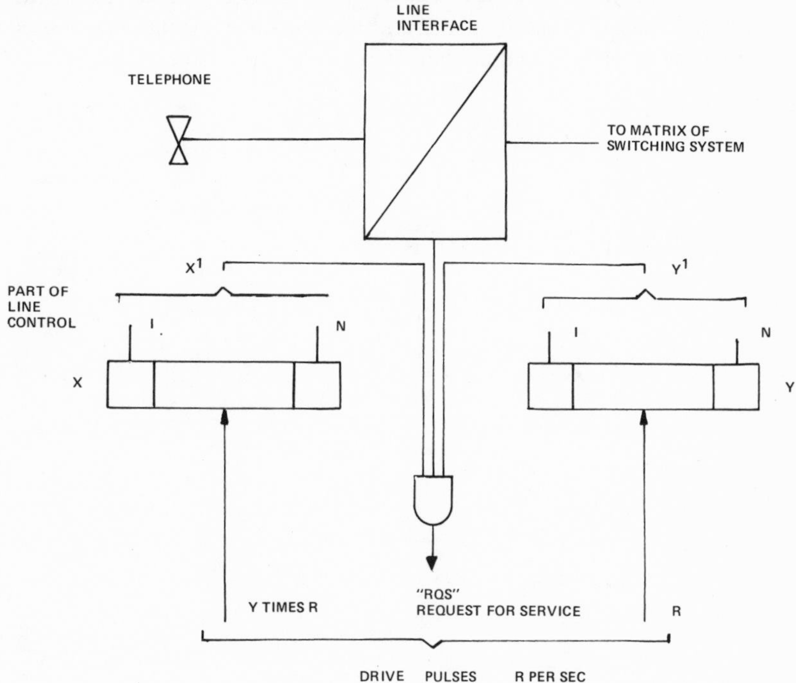

Figure 7.4. Principle of line scanning.

Line scanning. The state of a station loop is obtained by monitoring the line circuit interfaces at regular intervals. This is called *scanning* and the principle is shown in Figure 7.4.

A three-input "and" gate, which requires the presence of a signal on all three of the inputs at the same time, is provided for each interface. The input with the line sensor provides a request-for-service signal. The other two inputs represent the address of the interface. The example shows two inputs, x and y.

These have values depending on the size of the group being scanned. The size of the group depends on many factors, such as maximum time of scanning frames, yr, (which should not exceed 5 msec), modularity of system, etc. These must be decided on a system basis and include packaging considerations.

Generally speaking x and y should be of equal value since x should be the square root of xy for the minimum number of steps in the control counters which drive the gates. Excessive intervals in the successive interrogation of line interfaces can result in dial tone delays and, if scanning for dial pulse signals is used, mutilated signals.

7.3.3.2. Basic Called Line Interface Functions

Figure 7.5 shows the called line functions necessary in interfacing with a digital system via a concentration matrix. Since the same line interface is used for both originating and terminating calls it is necessary for the system control to

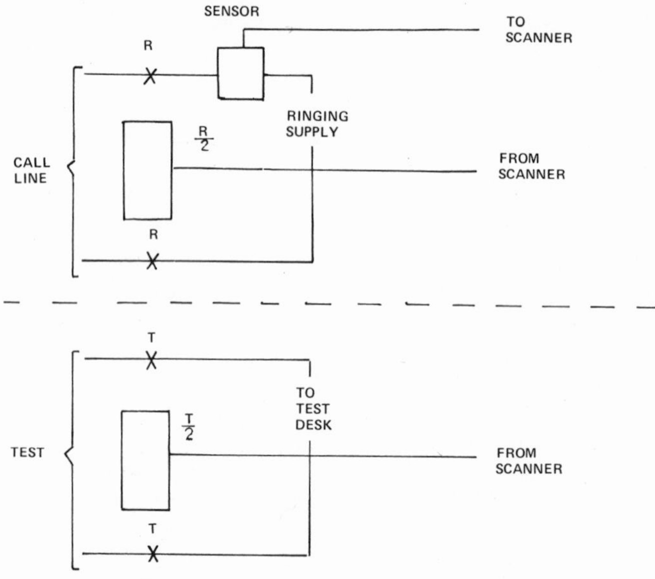

Figure 7.5. Basic line interfaces called functions.

remove the station loop sensor before a terminating call can be extended. This is accomplished by stopping the scanning action of the line group and setting the scanner to a position corresponding to the called line. This condition persists until the relevant instruction has been transmitted.

There are three basic functions used in extending a call to a line.

Ringing the station bell. This is done by superimposing the ringing signals on a direct current supply and extending them via a sensor. This is accomplished, in the figure, by the operation of relay R. An on-hook station has a telephone bell and condensor in series with the station loop. Hence, the sensor will not be activated. When the call is answered a loop for direct current replaces the previous condition and the sensor will be activated. It is important that the sensor be capable of detecting an off-hook condition during both the ringing and silent portion of the ringing cycle. If it does not, the answering party could well be the recipient of an acoustic shock.

Transmission bridge. There must be a means of providing the current for the called party's telephone; hence, the second half of transmission bridge must be substituted for the ringing circuit when the called party answers and the ring-trip sensor has recognized and signaled this fact to the system control.

Line testing. There must be a means for extending test circuits from a test desk or other facility to the customer's line. This is shown as the function of relay T in the figure which is operated as a result of an instruction from the system control via the scanner.

7.3.4. Types of Lines

The various types of lines are as follows.

7.3.4.1. Individual (Single-Party) Lines

Single-party residence and business lines are lines which have a main station and may have extensions. They use standard telephones or business telephones (which are multiline instruments with or without key telephone units), and other devices which use signals similar to telephones. These include such devices as automatic answering equipment and call rerouting equipment, teletype writer terminals, which use data-converting modems, telephone-company-provided connecting arrangements, etc.

Stations may be served by way of physical pairs, derived circuits, or concentrators; line circuits serving these lines recognize a loop closure (loop start) as a service request.

7.3.4.2. Foreign Exchange (FX)

Foreign exchange service provides for station equipment and services as described above except that the station equipment is situated in a remote exchange location which would normally provide it with telephone service. However, it is connected to and obtains service from an exchange other than its normal exchange. All features of the exchange serving the FX station are available to the served station.

FX service to a private branch exchange provides the same functions as local service to a private branch exchange. These lines usually serve businesses which anticipate a commercial advantage by such direct connection to the area. The lines will of course require special treatment to compensate for transmission and signaling loss.

7.3.4.3. Manual Lines

Manual service lines serve stations not equipped with a calling device. Such stations may or may not receive terminating calls automatically. Manual service lines receive no call progress tones except for ringback tone from the terminating point. The terminating point may be an attendant, operator, telephone instrument, or other device served by the same local exchange as the calling manual lines. Manual lines require a special request for service response and line lockout is only activated by the failure to disconnect from a call. The special response requires the automatic generation of the dedicated address of the only answering source.

7.3.4.4. Private Branch Exchange (PBX) Lines

PBX lines are lines which connect cordless manual PBXs, cord-type manual PBXs, and dial-type, automatic PABXs to a local exchange. Figure 7.6 shows the general arrangement of a PABX.

Its structure is similar to a local exchange and very large PABXs employ the same system architecture as local exchanges. Smaller PABXs use a scaled-down system or a special design. For example, a single unduplicated control could provide all the call-processing functions. They connect to a local exchanges via line or trunk interfaces depending on the traffic offered. They employ consoles for assistance in processing certain outgoing calls and routing incoming calls to the required PABX station. The PABX attendant may also be able to transfer or add parties to a call etc.

A PBX has a group of lines allocated to it which are accessed by a single directory number. Dialing this number (called the pilot number) results in testing each of the lines in the group until either a free line is encountered or, after all lines have been tested, a busy indication. Intermediate lines may be allocated their own directory numbers. Such lines are tested individually and if they are found to be busy no other lines are tested and busy tone is returned to the caller. The normal interface for a PBX line is a standard line interface.

In North America a request for service is made by the application of ground to one of the wires of the station loop pair (ground start).

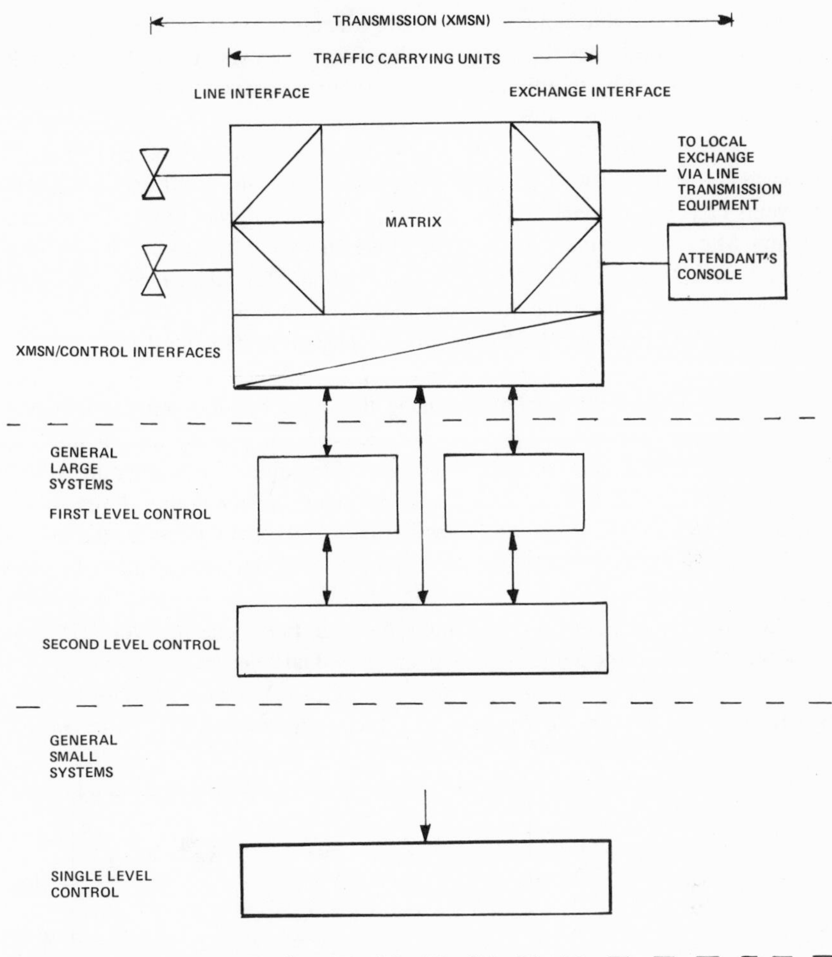

Figure 7.6. PABX system organization.

The following types of PBX services are provided from the local exchange.

Allocation of PBX directory number to line interfaces. All PABX lines have equipment numbers but usually only one directory number is assigned to the PBX. Other lines may be assigned directory numbers for "night service" or other purposes. Directory numbers are not required for the remaining lines, which can be reached only by dialing the pilot number. The number of lines in any PBX group may be two or more, the ultimate number depending on the system design. The pilot number may be any available directory number. It should be possible to segregate incoming from outgoing PBX lines since businesses often require unidirectional calling lines and sometimes called lines.

PBX night service. PBX night service is a feature which permits specified lines to be connected to stations at the PBX when the attendant is off duty. These night service lines may be assigned directory numbers other than the pilot number or reached by the pilot number after business hours. Each night service line is able to originate calls.

PBX night busying. PBX night busying is a feature which prevents calls from being routed to a PBX by busying out lines at the local exchange. This prevents calls being connected through to lines which do not have an answering facility when the attendant is off duty. This is usually effected by operating the battery cut-off key at the PBX attendant's switchboard, which causes the busying of the relevant lines at the local exchange. Night busying does not prevent outgoing calls from the PBX by using lines busied at the central office.

Direct inward dialing to a PABX (DID). Direct inward dialing (DID) permits calling to a PBX extension using the listed public telephone directory number to reach any listed PABX station without the intervention of the PBX attendant. The first three, four, or five digits of the directory number are used to select and route the call to the desired PBX with the remaining digits, which identify the station, sent (usually over a trunk) to the PBX to select the desired station there. The attendant can be reached by dialing a separate directory number.

Remote metering at PBX. Remote message rate metering over PBX lines is accomplished in the same way as over individual lines.

Central office (CO) centrex service. CO Centrex service provides for serving lines of one or more PBX subscribers directly from the central office instead

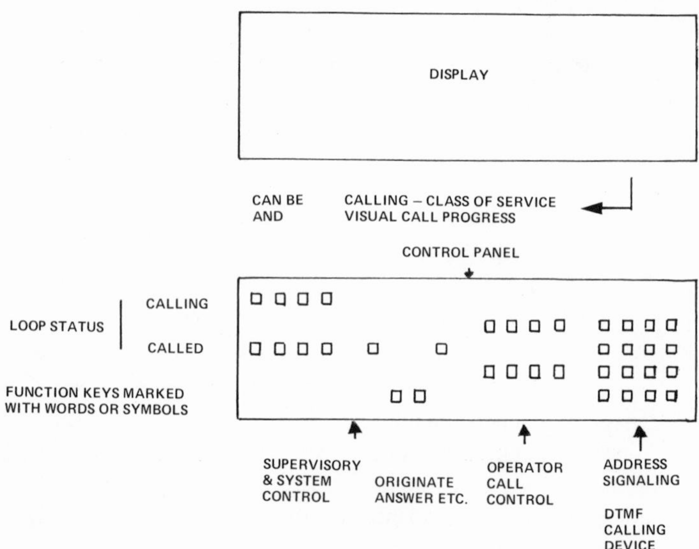

Figure 7.7. Layout of control display panel.

of a separate switching system on the subscriber premises. It requires very large numbers of classes of service since all the PBX stations appear as line interfaces, and each PBX requires separate grouping and service information.

Attendant's console. The PABX attendant's console has control keys, a calling device, and a display panel. Figure 7.7 shows some of the typical functions provided.

The display panel can display the number of a PABX station or that of an incoming transmission loop which has been extended to it. It can also display the called number, class of service, line state, call progress state, and so on.

The operation of the PABX is simplified by the use of a display panel with symbols. This enables it to be used by attendants speaking various languages; it was pioneered by the LM Ericsson Company.[31] Figure 7.8 shows a typical display panel layout. This provides a pictorial display of the progress of the call. The left-hand side shows whether the call is internal or external, if there are calls waiting, and if there is a request for operator intervention on a call (recall). The right-hand side shows the state of the call: idle, ring, answer, congestion, incoming blocked, outgoing blocked, etc. It also includes an alphanumeric display to provide the information discussed above.

7.3.4.5. Coin Box Lines

Coin box lines are lines serving coin boxes or, as they are called in North America, pay stations (see Figure 7.9).

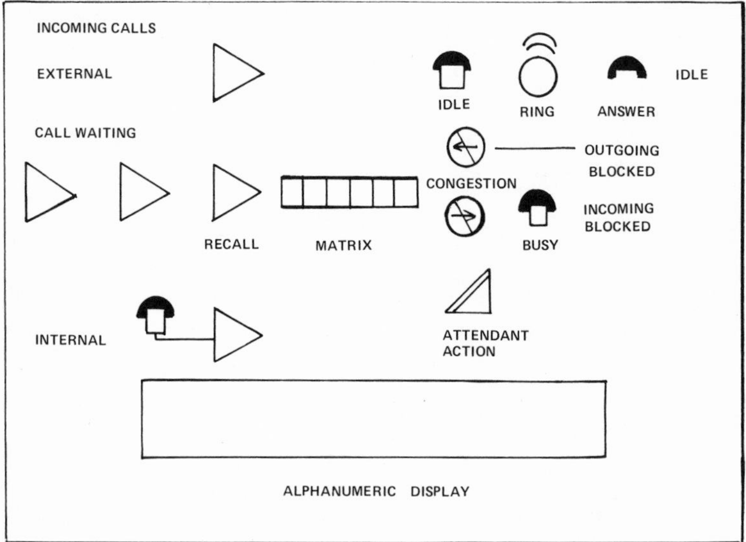

Figure 7.8. PABX attendant's display panel using symbols.

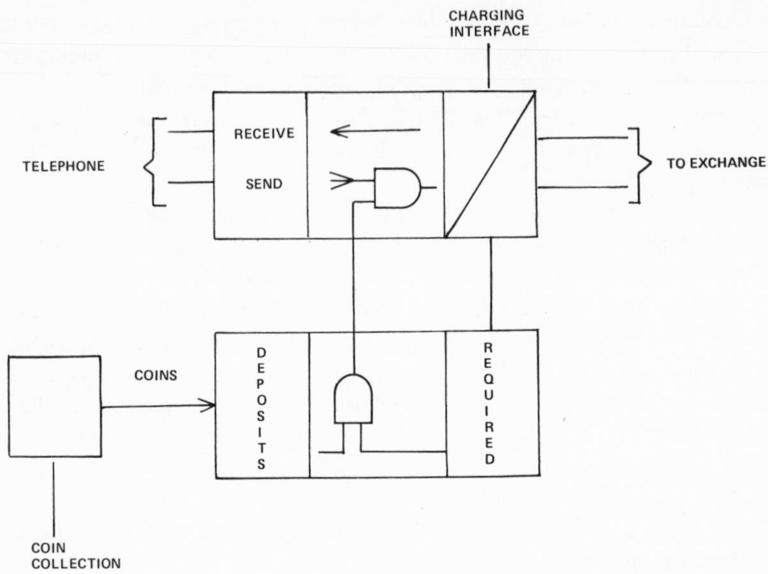

Figure 7.9. Principle of coin box.

A coin box may receive calls in addition to fulfilling its prime function of originating calls. It can operate in a number of different modes and requires the inclusion of the relevant control functions in the local exchange to which it connects.

7.3.4.6. Two-Party Lines

Two-party lines serve two separate subscribers. Each subscriber is alerted by ringing applied between one side of the line and ground. Each subscriber may have extensions.

In addition to two-party service there may be requirements for multiparty service which requires the provision of special ringing frequencies. A single line circuit interface with the switching center is used for all parties. A separate directory number is assigned for each main station.

7.3.4.7. Mobile Radio

Mobile radio customers are served in two ways: the original method, which required the intervention of an operator to originate or terminate calls, and an automatic method. The automatic method has two variants, low density (IMTS) and high density (HCMTS), which differ in the way that they interface with the telecommunications network.

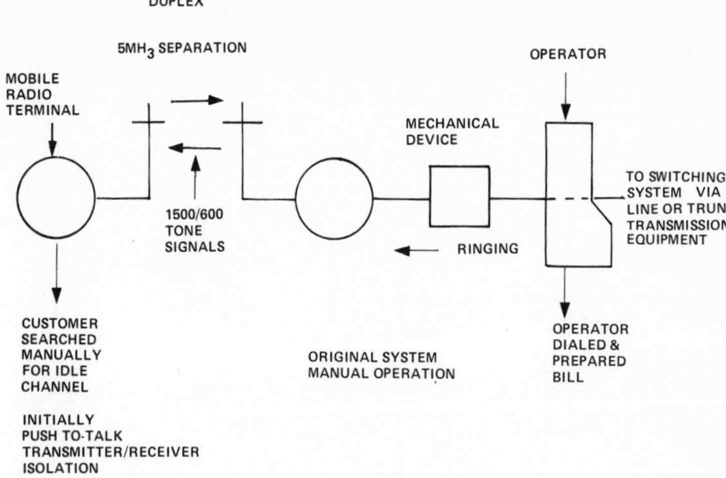

Figure 7.10. Mobile radio manual operation.

Manual operation. Figure 7.10 shows the general arrangements for manual operation. The customer searches manually for an idle radio channel. Tone signaling was used to activate the lamp signal to the operator who obtained the desired number, dialed the call and prepared the bill. Incoming calls terminating on the mobile radio terminal applied ringing to the called line, as in the case of any other terminating line call. This was converted into a lamp signal.*

*The system is now full duplex with 5 mHz channel separation. Originally it used "push-to-talk" operation.

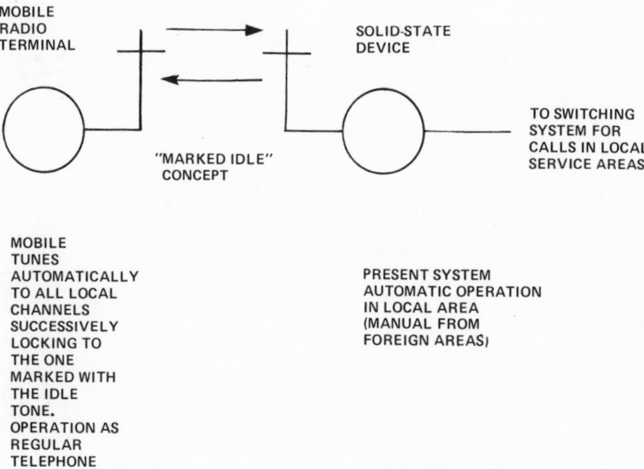

Figure 7.11. Mobile radio automatic operation.

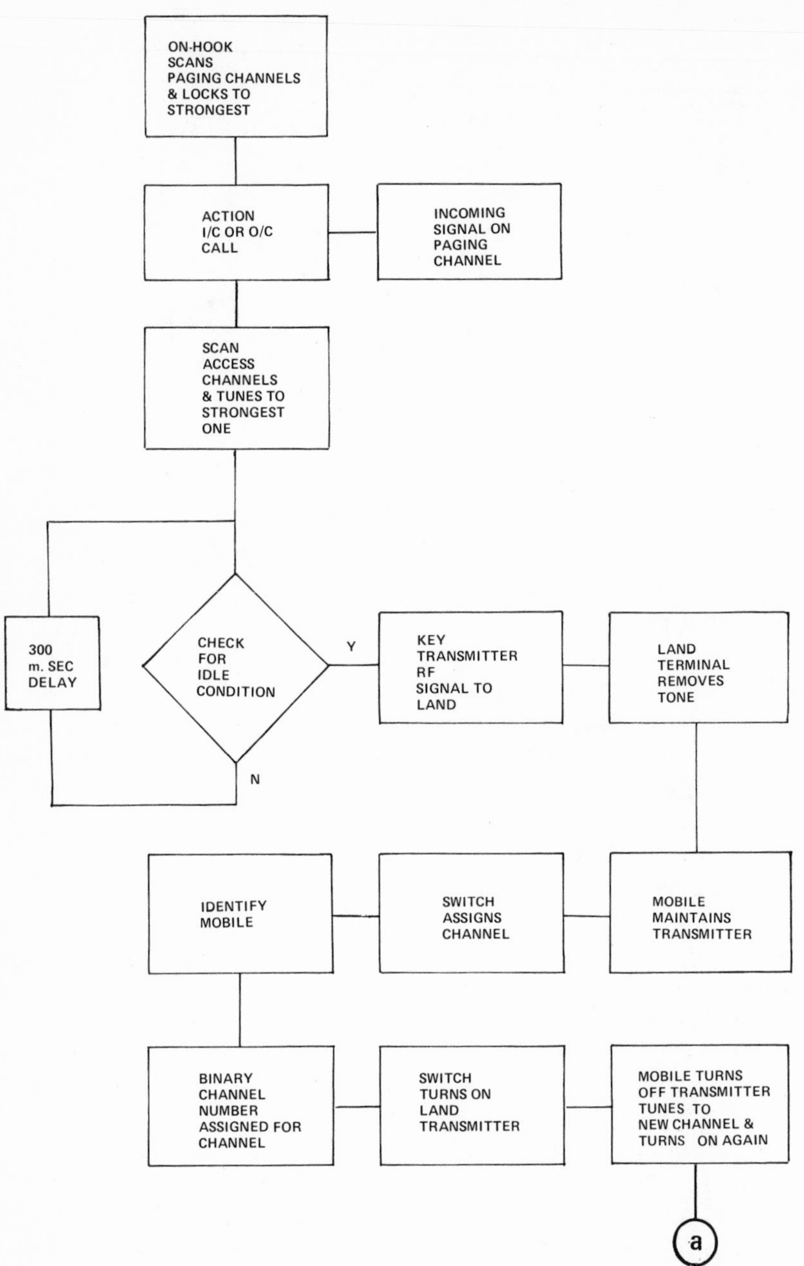

Figure 7.12. Part(i). Mobile radio flow chart.

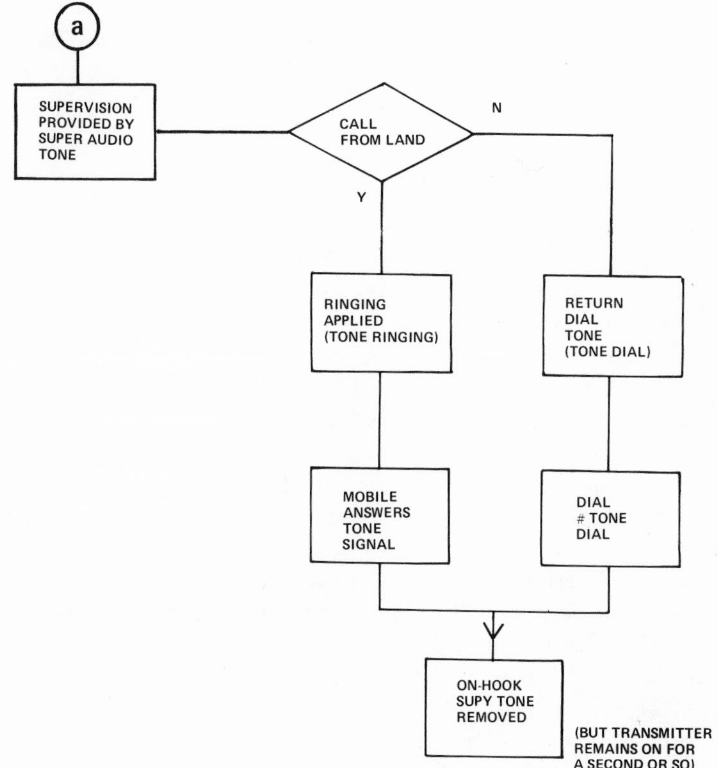

Figure 7.12. Part(ii). *See legend on previous page.*

Automatic operation (IMT). Figure 7.11 shows the central arrangements for low-density automatic operation.

In this case the mobile termination tunes automatically to each local channel successively, locking in to the channel marked with an idle tone. The system provides automatic operation in its local area with manual operation only used when in a distant area.

Figure 7.12 shows the flow chart of the sequence of functions. The operation of the mobile radio is the same as that of an automatic telephone, except a lamp indication of all channels busy is provided. The identifying number of the mobile station is automatically sent to the local exchange when the handset is removed. Dial tone is returned following the receipt of the calling number.

A mobile terminal unit (MTS) is used between the radio terminal and the local exchange. Standard line interfaces are used at the local exchange to which it connects via a two-wire transmission path which sends and receives normal telephone station signals. The mobile terminal unit provides the necessary interfaces between the local exchange and the mobile radio stations.

Automatic operation (HCMTS). This uses a special exchange function for its interface with the telecommunications network. At the time of going to press field trials of high-density systems are underway. The following is a summary of the original objectives of the ATT field trial. The system provides fully automatic working in a greatly increased area which is limited only by the availability of suitable exchanges.

The high capacity mobile telephone system (HCMTS) provides direct telephone network dialing service both to and from a vehicular telphone. The service is equivalent to that of a land subscriber in that dialing, dial tone, ringing, and full duplex operation (simultaneous transmission and reception) are provided to the mobile subscriber.

Calls to and from the nationwide telephone network are processed through a switching office for mobile service, which interfaces with radio equipment at strategically located base stations and in turn, to the mobile telephone. The system utilizes the ''cell'' coverage scheme in which a given metropolitan area is divided into a number of small areas or cells. Each cell is assigned a fraction of the available channels allowing for reuse of the same frequencies at spatially separated cells in the metropolitan coverage area. Calls in progress between the mobile and land networks may proceed indefinitely, being automatically reassigned to an available channel within a new cell as the vehicle moves between cells throughout the metropolitan area.

The mobile unit consists of three major sections: (1) an 850-mHz, approximately 600-channel transceiver section; (2) a supervisory or logic section; and (3) the handset and dialing control section. Radio operation is controlled by coded command from the land network. Commands are decoded within the supervisory section and its binary control data outputs perform the various functions necessary for uninterrupted land quality telephone service.

The new proposed high-density features include the ability to assign mobiles to radio channels optimally; charge for usage of radio facilities; allow same classes of service as fixed stations; associate specific groups of mobile units with their appropriate dispatchers; improve the grade of service and call-handling speed while serving much heavier traffic loads.

The system is capable of adaptively switching a call in progress as the relative geographic relationships of the station and processing centers change.

7.3.4.8. Mobile Line Capacity

Mobile line capacity requires a separate class of service for each group of lines served by a mobile terminal unit which can accommodate about 240 lines.

7.3.4.9. Volunteer Fire Alarm

A volunteer fire alarm circuit serves a group of firemen's lines which are single-party lines. These are reached by dialing a listed fire-reporting number.

Table 17. Some Telephone Dial Characteristics

Break/make (%)	Impulses per second	Masked pulse interval	Temperature humidity limits
62.5–66.7	10 ± 1	200 msec	−10 to +50°C 15% to 95% relative humidity
66.7	10 ± 0.8		
	20 ± 1.6	—	—
50–66	7–14	240 msec	—
50 ± 3	20 ± 1.6		
60 ± 3	16 ± 1		
67 ± 3	10 ± 1	320 msec	—
63 ± 2	10 ± 0.5	—	—
62–75	10 ± 1	200 msec	—

Ringback tone is returned until the first fireman answers. Idle lines are alerted by applying continuous ringing.

Busy lines have a special signal superimposed on the call. Such lines will be rung with continuous ringing when the called party hangs up. Ringing of unanswered lines and the signal to busy lines stops when the penultimate fireman line releases. This service alerting method is being displaced by the use of portable radios actuated from a fire station.

7.3.5. Telephones

The characteristics, features, and technical requirements for telephones used in the international market depend on the technical specification requirements of the country and the manufacturer. The following are typical characteristics.

7.3.5.1. Telephone Dial Characteristics

Telephone dial characteristics are shown in Table 17.

7.3.5.2. Ringer Characteristics

The ringer impedance may vary as follows:

- 4000 Ω at 25 Hz with 1 μF capacitor with a dc resistance of 1700 Ω.
- 9000 Ω at 16 Hz.
- 2000 Ω at 25 Hz and 20,000 Ω at 1000 Hz with a dc resistance of 100 Ω.
- 4000 Ω at 25 Hz with series capacitor.

7.3.5.3. Receiver Characteristics

Typical receiver characteristics are as follows:

- Receiving reference equivalent is 0–10.4 dB; impedance is approximately 300 Ω at 800 Hz.
- Specific response for receiver is approximately 71 dB; working impedance is between 65 and 280 Ω on the frequency range of 300–4000 Hz.
- Sensitivity is +46 dB relative to 1 dyn cm^{-2} $mV^{\frac{1}{2}}$ available power.

7.3.5.4. Transmitter Characteristics

Transmitter characteristics can vary as follows:

- Resistance during speech is 35–180 Ω; volume is 0.9–7.8 dB.
- Response range is −47 to −38 dB for frequency range of 300 Hz to 4 kHz; resistance is 35 Ω approximately.
- Sensitivity is +30 dB relative to 1 mV $dyn^{-1} cm^{-2}$ at 20 dyn cm^{-2} input.

7.3.5.5. Additional Telephone Features

Telephones may be equipped with keys which disconnect other parties, transfer calls, or hold calls. They may include a recall button and ringer volume control. They may have display lamps separate from or combined with the keys. However, in all cases they will interwork with standard line interfaces.

7.3.6. Coin Boxes

Many varieties of coin boxes will be encountered depending on the types used in the country of application.

Coin boxes use standard telephone calling devices and usually have standard telephone handsets and transmission circuits. They are equipped with ringers. They terminate on normal circuits but can generate heavy traffic.

Some coin boxes require the predepositing of coins before any action is taken; others, an audible signal before coins are deposited. Some collect coins on answer; others at the end of the call.

They vary in type from those having a completely manual operating mode to those having a fully automatic operating mode.

7.3.6.1. Fully Manual Operation

Fully manual coin boxes operate with two buttons. Except for calls to operator or to an emergency service (usually involving 0's and 9's in the code), a coin must be inserted before a call can be made. When the called person answers, the A button is depressed to collect the coin and conversation is possible. The operation of button B in lieu of button A returns the coins. This pay station handles single-fee local calls; other calls are handled by an operator. The manual control of coins by operators requires a means of generating the appropriate signals in the exchange under the control of signals from consoles or trunk positions.

7.3.6.2. Semimanual Operation

The operation is as above but the coin collection is achieved by a signal from the parent exchange. Sometimes coin refund is also handled in this way. Toll calls are handled by the operator. Some of the commonly encountered types are described below.

Semipostpay operation. This type of operation works as follows: lift the handset and dial following receipt of dial tone. The switching system reverses the loop current when the call is answered, if the call is chargeable. This reversal inhibits the transmitter. The insertion of a coin enables the transmitter. Coins once deposited cannot be returned. The method uses a standard line circuit but a reverse-battery function must be provided by the system.

Prepay operation. Prepay operation requires the deposit of a coin before or after the receipt of dial tone. The first case is called *coin first;* the second case is called *dial tone first*. The deposited coins are collected by a signal from the exchange. The coin box may include a means of automatically detecting that coins have been deposited.

Local prepay operation (North America). This dial-tone-first service, which is not compatible with coin-free emergency working, requires the deposit of a coin before dialing can be accomplished. The interface requirements require a means of applying a dc pulse of negative or positive polarity. The duration of the pulse is 600 plus or minus 50 msec with a minimum interval of 500 msec between successive pulses. It is necessary to temporarily apply a line discharge circuit to the line after the application of a pulse in order to reduce audible line transients to an acceptable value.

The direct current signals are made by applying a pulse of 130 V plus or minus 5 V. A positive voltage of 130 V with reference to ground is used to collect coins and a negative voltage of 130 V with reference to ground is used for refunding coins. The depositing of a coin causes the application of ground connected through a coin control magnet, having a nominal resistance of 1000 Ω, to the "tip" leg of the line transmission wire pair.

The tip conductor is the leg of loop which normally has ground connected through the transmission bridge to it. The coin magnet usually requires a minimum current of 65 mA for it to operate. Dialing is inhibited until a coin, or coins, is deposited. The system must apply a refund signal to the coin box when a disconnect signal is encountered.

Single-slot prepay pay stations (North America). This type of pay stations can provide either service without an initial coin deposit (dial tone first) or with an initial coin deposit depending upon whether emergency services are reached without first depositing a coin. Reversing the battery on answer sets the coin mechanism to its collect state. The local exchange examines the first one- to three-digit codes for free service listing and blocks unauthorized codes from further call processing.

Calls outside local area. These require the intervention of an operator. If the system requires the deposit of a coin to reach the operator then the system must refund the coin when the outgoing trunk is seized. When the operator answers, the reversal of battery at the relevant interface causes the immediate collection of any subsequent coins deposited, in case of operation from a local prepay pay station, by rerouting the call to busy. The operator determines the fee for the call and requests the deposit of the relevant coins.

The pay station is arranged so that the dropping of the coins in the coin slots produces audible signals to the operator. These may consist of tones acoustically generated by gongs and detected by the telephone transmitter or electrical tone signals. The latter tones are designed to permit the extension of direct dialing by providing "machine-readable" tones. This will ultimately allow coin boxes the same direct dialing features as a normal customer station. Calls routed to an operator from a coin box are usually identified by a burst of tone when the operator answers.

Coins are disposed of by the operator actuating special keys. These keys initiate the relevant "collect" or "refund" signal, which must be produced automatically and hence have the appropriate timing and other parameters at the local exchange to which the pay station is connected. Signaling through the intervening network from the operator's console can be by one of the following methods.

In-band coin control. This uses a pulse of 700 plus 1100 Hz for collect, a pulse of 1100 + 1700 Hz for refund, and a pulse of 700 + 1700 Hz for re-ring. These frequencies must be within ±1.5% of their normal value and have an output value of −6 dBm per frequency. The pulse must have a duration of 100±30 msec at the point of origin. This is preceded by a momentary "on-hook" signal.

Tone receivers at the point of conversion to dc coin control signals must respond to frequencies having a power of −22 dBm per frequency and to pulse durations of 100±40 msec.

Wink pulse coin control. This uses a series of pulses 100±30 msec with intervals of 125±25 msec.

These are decoded in the following way: one pulse for operator disconnect, two pulses for operator connect, three for coin collect, four pulses for coin return, and five pulses for re-ring.

7.3.6.3. Fully Automatic Operation

The collection of multifee coins can be made by dc signaling such as reverse battery for 400 msec to 1000 msec or by ac signaling as described below.

Local service. Collection and return is automatic for single-fee local calls. The replacement of the handset before the coin has been collected results in a refund.

Other service. Other pay stations include the means to control multifee calls and hence have elaborate circuitry. The operation of a typical pay station is described in the text below. Only calls requiring special service are handled by operators.

7.3.6.4. Operation of a Typical Single-Fee Coin Box

Typical single-fee coin boxes extend a loop to the exchange. A reversal of battery toward the station takes place at the time of answer supervision. Coin deposits are required initially to enable dialing. Such coins enable the calling device until they are collected. Coins are collected automatically when the call is answered by receiving reverse polarity answer signal. The pay station may be arranged to allow dialing 0 and 9 codes and other codes without coin deposits. Calls to free services will not receive reverse polarity and hence the coins will be returned upon release.

7.3.6.5. Typical Multifee Coin Boxes

Typical multifee coin boxes are coin-operated telephones which provide facilities for local, trunk, and international calls. The Sodeco Phonotaxe TE50 is typical of modern coin-operated telephones for the international market which are controlled from the telephone exchange. It can be connected to any line circuit of a switching system that has equipment for transmitting metering pulses to the subscriber.

Figure 7.13 shows the general arrangement. It does not require answer supervision and uses metering signals of the following types.

Metering signals (loop). These consist of pulses of 12 kHz ± 1% or 16 kHz ± 1% with a minimum duration of 50 msec and a minimum impulse interval of 50 msec receiving level; minimum 105 mV, maximum 2400 mV.

Metering signals (simplexed to ground). These consist of 50-Hz pulses between the a and b lines in parallel and return via ground 80 V ± 10% with a minimum pulse duration of 50 msec and a minimum impulse interval of 50 msec.

7.3.6.6. Operation of Phonotaxe Coin Box

Coins are deposited after the handset has been lifted (dial tone first). The cost of a call depends on the number of metering pulses received from the parent exchange.

The value of the inserted coins is shown on the reserve indicator on the coin box.

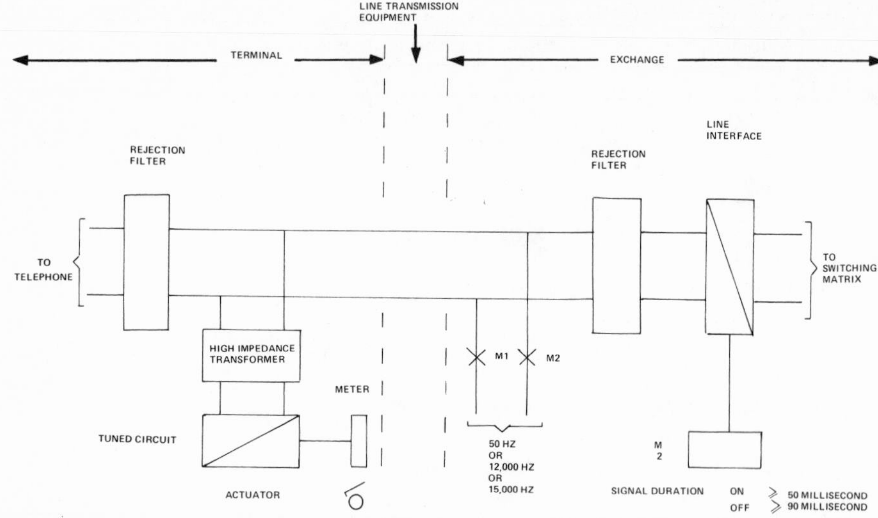

Figure 7.13. Remote metering principle.

The principle of operation of a coin box is shown in Figure 7.9.

It includes a telephone with a ringer and calling device and a means of coin control. The latter consists of a charging interface which has a means of indicating when coins are deposited including the value of the coin and a means of collecting coins.

The means of accomplishing the coin collecting and refunding functions depends on the operating mode and whether the coin box is semiautomatic or fully automatic, In any case it connects to a line circuit at the local exchange.

Some of the operating modes are described below.

- During the course of a call, the balance shown in the reserve indicator is reduced depending on the incoming metering pulses.
- A warning signal to insert another coin is given when the reserve has been emptied.
- Any coins inserted during the course of a call are added to the balance.
- When the receiver is replaced, the circuitry determines the best combination among the inserted coins for refunding the amount due.
- When these coins have been returned, the remainder are collected.

7.4. Trunk Interfaces

Trunk interfaces can be required for analog and/or digital transmission and be two wire or four wire. Transmission hybrid circuits will be provided as required.

7.4.1. Analog Interfaces

Analog interfaces may be derived or nonderived. Derived channels have a bandwidth of 4 kHz and will normally use E&M line signaling; they will be four-wire inputs. Non-derived channels will usually be two wire and normally use loop–disconnect line signaling.

7.4.1.1. E&M Signaling Interface (Continuous Signals)

Figure 7.14 shows a typical circuit arrangement for interfacing a trunk with a multiplexed carrier system used in North America.

It includes the two-wire to four-wire hybrid for the signaling path and the means of converting the single frequency line signals to E&M leads. The trunk termination must include surge suppression if an inductive sensor is used. The original E&M lead signaling circuit used only one lead with a common ground return for each direction of signaling. This means that the signaling leads have a greater noise influence than if a pair of balanced leads were used similar to transmission circuits. This is satisfactory for electromechanical systems, but not satisfactory for electronic systems. As a result, new E&M lead interfaces were designed by AT&T for electronic switching systems.

The preferred arrangement for trunk circuits in electronic switching environments is a four-wire, fully looped arrangement in which open and closure signals are used in each direction.

7.4.1.2. E&M Signaling Interface (Discontinuous Signals)

There is an alternative method of E&M signaling which employs pulsed signaling to increase the number of possible signals. When used with single (nonlooped) E&M leads it uses a ground signal, instead of the battery signal, for the M lead.

7.4.2. Interfaces with Signaling Systems

Interfaces with channel-associated signaling systems are provided by appropriate trunk circuits. These match the switching system to the outside plant.

The two basic types of signaling interfaces are those for derived trunks which use E&M signaling, and those for direct connected trunks, which normally use loop signaling.

In the case of common channeling signaling systems there is a separate, common channel used exclusively for signaling which is independent of the speech circuits.

The transmission of signal information on a separate channel simplifies the interface.

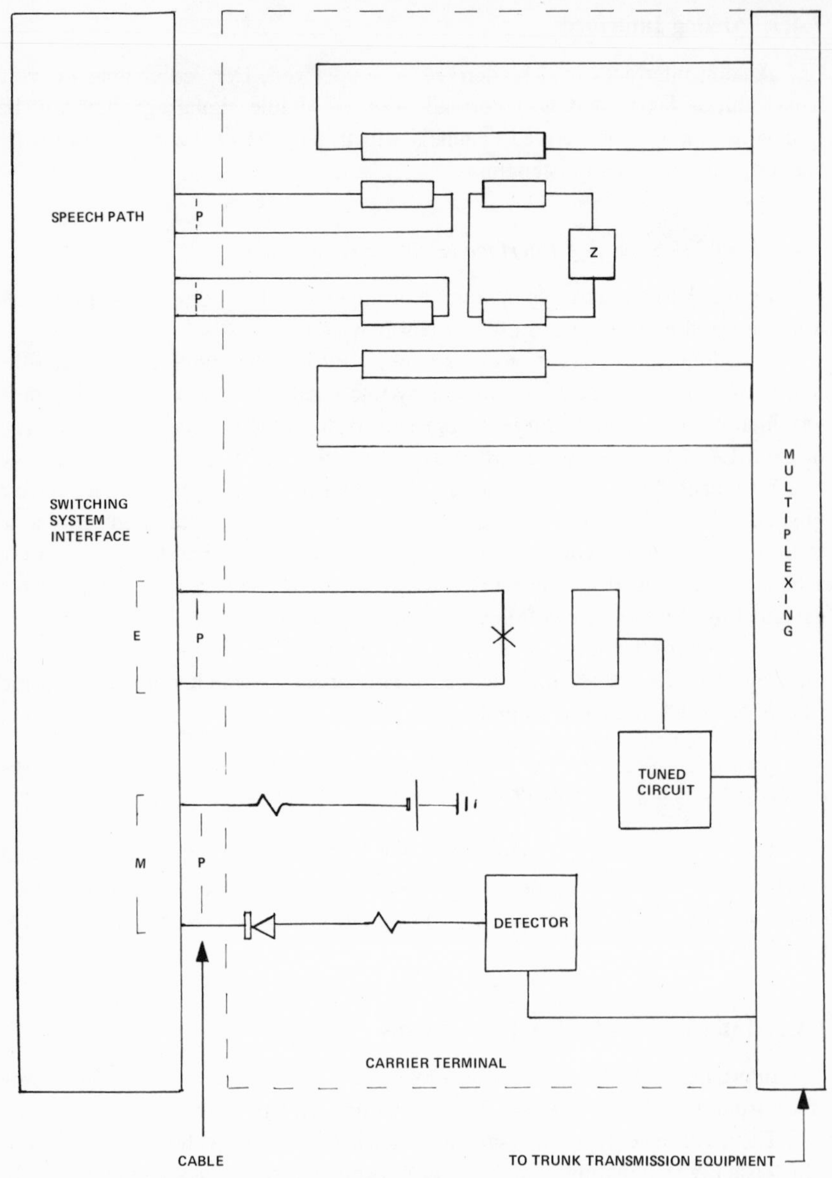

Figure 7.14. Conversion of E and ME to analog.

7.4.2.1. Interfacing with CCITT *No. 6 Signaling System*

Figure 7.15 shows the general arrangements. A universal signaling terminal can work with frequency division multiplex carrier systems via a modem at a rate of 2400 bits/sec. It can also utilize PCM carrier system channels by the use of a sychronous interface which operates at a rate of 2400 bits/sec. It is also possible

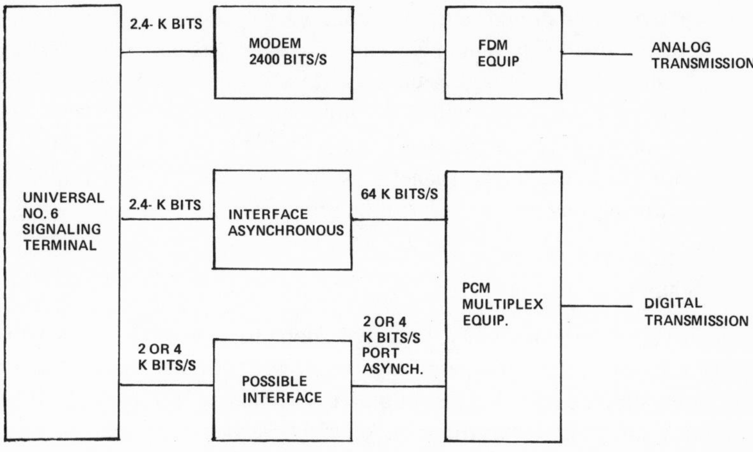

Figure 7.15. General interface arrangements for CCITT signaling system No. 6.

to work at bit rates of 2000 bit/sec or 4000 bits/sec sychronously with the port of a PCM transmission system.

7.4.2.2. Three-Wire Trunks

Interexchange trunks using three wires may be required in some networks. The third wire connects directly to the distant exchange and is used for line signaling and/or metering.

7.4.3. Features Common to All Types of Trunk Interfaces

Common features include the following:

- Maintenance access should be provided for monitoring and testing.
- Visual indications should be given when a trunk is in use, defective, or has been taken out of service by maintenance.
- Idle circuit termination (900 Ω standard, 600 Ω optional) should be provided toward remote end when trunk is in idle state.
- Optional on-demand local and/or remote readout of locked-out, defective, and maintenance out-of-service trunks should be provided.
- A traffic-measuring means should be provided, as well as a balanced transmission circuit toward the remote end.
- A means of repeating any pulsing and line signals between the system control and the associated trunk should be provided.
- Protection should be provided against false operation due to differences in battery or in ground potentials at the distant end the originating end of the trunk.

- The interface should provide a trunk "make-busy" feature controlled from the trunk circuit directly or remotely. The automatic "make-busy" feature is required when the trunk card is unplugged, when the power fuse is blown, when there is power failure within the trunk or outside it, when the trunk is under test, or when any circuit associated with the trunk is faulty. Busy fault indication to maintenance personnel must be provided.

  The following features should also be provided:

  - Listening and test jacks.
  - Lightning protection.

The means, where necessary, to accept loop-disconnect (dial signals) from the trunk with an open period percentage of 60 ± 2% at any required speed.

Additionally, the insertion loss of the trunk should be less than 0.3 dB and it should satisfy all requirements for noise, cross talk, echo return loss, longitudinal unbalance, and other transmission requirements.

At call release, the outgoing trunk should be momentarily marked as busy after call release to allow for the release of the associated equipment in the other exchange.

Message metering pulses should be received and repeated. Also such pulses should be generated and sent as required to the associated equipment.

The system should be programmed to hold the trunk under the control of the calling party, except in case of trunks connecting to manual service and when the called party is under malicious call supervision. In such cases, the call is held under the control of both the called and calling parties. It is also necessary under some conditions to hold the call under the control of an operator.

The system should provide an indication with other trunks in its group of an all-trunk-busy (ATB) condition of a trunk group.

The system should be adaptable to connecting a variable attentuator pad of 0–7 dB.

The system should inject progress signals into the transmission path using a balanced method. An application of tones when an analog tone distribution is used is shown in Figure 7.16.

7.4.3.1. Outgoing Loop Signaling Trunks

The following additional basic features are necessary:

- Seizure by providing a dc loop with no greater resistance than 300 Ω.
- Immediate busy to outgoing traffic upon seizure.
- Continuous monitoring of the loop current for reversal of polarity and for the presence of a loop current of a specified level.
- Repeat dialed digits received from the sender function of the system without distortion toward the remote end as interruptions of the dc loop. Use two-stage

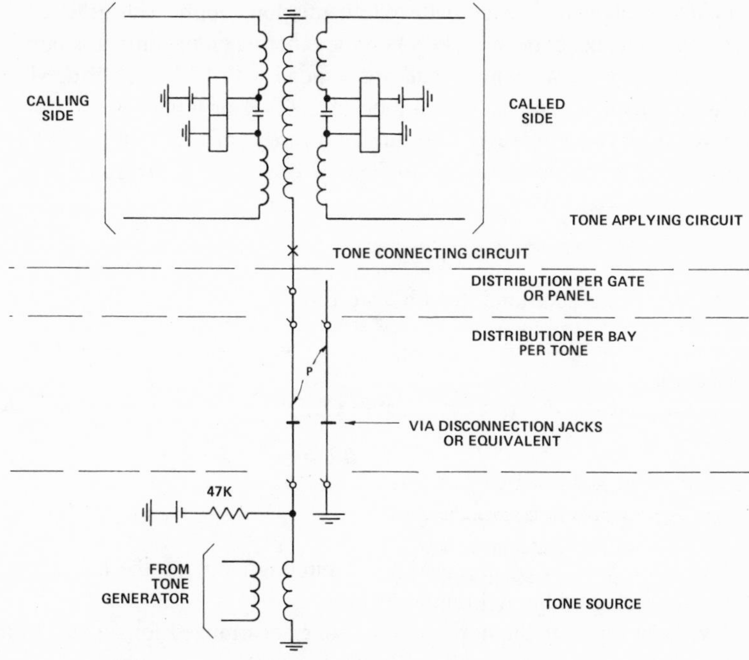

Figure 7.16. Application of balanced tones.

reversion of a circuit to its inductive state of originating end by providing a temporary intermediate connection such as bridging the loop by an 800 ohm non-inductive resistor momentarily.

- Conduct channel associated signals without distortion.
- Provide for remote busying by reversing the loop current.

7.4.3.2. *Outgoing Trunks, E&M Signaling*

The following additional basic features are provided:

- Seizure by completing the M loop.
- Immediate busy upon seizure to outgoing traffic.
- Continuous monitoring of E loop for signals.
- Conduct channel-associated signals received without distortion.
- Provide a guard interval following release during which the trunk is maintained busy.
- Provide for remote busying.

7.4.3.3. *Incoming Loop Signaling Trunks*

The following additional features are provided:

- A balanced high-impedance battery transmission supply with ground on one side and exchange battery (−48 V) on the other when the circuit is normal. A current-sensing device which shall not respond with 15,000 Ω bridged across the loop but shall respond with a loop of 2000 Ω or less.
- Answer supervision by reversal of the loop current and recognition of a reduction in the trunk loop resistance to 2000 Ω or less as a seizure.

7.4.3.4. Incoming Trunks E&M Signaling

The following additional features are provided:

- Continuously monitoring the E loop for closed (off-hook) and open (on-hook) conditions.
- Repeat the answer supervision signal toward the trunk by looping the MA and MB leads.

7.4.3.5. Two-Way Trunk Features

Two-way dial-to-dial or E&M trunk circuits combine the above features with the following additional requirements:

Two-way trunks in the normal (free) state are arranged for seizure from the remote end for an incoming call or by the local switching system for outgoing calls. They provide an immediate busy condition to outgoing calls on seizure either incoming or outgoing.

8

Stored-Program Control

8.1. Introduction

All modern switching systems operate in the stored-program mode. This is because it is a technique which allows the relatively inexpensive accomplishment of complex call processing. The cost of the equipment to provide stored-program control is reducing due to the availability of low-cost electronic components. This is due to the development of techniques for large-scale integrated circuits. Unfortunately the cost of writing programs is not a function of material costs and hence this remains a major cost in the development, manufacturing, installing, and operating of stored-program-control switching systems.

A stored-program system operates from instructions extracted from memory according to predetermined sequences stored in memory. This determines what is to be done. The stored logic determines how it is to be done and componentry, hardware, accomplishes it. This means that the functions necessary in effecting the physical results of the system are controlled by hardware. Typical hardware functions are accomplished at the line and trunk interfaces and switching matrices. Hardware is used for logic processing.

The application of stored-program techniques to automatic telephone switching offers tremendous possibilities in standardization of hardware at the expense of variety of programming. It simply means that the work is performed in a different part of the system. This means that the variations in operation are the results of different orders, or sequences of references to the memory.

8.2. Stored-Program Control

Stored-program control is based on the computer technique of storing the computation program in the same memory that stores data. This enables the computer to automatically modify a program as it is being executed.

8.3. Application of Stored-Program to Telephone Switching Systems

The viable application of the stored-program principle in telephone switching is due to the successful exploitation of the technique by the Bell Telephone Laboratories in the 1960s.[32] This culminated in the production of a number of electronic switching systems. These used space division or digital switching matrices.

The stored-program technique is a means of accomplishing the controls of all, or part, of a telephone system by using a memory and processing circuits. The logical operations are performed in a similar way to a commercial computer.

A program is a series of instructions which defines what is to be done (instruction), how it has to be done (rules), and what to do it to (data). A system can process the steps in the program (the operation) in two basically different ways.

8.3.1. Wired Program

A wired program is achieved by storing all the rules permanently either by a wired circuit or by a read-only memory. This mode can apply to well-defined routines such as line and trunk scanning and to special interfaces which occur in small quantities.

8.3.2. Stored Program

A stored program is achieved by storing only the basic rules in a permanent form and storing the various combinations of the rules, as required, to provide a particular call process, in portions of a memory, from which the results can be extracted when required.

When the rules are stored in a memory, only those combinations of rules applicable to the task in hand are used. New combinations of the rules can be carried out by changing the program (the sequence of operations is variable). Such a system having the ability to synthesize rules out of any combinations of basic wired rules is a stored-program system. The application of the program needs a memory to store the data and instructions and a means of locating any desired memory area to read or, sometimes, write in new data.

It needs a processor to interpret and process instructions. The processor functions may be spread over a number of controls. Interfaces are necessary to allow for speed or other differences between the inputs and outputs to the processor.

System organization. System organization includes the organization of the memory. The memory has areas which store groups of bits, organized into bytes and words. These can be of various sizes. The word size varies; typically 16 to 44 bit words have been used for stored-program switching systems. The memory is usually modular so that it can be increased when new lines or traffic sources are added. The memory areas are organized according to the operating program and hence the memory contains specific instructions as to how a call should be processed.

The program sequence relates both to instructions from the memory and to information obtained from the system interfaces, such as lines or trunks. Hence the things that are to be done in a stored-program system are determined by addressing and obtaining information from memory and using it with information obtained from other sources, in a predetermined way, to decide what has to be undertaken.

This information may have to be located at different places in the memory during its collection and modified after processing. It may be necessary to store some, or all of, the information for future use (such as statistics or call charging).

The logic functions accomplished by the processor typically consist of obtaining data from interfaces addressing and extracting or writing into the memory, checking for error, and comparing data. The actual functions depend upon the services and features that the system requires.

The program uses combinations of the logic functions to perform the relevant tasks. The tasks may be a command to an interface, such as, apply ringing to a specified line. The task may also result only in the modification of data or the invoking of an alternate procedure, such as, when a called line is found busy, modifying the program by indicating a different next step.

8.4. Programming the System

It is necessary that the memory contain all the relevant data to operate the system. The relevant information is located in assigned memory areas so that the sequence of operations, the combination of the various logic devices, etc. can be established. Hence it follows that a stored program system will only perform its operations according to its rules in the exact way defined by the programmer.

The processor operates on a bit-by-bit basis. That is, it makes all decisions based upon the state of the individual bits in the program. Therefore, it is necessary to assign values to each of the bits. The values assigned to the bits determine what information is stored in the computer. In addition to this, it is necessary to define where they are stored and how they are to be used.

Before a stored-program system can operate, it is necessary for a program to be written into its memory. This establishes the basic operating procedures. Later it has to be programmed for the dimensions and services of the switching system that it serves. This information is encoded and written into the memory from a magnetic tape or disk.

8.4.1. Programming Language

The writing of a program on a bit-by-bit basis is impractical except for very small programs. In order to permit the writing of program in a straightforward manner, programming languages have been devised. These programming languages enable the programmer to write out his program in a conversational manner. These instructions must be converted to machine language for the processor. This is accomplished by the use of programs which convert the basic programs into programs suitable for the processor.

Typical of such programs is the program that translates the instructions into the actual bit-by-bit machine code. This is called an assembler and a program to be translated in this way is written in assembly language. An additional program is written in a higher-level language which takes the detailed job of writing each individual instruction from the programmer, enabling the programmer to write the program in a statement that reads much like English or mathematical equations.

8.4.2. Basis for Programming Language

When stored-program systems were first designed the manufacturer either used a variant of a commercial programming language, or developed his own language, or wrote programs in lower-level languages such as assembly language. This resulted in a number of noncompatible languages, sometimes from the same manufacturer. The need for a standard language applicable by any manufactuer in any location caused the CCITT to establish study group II. This study group produced the basis for the three basic languages. These are as follows.

Functional specification and description language (SDL). SDL is the language that was discussed in the first chapter, which is used for documentation and description of a system.

Man–machine language (MML). MML is used in the operation and maintenance of stored-program-controlled switching systems. It is basically intended to apply to interfaces with the level-1 and level-2 controls for both local and remote maintenance centers. It is not intended to apply to level-3 control administrative functions or to any associated control functions provided at a remote administrative center. At the time of going to press the recommendation had not been issued. Its general objectives were to use 72 characters per line and 66 lines per form (F1).

A second format (F2) would use 120 characters per line and would be used for typewriters and line printers; it also would use 66 lines per form. The character set used is a subset of the International Alphabet No. 5, which is described in the CCITT *Green Book.*[33] A character in the table has a position which is indicated by a row identity and a column identity. The row is indicated by one of the 16 combinations of the first four binary digits where bit one is the least significant bit. The column is indicated by one of the eight combinations of the last three binary digits. For example position 3/1 is identified by the binary code 1000110 (bit 7 is the left-most bit, e.g., 1). Position 3/2 is identified by 0100110. These correspond in the alphabet to digits one and two, respectively. The allocation of values to the subset will be the subject of a recommendation by the CCITT.

High-level language (HLL). The basic trouble with standardization after the fact is that it is not always possible to apply it retroactively. A new international telecommunications programming language will face this problem. Thousands of man years of programming have been applied to develop the SPC systems that are going into service in increasing numbers. Their operating diagnostic and support programs are written and working. This means that the scope of application of the yet-to-be-recommended CCITT HLL will inevitably be restricted—probably to the next generation of telecommunications switching systems!

The objectives for the language were that it be applicable to call processing, diagnostics, and operating programs including initialization and recovery. It should also be able to implement instructions received in the man–machine language. In addition it should be suitable for on-line support programs or off-line programs and installation and acceptance testing programs. The language should be structured[(34)] so that modular programs can be developed.

8.5. Software Design Principles

Software design must comply with software standards. All programs must be reliable, reusable, maintainable, extensible, and as efficient as possible. This requires adequate documentation for methods of specifying what the program has to do. It requires a method of writing programs and of controlling and recording modifications.

The method of program organization for a specific application must include all required functions and a means of testing compliance with them. Operating manuals are necessary to telecommunications switching system both from the standpoint of the overall system operation and from the standpoint of personnel concerned with administrative functions.

Programs should consist of groups of instructions in modules. The program modules should be assembled to provide all the required data for an installation. The program should be recorded magnetically in machine-readable data which will be used to check the validity of the program prior to its use in the intended system. This test checks that the programming rules of the system have been adhered to and that the association of the software modules has produced the desired result.

8.5.1. Organization of the Program

The program should have a number of independent modules, which are the smallest programming units, grouped functionally to form clusters, with the clusters arranged in hierarchical levels. This segregates each function and any associated data base from other functions and their data bases. It has well-defined interfaces between functions. This structure provides a varying degree of detail at each level, the detail increasing at each successive lower level.

The first level of clusters should consist of the minimum number needed to meet the system requirements. Hence each cluster handles some specific requirement, or parts of several closely related requirements. The next hierarchical level of clusters are those needed to support the first level and so on.

8.5.2. Clusters

A cluster is made up of a number of related modules which together provide a function without requiring the user to know how it performs the function or how it is organized. If the cluster provides access to a data base, the logic required to access, write, or read that table exists in the cluster modules, and the table users need only know how to refer to the appropriate modules. It will include hardware which is only accessible through it.

8.5.3. Modules

Modules are the basic program building blocks. A module performs a single function. Modules in the higher-level clusters indicate what is to be done and where it will be done. Modules in the lower-level clusters carry out defined tasks.

Modules are designed to facilitate the design, understanding, and maintenance of programs, hence the maximum size of a module should be such that it can be displayed on a single printer page or no more than two video screen displays, exclusive of headings and other general information not related to the performance of the function. The interfacing of the modules must only be possible by following the relevant interfacing standard. This normally requires a single entry point and a single exit point to the module for interlinking modules.

8.6. Implementation of the Program

The implementation of the program is the part of the text that should illustrate how programming is implemented, but this requires the use of a programming language. There are 27 existing programming languages, maybe more, but in spite of this the CCITT working group decided none was acceptable either in its existing or modified form. Hence the decision to develop a new language (HLL).

It is hoped that subsequent editions of this book will be able to describe the new HLL language. In the meantime designers of new systems will have to continue with the use of the programming language of the manufacturer or administration. Documentation for this will have to be sought elsewhere.

8.7. Processors

The stored-program control processors are basically special-purpose computers. The earliest stored-program switching systems, and some of the newest, were based upon processors designed by the switching manufacturer. In the case of the very large scale manufacturers, such as Western Electric, the processors were based on components also manufactured by them. The costs of such developments needed a large volume of sales to justify them. They also required the availability of in-depth research and development resources to maintain and expand the technology on which the processors were based. This enabled the Bell System to double the capacity of its processors while substantially reducing their space and power requirements. However, some other manufacturers have chosen to adapt a general-purpose computer to their processor requirements.

The selection of such a computer must be based on meeting the requirements of providing service 24 hours a day, with no scheduled down time. Such computers may also have projected shorter lifetimes with correspondingly shorter depreciation times than that required for (and allowed for) in telecommunication systems. The basis for the decision to develop a special processor or to modify an existing one may not be an engineering one but rather be based on marketing or executive decision. It depends on many factors and requirements as is brought out in the following discussion on stored-program control.

8.7.1. Computer and Telephone Processors

A computer is designed to perform services in response to the request of the user. This normally occurs during specified periods so that there are periods which can be devoted to maintenance during which no services are provided. Such an arrangement, of course, is inadmissible for a telephone switching system in which service must always be maintained. However, the application of computers to telephony switching system controls means that they must meet the rigorous reliability and differing service requirements of telephone switching systems. A general-purpose computer provides a range of possible uses, some of which do not apply to telephony. The special-purpose computer is designed to provide only the necessary telecommunications functions.

Telephone switching systems vary in size and hence require a family of processors if they use a single, but redundant level-2 control. Switching computers are designed for specific applications and hence do not need to have all the facilities necessary for data processing. They also do not need facilities for arithmetical functions.

A major point of difference between a computer and a processor is that the computer produces answers to specific questions in a printed or graphical readout form or other data form. The output of the computer shows the answer to the questions that were raised, whether this be reference information or the computation of a mathematical expression.

In the case of a processor for a telephone system, the output from the processor occurs in a similar way to the computer. However, the interface to which it is applied has to perform the delegated task in order that the requisite results be obtained. For instance an instruction in a stored program system to apply a ringing signal to a called line may be connected to a ringing functional unit and ringing applied to the line.

8.7.2. Operation of Stored-Program Control

The processing of calls by the various controls is effected by the program. The memory which contains all the relevant data necessary to process a call is associated with the appropriate controls. The control directs the procurement of the information from its source as referred to in its instruction. If necessary it transfers this information into the memory to replace existing information or transfers it into the memory for temporary storage. The instruction being processed is analyzed by the controls which directs the programmed information to its destination. The input and data output together with any relevant information are then analyzed. This information can either have been collected previously or obtained when required. The result of the processing either is used to issue an instruction to perform a task or is stored for further action depending on the analysis of the instruction which is being processed.

8.7.2.1. Operating Procedure

A typical operating procedure for a stored-program control is as follows.

An instruction word is stored after being extracted from memory. The instruction part of the word is decoded. This chooses a function, makes access to memory, etc., and the information obtained is directed by the decoding information.

The location of the next order is then determined.

It could be seen therefore that programmed logic requires the following:

- A data and instruction store and a means of location of information relating to designated addresses.
- An overall control which relates the various actions and directs them to their destinations and if necessary stores them for future use.

- Logic circuits for performing such functions as comparison, addition, subtraction, decoding, addressing, etc.
- A means of input and a means of output.

8.7.2.2. Input and Output Means

The output means may consist of such devices as teletypewriters, magnetic disks, magnetic tape, etc. The input in the case of a telephone system processor will also include sensing devices to collect information relating to conditions occurring during the progress of a call. A processor requires a means of instruction to program the machine to perform its tasks and a means of collecting additional information according to the rules of operation of the system which it controls.

Output devices can also consist of magnetic tape, printed pages, drawing machines, or cathode ray tubes. Printed output is necessary in order that records can be obtained when required. Such a case occurs if the maintenance personnel need to locate data associated with a given directory number. A processor output or outputs to the system control is necessary in order to issue the necessary commands for the control of calls.

8.7.2.3. Memory Types

Storage requirements include two types of memory. Processing a call requires the use of temporary storage of information of input data which is only relevant to a specific call and permanently stored information (which relates to the processing of all calls). The permanently stored data includes the operating programs and exchange line and trunk data.

A memory area is necessary for the diagnostic program. This stores the data to check that the processor is carrying out its task correctly. This requires elaborate programs which check the operation of the program to see if performing inaccurately is due to the failure of equipment, etc.

8.8. Security of System Operation

The operation of the control should not be affected by spurious input information which could affect the entire operation of the system and result in the mutilation or loss of calls. The information must therefore be stored in duplicate at least. Hence processors have to be provided on at least a duplicated basis and at best on a majority basis. In the case of a majority basis, three or more computers or processors, each containing identical call-processing information, would be used, and only if the output of at least two of them were identical would the information be allowed to control the system. Such an arrangement has its economical penalities and hence in general processors are usually duplicated or provided on a load-sharing basis.

If two or more controls are capable of acting on the same information a means of using only one of them to operate the system is necessary. Hence the possibility of failure at the point at which the controls are interconnected arises. This control selection means may consist of a series of simple "make" contacts (which may be sealed for additional reliability). Alternatively it may consist of a number of capacitors or transformers which are coupled into the requisite output.

Regular or standby operation. The use of regular and standby processor operation, as opposed to sharing the system load over a number of independent but similar controls, has pros and cons. The biggest advantage of the multiple independent processors is that only that portion of the equipment which it controls is unusable when it fails. There is no possibility of it affecting other similar controls. However, the removal of such a control from service does have a problem in that the traffic-carrying capacity of the system is reduced when it is inoperative. There is some degree of reduction of such an event due to the fact that the failure may not occur during the busy hour. Hence the resultant effect on the traffic being carried may not be as great. Such a control can be designed to have a failure rate of once in ten years if the components are properly derated. A fault could occur during a nonbusy hour, since there is no reason to suppose the control more likely to fail at any one time rather than at another.

If we assume for the sake of argument that there are four major traffic periods during the day (the four busiest hours), then during the remaining 20 hr the effect of a faulty control will be minimized. This will be especially true if it occurs during the period of night when the traffic is usually very low and the impact of such a failure would be minimal.

It should be remembered that even if complete redundancy of processors is provided components have a possibility of random failure. Hence, faults will occur in the system and it is not possible to isolate such fault occurrence from the point where redundant equipment is switched in or put.

8.9. Use of Processor for Nontelephone Switching Tasks

It can be shown that a processor will carry *a* Erlangs of traffic with a predictable average waiting time for obtaining access to it.[35] This traffic is the product of the number of calls processed and the average time for the control to perform its task.

This traffic must include an allowance for false starts, i.e., calls which do not mature, and for peak traffic.

In general this objective can be met (when only level-2 controls are used) by loading the processor up to 0.95 Erlangs due to all such traffic.

Table 18. Some Delays for Single Server Systems

	Operating times of processor for complete call				
	50 msec	25 msec	10 msec	5 msec	1 msec
Number of calls per busy hour	54,000	10,000	270,000	540,000	2,700,000
Percent of calls delayed	75	75	75	75	75
Average delay on calls delayed (msec)	100	50	20	10	2
Average delay on all calls (msec)	75	37.5	15	7.5	1.5

Now it can be argued that by making the control operate very rapidly and by speed buffering so that control of slower operations can be made off line such a processor can also carry out computation through the association of suitable equipment. The amount of time available can be shown by using the following type of calculation.

The average delay on all calls is equal to the average delay on calls delayed multiplied by the percent of occupancy of control (traffic in Erlangs).

The processing time must be equal to or greater than the time to carry out the instructions or the instruction must be stored at its point of application for future use. If the control is loaded to 0.5 Erlangs then the average delay on calls delayed is ½ (holding time in seconds)/ (1−traffic in Erlangs) Thus if the holding time, i.e., the time that the control requires to perform its task, is 10 msec, then if it is to process 40,000 call attempts the traffic that it carries is 40,000 multiplied by 0.01/3600 (we divide by 3600 since Erlangs are expressed in terms of hours), which is equal to 0.111 Erlangs.

This means that the control is in use for only 11% of the time and the average delay on reaching the control would be 5 msec for calls delayed and for all calls will be 450 μsec. The remainder of the time could be used for other purposes.

Table 18 shows the delays that can be anticipated for various processor operating times when the processor carries 0.75 Erlangs of traffic. This traffic loading will require level-1 controls to absorb the false traffic.

8.10. Some Applications of Stored Program to System Functions

Typical functions are the control of certain line functions and the use of memory in call routing and transmission path control.

8.10.1. Line Lockout Function

Line lockout can be indicated by the state of the relevant memory word (usually a two-bit word is provided per line interface), so that requests for service are ignored until the normal on-hook state is resumed. The line circuit need not have special hardware functions as would be necessary with a wired-logic circuit and hence is simplified. The circuit may be arranged so that an alarm is only given if a group of lines are locked out of service. This may be the result of a cable failure and hence need immediate action. The wired logic function is usually provided by a means of measuring the current from a number of circuits to see if it represents an alarm. The equivalent stored-program solution is to use a counter to add up to the requisite number of permanents to establish the alarm.

8.10.2. Line Load Control

Line load control is needed because of heavy traffic conditions or an emergency which can restrict the certain lines from originating calls (terminating service is still provided). This arrangement can be accomplished readily by removing the potential from the line relay in the case of a conventional line circuit. Such a removal can be accomplished on a group basis by throwing keys. In the case of stored program it can be accomplished by using an appropriate class of service. The use of a class of service rather than a line state reduces the dedicated memory size perline to two bits. (Two bits provide four states: quiescent, request for service, call in progress, and lockout.)

8.11. Map in Memory

A very important aspect of a stored-program system is the ability to keep a record of the state of the paths and traffic-carrying devices in memory. This means that a call can be established through the system simply by reference to the memory without examining the paths to the required destination.

This allows the use of crosspoints without a holding path. Crossbar and some other space division systems used a separate, switched, holding path or paths. Hence the number of switched leads in the system can be reduced since the holding function is removed. This effects economies in cabling and reduces the cost of the crosspoints; however, this is at the expense of preserving an accurate record of the condition of all the links in the memory.

A map in memory is essential in a digital system which employs intervals of time as transmission paths.

The memory can be centralized or distributed over the system controls depending on the technique employed in the switching system.

Path reservation. There is an exceedingly important advantage of the use of map in memories in that a switching path can be interrupted without releasing the connection. It is possible for instance to temporarily connect circuits in parallel with an existing line or trunk connection to perform a certain function. For instance it is possible to establish a path to a traffic-carrying unit and then to disconnect the path temporarily, without affecting the busy status, and then connect it to another function. The traffic-carrying unit and its path are still reserved for future use in the connection.

When the second function has been completed it is disassociated with the call and the reserved path and traffic-carrying unit connected, instead of it. There is a disadvantage in such an arrangement in that one has to be very sure that the memory does, in fact, reflect in complete accuracy the state of each of the established connections, as otherwise double connections could be possible. There is also the added requirement that, unless the arrangements are such that the crosspoints are released automatically at the end of the call, it is necessary to refer to the control to take down the connection. An exception to this occurs in systems which employ crosspoints which are actuated by energizing both of its two windings but released if only one coil is energized.[36] The coils are connected in a coordinate manner using rows and columns of wiring. Hence the actuation of any crosspoint releases any actuated crosspoints in either the row or column of its coordinates. Such an arrangement only allows one call at a time in such a matrix.

8.12. Dependence on Hardware

A stored-program system has the facility of ready modification of instructions on what the equipment has to do, but the realization of a command may require the availability of the hardware to carry it out. For example suppose that it is necessary to provide multiparty lines which necessitate special ringing conditions. This requires a functional unit which will apply the relevant ringing frequencies to the appropriate line side or to the loop. If there is no such circuit in the system then the instruction cannot be carried out, and if the feature is to be added both hardware and software changes or additions must be made.

The manipulation of data relating to code interpretation, class of service analysis, etc., can be readily accomplished by changes of program if the memory capacity is large enough. Again if the feature change extends the memory size beyond its capacity then once again hardware additions must be made to the memory.

8.13. Routing Facilities

All routing in a stored-program digital exchange is under the control of software programs. Routing instructions are entered in the memory either via a typewriter entry for updating or from tape or disk recordings. These instructions are part of the data base of each exchange and depend on the specific exchange requirement.

Space in a dedicated memory area is used. The instructions record a series of tables in this memory area. These tables provide the data to route each call by using the address data, class of service, and other data associated with each call. The information is identified by a matching technique.

One such table is the code point table which is arranged in a tree-structured format. This allows the examination of every digit combination from 1 to 999. Generally any digit combination beyond three digits which requires examination for routing purpose is entered into code point expansion tables.

When, during the course of dialed digit entry, a match in the code point table is found, a code point screening table is addressed. This adds the appropriate screen class which relates to the class of service of the originating subscriber or incoming trunk circuit.

These classes of service differentiate between regular subscribers, coin boxes, priority subscribers, test equipment, etc., or indicate special treatments such as blockage to incoming calls or permitting outgoing local calls only.

The number of variations that can be handled can be very large because 500 or more different classes of service are usually available. Hence, the code point screening table addresses a route treatment table (which can be programmed to contain the appropriate routing information for each kind of call). This reduces the amount of information contained in the first screening to the address of the relevant route treatment point. This allows the call to be directed to a particular trunk or trunk group or to a recorded announcement or to an operator, and to different trunk groups if the previous trunk group was found to be busy (alternate routing).

Modification to the office data base including the routing instructions and trunk group sizing are usually performed locally and also remotely.

The latter is used for centralized office administration and network management.

9

Associated-Channel Signaling

9.1. Introduction

The majority of the signaling systems in use in telecommunications use channel-associated signaling. The repertoire of signals has increased over the years with the most modern signaling systems including signals which transfer information backwards from the called side. Each successive system has built on the experience of its predecessors, and recent systems have included network and administrative signals. The earlier signaling systems, most of which will be in use for some time, may be augmented by the common channel systems described in the next chapter for new applications. Associated-channel systems have reached the peak of their development. The systems are described in the order of their development. The line signaling and register signaling portions of the system are described using their CCITT designations.

9.2. CCITT System No. 1

The CCITT system No. 1 is used for manual international circuits, often for radio links. It uses a signaling frequency of 500 Hz interrupted at a frequency of 20 Hz. The general arrangements are shown in Figure 9.1.

It is a ringdown system in which seizure and release signals are sent as pulses of 500/20 Hz.

The signals may be sent over wide-band carrier systems.

When it used over short two-wire circuits a low-frequency signal (16, 25, or 50 Hz) may be used in lieu of 500/20.

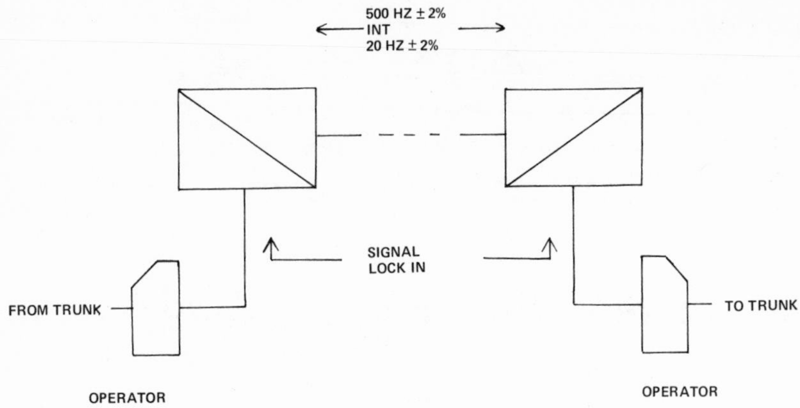

Figure 9.1. CCITT system No. 1 general arrangements.

9.2.1. Receiver Limits

Level. The limit level is minus 8.5 plus $n \leqslant N \leqslant$ plus 2.5 plus n decibels, where n is the relative power level at the point of the circuit at which the signal receiver is connected and N is the absolute power level.

Frequency tolerance. This is 500 Hz plus or minus 2% at an interrupting frequency of 20 Hz plus or minus 2%.

Reaction time (guard time). This is 100–1200 msec.

Insertion loss. This is less than 0.3 dB for circuit bandwidth.

The receiver circuitry is usually arranged to store the signal until it is answered at its point of reception. The pulse signal is converted to a visual indication, usually a lamp per channel.

9.2.2. Sender Limits

Applications. The two frequencies must be applied as nearly simultaneously as possible in a balanced mode.

Effective power. The effective power is, interrupted, 0.5 mW plus or minus 1 dB, or an absolute power level (ref. 1 mW) or −3 dB plus or minus 1 dBO.

Signal duration. The outgoing signal must be limited automatically to a duration of 2 sec, even if the signal is from a key-operated source.

9.2.3. Scope

This system provided line signals only.

9.3. CCITT System No. 2

This system, which is described in the CCITT *White Book*,[37] was intended for two-wire semiautomatic operation but was never used in international service.

9.4. System No. 3

This system is described in the CCITT *Red Book*.[38]

The system is applicable in semiautomatic and automatic working and was only used either for terminating or transit traffic operation in Europe. It will not be used for new international connections. It uses a single frequency, 2280 ± 6 Hz, and provides for one-way circuits only. A digit requires a start pulse (1) four successive pulses, without a space between them, to convey the value of the digit, and a stop pulse (0).

The register signaling code is shown in Figure 9.2. Signals are sent without interruption: tone on (shown as —— in the figure) is equal to 1: tone off is equal to 0 and is shown as a blank in the figure.

9.4.1. Receiver Limits

Level. The operate level is $(-15 + n)$ to $(3 + n)$ dBm; the nonoperate level is $\leqslant (-32 + n)$ dBm.

Frequency tolerance. The frequency tolerance is 2280 ± 15 Hz.

Recognition time. The recognition time is 50 ⩽ 8 msec for tone pulses and for silent periods is ±8 msec of nonoccupied time slot at sending end.

9.4.2. Sender Limits

Frequency tolerance. The frequency tolerance is 2280 ± 6 Hz.

Level. This is −6 ±1 dBmO.

Duration of six time slot frame. This is 300 ±3 msec. The sender uses 50-msec time slots with ±10 msec at beginning and end of each nonoccupied time slot.

9.4.3. Scope

This system provided in-band pulsed register, address, signals only.

ARITHMETIC VALUE			8	4	2	1	
SIGNAL MEANING		START	ELEMENTS 2280 ± 6 HZ				STOP
			1	2	3	4	
DIGIT	1	—				—	
	2	—			—		
	3	—			—	—	
	4	—		—			
	5	—		—		—	
	6	—		—	—		
	7	—		—	—	—	
	8	—	—				
	9	—	—			—	
DIGIT	0	—	—		—		
CALL OPERATOR CODE 11	11	—	—		—	—	
CALL OPERATOR CODE 12	12	—	—	—			
SPARE CODE	13	—	—	—		—	
SPARE CODE	14	—	—	—	—		
END OF PULSING	15	—	—	—	—	—	
SPARE CODE	16	—					
MILLISECONDS		0	50	100	150	200	250 300

Figure 9.2. CCITT system No. 3 signaling code.

9.5. CCITT System No. 4[(39)]

During 1953 and 1954, tests were made on services operating with system No. 4; the results were satisfactory and the system was standardized by the CCITT. At the time of going to press it is the most widely used system in Europe; its development has been rapid and in about 15 yr its use grew from a small number of experimental circuits to several thousand circuits.

The frequencies 2040 and 2400 Hz are used for register signals with a pulsed code. A digit requires four successive intervals—each interval has a pulse period and a silent period.

The code is shown in Figure 9.3.

System No. 4 uses unidirectional circuits only and can be applied to any type of cable or radio-link circuit, but is not compatible with intercontinental circuits or circuits using time-assigned speech interpolation (TASI). (TASI is described below.)

9.5.1. Receiver Limits

Frequency tolerance. x is equal to 2400 ±15 Hz; y is equal to 2400 ±15 Hz.

ARITHMETIC VALUE		8	4	2	
SIGNAL MEANING		X	2040 HZ & 2400 HZ		Y
		1	2	3	4
DIGIT	1	Y	Y	Y	X
	2	Y	Y	X	Y
	3	Y	Y	X	X
	4	Y	X	Y	Y
	5	Y	X	Y	X
DIGIT	6	Y	X	X	Y
	7	Y	X	X	X
	8	X	Y	Y	Y
	9	X	Y	Y	X
DIGIT	0	X	Y	X	Y
CALL OPERATOR CODE 11	11	X	Y	X	X
CALL OPERATOR CODE 12	12	X	X	Y	Y
SPARE CODE	13	X	X	Y	X
SPARE CODE	14	X	X	X	Y
END OF PULSING	15	X	X	X	X
SPARE CODE	16	Y	Y	Y	Y

Figure 9.3. CCITT system No. 4 signaling code.

Level. The operate level is $(-18 + n)$ to $(0 + n)$ dBm; the nonoperate level is $\leqslant (-35 + n)$ dBm.

Recognition time. The recognition time is $\leqslant$ 20 msec.

9.5.2. Sender Limits

Frequency tolerance. x is equal to 2040 $\pm$6 Hz; y is equal to 2400 $\pm$6 Hz.

Level. This is -9 ± 1 dBmO.

Maximum difference in levels. This is 0.5 dB.

Transmit signal duration. This is 35 $\pm$7 msec.

Interval between signal elements. This is 35 $\pm$7 msec.

9.5.3. Scope

This system provided in-band, pulsed, register, address, signals only.

Use of systems Nos. 3 and 4. In Europe, comparative tests were made in Europe from 1949 to 1954 of systems Nos. 3 and 4. However, neither the results of the tests nor the evaluation by experts showed a preference for one over the other. Hence both systems were adopted for use between European countries. In practice, however, system No. 4 has spread much faster and wider than system No. 3.

9.6. CCITT System No. 5

This system was standardized by the CCITT in 1964[40] and is used for intercontinental traffic, both terminating and transit. It can be used on underground and submarine-cable circuits, as well as on radio links.

The system uses six frequencies, separated by 200 Hz from each other, between 700 and 1700 Hz in the code shown in Figure 9.4; register signals are sent in pulses, each of which is a combination of two of the frequencies. One pulse represents one digit, and there is a silent interval between pulses.

The main difference between earlier systems and system No. 5 is the use of compelled line signals. The biggest advantage with line signals that are compelled continuously is found in their use on submarine-cable circuits with TASI. The continuity of the signal ensures circuit–TASI channel association even under high-traffic load conditions when the assignment time can reach 500 msec.

The No. 5 system uses link-by-link in-band, compelled two-frequency (2400 and 2600 Hz) signaling for all line signals, except the forward-transfer signal.

Line signaling is either applied at the trunk interface or it may be built into a carrier system. In the latter case, it will be controlled by the E&M signals described above. If signaling is applied from the outgoing trunk over audio (wire)

SIGNAL	DIRECTION		SIGNALING FREQUENCIES		
	FORWARD	BACKWARD	2400	2600	
SEIZE			—		C
PROCEED TO SEND				—	
BUSY FLASH		—		—	C
BUSY FLASH ACK	—		—		
ANSWER		—	—		C
ANSWER ACK	—		—		
FORWARD TRANSFER	—			—	850 ± 200
CLEAR BACK		—		—	C
CLEAR BACK ACK	—		—		
CLEAR FORWARD	—		—	—	C
RELEASE GUARD		—	—	—	

LINE SIGNALING

Figure 9.4. Part(i). CCITT system No. 5 signaling code. C = compelled pair.

circuits, then converters are inserted between the outgoing E&M trunk and the outside conductors.

This system also differs from previous systems in its use of two-way trunks. The high cost of very long trunk circuits and the variations in traffic arising from differences in various parts of the world between traffic passing in opposite directions can easily justify the use of two-way circuits. However, such circuits present the problem of simultaneous seizure of the circuit from both ends.

In system No. 5 the double seizure is automatically indicated by the received frequency being the same as the sent frequency, that is, 2400 Hz rather than the 2600 which is the frequency of the proceed-to-send signal that is the normal response to a seizing signal.

9.6.1. Receiver Limits

Frequency tolerances. The frequency tolerances, in hertz, are as follows:

700 ±15	1300 ±15
900 ±15	1500 ±15
1100 ±15	1700 ±15

SIGNAL		FREQUENCIES						DURATION
		700	900	1100	1300	1500	1700	MSEC
SEIZE PROCEED TO SEND								
60-100 MSEC LATER	KP1			—			—	100 ±10
	KP2				—		—	
DIGIT	1	—	—					
	2	—		—				
	3		—	—				
	4	—			—			
	5		—		—			
DIGIT	6			—	—			
	7	—				—		55 ±5
	8		—			—		
	9			—		—		
DIGIT	10				—	—		
CODE 11		—					—	
CODE 12			—				—	
ST						—	—	

REGISTER SIGNALING

Figure 9.4. Part(ii). *See legend on previous page.*

Level. The operate level is $(-14+n)$ to $(0+n)$ dBm; the nonoperate level is $\leqslant (-24+n)$ dBm.

Recognition time. The recognition time is $\geqslant$ 30 msec.

Silent period. The silent period is $\geqslant$ 30 msec.

Transmitted tone coincidence. The transmitted tone coincidence must be $<$ 1 msec.

9.6.2. Sender Limits

Frequency tolerances. The frequency tolerances, in hertz, are as follows:

700 ± 6	1300 ± 6
900 ± 6	1500 ± 6
1100 ± 6	1700 ± 6

Level. The level is −7 ± 1 dBmO. The maximum difference in transmit levels is 1.0 dB.

Duration of signal. For KP1 the signal duration is 100 ± 10 msec. For KP2 it is 100 ± 10 msec. For all others it is 55 ± 5 msec. Between the end of the seizing signal and the start of KP1 and KP2 tones there must be a silent period of 80 ± 20 msec. There must be an interval between all other signals of 55 ± 5 msec.

9.6.3. Scope

This system provided compelled in-band line signals and pulsed in-band, link-by-link, register and address signals.

Signaling from the register commenced with a continuous "seize" line signal which was acknowledged by a return "proceed-to-send" line signal.

This is followed by the pulsed register signals which indicate the start of pulsing: KP1, which indicates that the call will terminate in the country of origin, or KP2, which indicates that the address will include a country code of one to three digits. The called-address information is sent next, followed by the end-of-pulsing signal ST.

9.6.4. Signaling over Cables Using TASI

Both the line and interregister signaling arrangements for System No. 5 are designed to work together and are compatible with time assignment speech interpolation, TASI, which is used extensively on long overseas cables.

TASI connects an "active" trunk to an "inactive" transmission channel for the duration of a "talk-spurt" rather than for the duration of a whole conversation. Thus, during any given telephone conversation, TASI may have switched the conversation over several available channels. When all channels are "active" any delay in securing a free channel, although not noticeably affecting speech, could cause an intolerable shortening of the signaling tones unless special signaling procedures are used. Compelled line signaling in conjunction with pulsed register signaling overcomes this problem.

The line signal used as a seizing signal looks like a talk-spurt, so the equipment connects the trunk to a free TASI channel. The signal will persist until the distant exchange has recognized and acknowledged the signal. The distant exchange recognizes the seizing signal, and connects a register to the trunk. It then returns a proceed-to-send signal as an acknowledgment signal and as an indication to the originating exchange that is ready to receive address information. Upon recognizing this signal, the originating exchange orders the removal of the seizing signal and before the TASI channel can be released directs the interregister equipment to send the complete address in a continuous uninterrupted stream of closely spaced multifrequency pulses (en bloc transmission). Since these closely spaced pulses continue to hold the TASI channel connected, their lengths are not affected by TASI. The forward-transfer signal is made long enough (850 msec) to ensure that it will be recognized despite an occasional reduction of its duration by TASI.

9.7. Commonality of Signals

A switching system must interwork with whatever signaling systems are in use at its location, hence it is important to identify differences between them.

9.7.1. Commonality of Signals with AT&T R1 Signaling

Some of the No. 5 signals, or their equivalents, are used within the North American network. Line signals that are not used in North america are the busy-flash and release-guard signals. A busy-flash signal indicates congestion; that is, either all trunks are busy or a trunk exchange is unable to handle the call because of congestion. When a busy-flash signal is received at the originating international exchange, the trunk is released and busy tone is returned to the caller. The release-guard signal acknowledges the clear-forward (disconnect) signal and indicates to the originating exchange that the trunk is available for a new call.

Use of KP signals. Two signal codes, forming the first multifrequency pulse of the address, are used to indicate whether or not the address contains a country code. As explained above, one signal code (KP2), indicates that a country code consisting of one to three digits will follow. The other signal code (KP1) indicates that the call will terminate in the national network. A language-discriminating digit follows the country code (or the KP1 signal if the address does not contain a country code). This digit specifies the language to be spoken by the operators in the case of operator-handled calls.

In the case of unassisted direct-dialed calls, the prefix digit indicates this fact and hence no operators are involved. The langauge-discriminating digit is followed by the national telephone number of the called station.

Use of ST signals. The R1 system uses three ST signals which are explained later.

Coin-control signals. The R1 system uses three coin-control signals.

9.7.2. Signals Common to Systems Nos. 3, 4, and 5

The characteristic signals common to systems Nos. 3, 4, and 5 are as follows.

Forward-transfer signal. This is sent by the operator of the originating center to initiate the intervention of the operator at the incoming end for automatic switching of international calls in that center.

Busy-flash signal. This is sent by the international transit or incoming center to indicate the busy condition to the outgoing-call operator. In the case of automatic calls the signal releases the international connection and returns busy tone from the originating exchange.

Other signal combinations. The number of numerical combinations of the signaling code may be greater than 10, e.g., 16 for systems Nos. 3 and 4 and 15 for system No. 5. Combinations above 10 are used in the semiautomatic service for calls between international operators (code 11 and 12 in all systems), or to provide access to test facilities (code 13 in systems Nos. 3 and 4). Code 15 (end of pulsing or ST) is used in all systems to indicate to the next center that no further digits are to be transmitted.

9.7.3. Comparison of Line Signals

Figure 9.5 shows a comparison of the line signaling codes of CCITT signaling systems 3, 4, and 5. The following symbols are used in the figure.

System No. 3. For signaling frequencies of 2280 Hz the following symbols are used:

X Short signal element (150 ± 30 msec)
XX Long signal element (600 ±120 msec)
S Silent interval (100 ±20 msec)

System No. 4. For signaling frequencies of 2040 Hz (frequencies X) and 2400 (frequencies Y) the following symbols are used:

P Prefix signal constituted by two frequencies X and Y (150 ±30 msecs)
X Short signal element (100 ±20 msec)
Y Short signal element (100 ±20 msec)
XX Long signal element (350 ±70 msec)
YY Long signal element (350 ±70 msec)

System No. 5. For signaling frequencies of 2400 Hz, f1 is the symbol used, and for signaling frequencies of 2600 Hz, f2 is the symbol used.

9.8. System No. 5 Bis

System No. 5 Bis was standardized in 1968 by the CCITT but its development has been overshadowed by CCITT No. 6 and hence it is unlikely to be encountered. It is an adaptation of CCITT system No. 5 using the same signaling frequen-

SIGNAL	DIRECTION	SYSTEM NO. 3	SYSTEM NO. 4	SYSTEM NO. 5
SEIZING	—	TERMINAL X TRANSIT XX	TERMINAL PX TRANSIT PY	F 1 CONTINUOUS
PROCEED TO SEND	—	X	TERMINAL X TRANSIT Y	F 2 CONTINUOUS
NUMBER RECEIVED ACKNOWLEDGMENT	— —	X	P	
BUSY FLASH ACKNOWLEDGMENT	— —	XX	PX	F 2 CONTINUOUS F 1 CONTINUOUS
ANSWER ACKNOWLEDGMENT	— —	XSX	PY	F 1 CONTINUOUS F 1 CONTINUOUS
CLEAR BACK ACKNOWLEDGMENT	— —	XX	PX	F 2 CONTINUOUS F 1 CONTINUOUS
FORWARD TRANSFER	— —	XSX	PYY	F 2 850 ± 200 MSEC
CLEAR FORWARD RELEASE GUARD	— —	XXSXX XXSXX	PXX PYY	F 1/ F 2 CONTINUOUS F 1 / F 2 CONTINUOUS
BLOCKING UNBLOCKING	— —	CONTINUOUS	PX PYY	

Figure 9.5. Line signals CCITT signaling systems 3, 4, and 5. Significance of symbols: System No. 3, signaling frequencies 2280 Hz X, Short signal element (150±30msec); XX, Long signal element (600±120 msec); S, Silent interval (100±20 msec). System No. 4 signaling frequencies 2040 Hz (frequencies X), 2400 Hz (frequencies Y); P, prefix signal constituted by two frequencies X and Y (150±30 msec), X, short signal element (100±20 msec); Y, short signal element (100±20 msec); XX, long signal element (350±70 msec); YY, long signal element (350±70 msec). System No. 5 signaling frequencies 2400 Hz—F 1, 2600 Hz—F 2.

cies but using forward and backward interlocked signaling to provide more facilities. It is fully described in the CCITT *Green Book.*[41]

This system is suitable for terminal and transit traffic and may be used over submarine or land cable circuits and microwave radio circuits (whether TASI is used or not) with certain reservations. It is a link-by-link system using an initial address block followed by overlap signaling. It includes TASI locking signals.

9.9. CCITT Regional Standard No. R1

The CCITT regional standard No. R1 uses multifrequency register signaling with pulses in 2-out-of-6 code. It is a link-by-link system employing forward signals only. It uses in-band line signaling.

Signaling speed is higher than that of the R2 system, the other CCITT Regional standard, but the information content is lower as each frequency combination has only one meaning.

The system is described in the CCITT *Green Book,*[42] and the AT&T *Blue Book.*[43]

9.9.1. R1 Line Signaling (Analog Version)

The R1 line signaling system (analog version) uses line signaling based on the application and removal of only one frequency, 2600 ±5 Hz in both directions on the transmission facility. Signals are transmitted at a level of −8 ±1 dBmO for the duration of the signal or a minimum of 300 msec (whichever is shorter) for a maximum of 550 msec, after which the level of the signal is reduced to −20 ±1 dBmO.

E&M signals in the terminating trunk interface control the carrier signals. The signaling code is shown in Figure 9.6.

OPERATING CONDITION OF THE CIRCUIT	SIGNALING CONDITION			
	FORWARD		BACKWARD	
	TONE ON	TONE OFF	TONE ON	TONE OFF
IDLE	——		——	
SEIZED		——	——	
DELAY DIALING		——	——	
START DIALING		——	70 to 130 Milli Seconds	——
RING FORWARD	65-135 Milli Seconds			
ANSWERED		——		——
CLEAR BACK		——	——	
RELEASE	——		EITHER	EITHER
BLOCKED	——			——

Figure 9.6. R1 line signaling code (analog).

9.9.2. R1 Line Signaling (PCM Version)

The R1 line signaling system (PCM version) uses the A and B signaling bits of the T1 carrier system described in Chapter 6. E&M signals in the terminating trunk interface control the carrier signals.

The signaling code is shown in Figure 9.7.

9.9.3. CCITT Regional System R1 Register Signals

The register signals are sent in pulses consisting of a combination of two frequencies. The system checks for the presence of only two frequencies. A silent interval separates the pulses. Each pulse corresponds to a digit or other signal. The signaling code is shown in Figure 9.8. The signaling sequence is similar to CCITT system No. 5.[(44)]

9.9.3.1. Register Limits

Frequency tolerances. The frequency tolerances, in hertz, are as follows:

700 ±1% ±10	1300 ±1% ±10
900 ±1% ±10	1500 ±1% ±10
1100 ±1% ±10	1700 ±1% ±10

Levels. The operate level is $(-14 + n)$ to $(0 + n)$ dBm; the nonoperate level is $\leqslant (-23 + n)$ dBm.

Duration. The duration must be $\geqslant$ 30 msec. The silent period must be $\geqslant$ 20 msec.

OPERATING CONDITION OF THE CIRCUIT	TIME SLOT CODE			
	FORWARD		BACKWARD	
	A	B	A	B
IDLE	0	0	0	0
SEIZING	1	1	0	0
SEIZING ACKNOWLEDGED	1	0	1	1
ANSWERING	1	0	1	1
CLEAR BACK	1	0	0	0
CLEAR FORWARD	0	0	0 or 1	0 or 1
RELEASE GUARD = IDLE	0	0	0	0

Figure 9.7. R1 line signaling code (PCM).

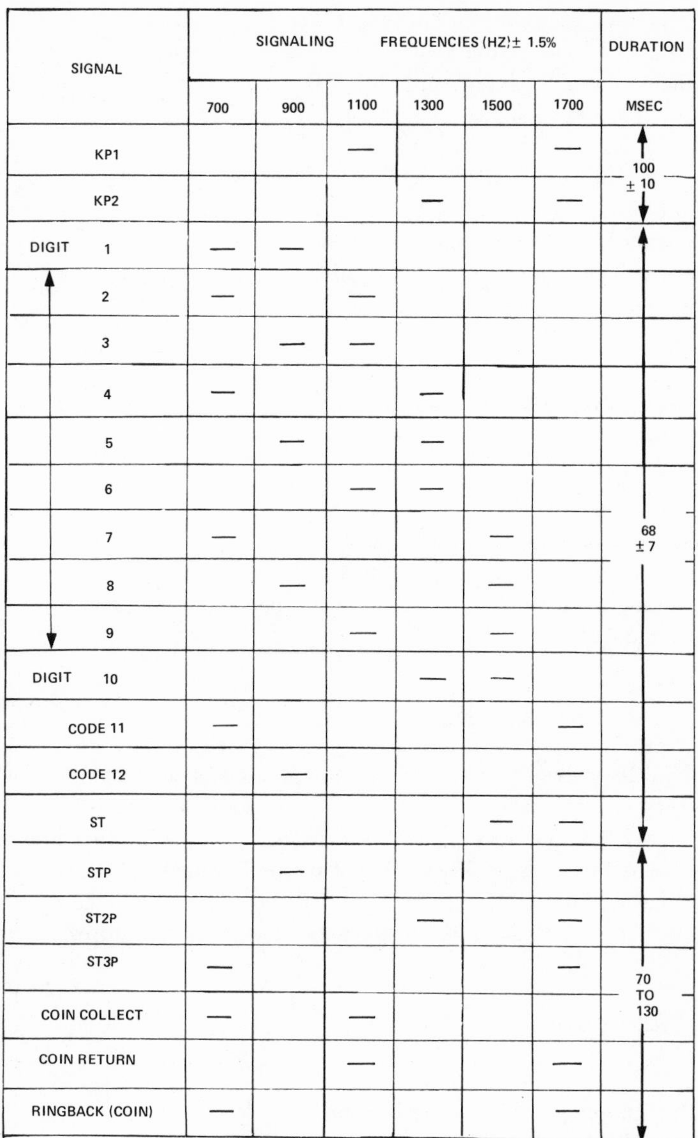

SIGNAL	SIGNALING FREQUENCIES (HZ) ± 1.5%						DURATION
	700	900	1100	1300	1500	1700	MSEC
KP1			—			—	100 ± 10
KP2				—		—	
DIGIT 1	—	—					68 ± 7
2	—		—				
3		—	—				
4	—			—			
5		—		—			
6			—	—			
7	—				—		
8		—			—		
9			—		—		
DIGIT 10				—	—		
CODE 11	—					—	
CODE 12		—				—	
ST					—	—	
STP		—				—	70 TO 130
ST2P				—		—	
ST3P	—					—	
COIN COLLECT	—		—				
COIN RETURN			—			—	
RINGBACK (COIN)	—					—	

Figure 9.8. R1 register signals.

9.9.3.2. *Sender Limits*

Frequency tolerance. The frequency variation should not exceed ±1.5% of each nominal frequency.

Level. The transmitted signal level is −7 ±1 dBmO frequency. The difference in transmitted level between the frequencies comprising a signal must not exceed 0.5 dB.

Frequency leak and modulation products. The level of the signal leak current transmitted to the line should be at least 50 dB below the signal frequency level when a multifrequency signal is not being transmitted. It should be 30 dB below the transmitted signal level of either of the two frequencies when a multifrequency signal is being transmitted.

Signal duration. KP signal is 100 ±10 msec. All other signals have a duration of 68 ±7 msec. The interval between signals should be 68 ±7 msec.

Compound signal tolerance. Frequencies in a signal must be applied or removed with a difference in the application or removal of 1 msec or less between them.

9.10. CCITT Regional Standard No. R2

The CCITT regional standard No. R2 signaling system is now used in many European, Latin American, and developing countries for national and international calls. It meets the various requirements for semiautomatic and fully automatic service. The characteristics of the system were outlined at a conference held at Berne in November, 1962 and it has been recommended by CCITT as a regional standard.

9.10.1. Compelled Signaling

The R2 signaling system uses end-to-end compelled signaling. It uses signals having two frequencies in a two-out-of-six code (540 to 1140 Hz for backward signals and 1380 to 1980 Hz for forward signals). The signaling code for register signals is shown in Figure 9.9. It should be noted that the register signal frequencies used in the R2 system differ from CCITT No. 5 and the R1 system as shown in Figure 9.10 and that the signaling tolerances also differ.

9.10.2. R2 Line Signaling

9.10.2.1. Line Signaling Analog Version

The signals are transmitted link by link. It is a "tone-on-idle" signaling method. The signaling channel is outside the speech frequency band.

The recognition time for a changed condition is 20 ±7 msec at the output of the signal receiver.

If seizure is immediately followed by release, the tone must be removed for at least 100 msec.

Figure 9.11 shows the signaling code.

SIGNALS	FREQUENCIES (HZ)					
NUMERICAL VALUE	FORWARD DIRECTION					
	1380	1500	1620	1740	1860	1980
	BACKWARD DIRECTION					
	1140	1020	900	780	660	540
1	—	—				
2	—		—			
3		—	—			
4	—			—		
5		—		—		
6			—	—		
7	—				—	
8		—			—	
9			—		—	
10				—	—	
11	—					—
12		—				—
13			—			—
14				—		—
15					—	—

Figure 9.9. R1 register signals.

9.10.2.2. R1 PCM Version

The PCM version uses the sixteenth time slot of the 2.048 Mbit primary multiplex system described in Chapter 6. The 8 bits in the time slot serve a pair of circuits. Only two of the bits, a, b, c, and d, serving a channel are used; Figure 9.12 shows the signaling code.

9.10.3. Register Signaling

Figure 9.13 shows the principle of signaling. An originating exchange signals to a controlling exchange, which at certain stages of the call will utilize two

FREQUENCY	USED IN CODE R2	USED IN CODE #5 & R1
540	—	
660	—	
700		—
780	—	
900	—	—
1020	—	
1100		—
1140	—	
1300		—
1380	—	
1500	—	—
1620	—	
1700		—
1740	—	
1860	—	
1960	—	

Figure 9.10. Comparison of frequencies used in R1 and R2.

OPERATING CONDITION OF THE CIRCUIT	SIGNALING CONDITION			
	FORWARD		BACKWARD	
	TONE ON	TONE OFF	TONE ON	TONE OFF
IDLE	—		—	
SEIZING		—	—	
ANSWERED		—		—
CLEAR BACK		—	—	
RELEASE	—		EITHER	EITHER
BLOCKED	—			—

Figure 9.11. R2 line signaling code (analog).

OPERATION CONDITION OF THE CIRCUIT	TIME SLOT 16 CODE			
	FORWARD		BACKWARD	
	A_F	B_F	A_B	B_B
IDLE	1	0	1	0
SEIZING	0	0	1	0
SEIZING ACKNOWLEDGED	0	0	1	1
ANSWERING	0	0	0 OR 1	1
CLEAR BACK	0	0	1	1
CLEAR FORWARD	1	0	0 OR 1	1
RELEASE GUARD = IDLE	1	0	1	0
BLOCKED	1	0	1	1

Figure 9.12. R2 line signaling code (PCM).

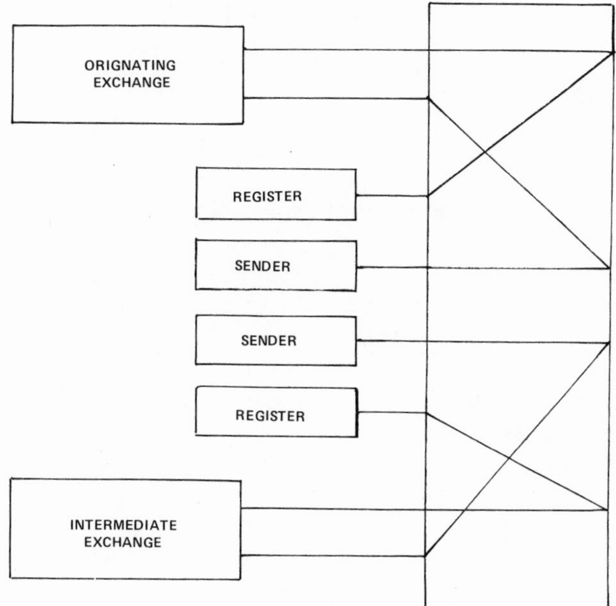

Figure 9.13. Principle of compelled signaling.

pairs of register senders. The system normally operates on an end-to-end signaling basis. It provides for overlapping sending and receiving by permitting sending and receiving in both directions at the same time. It is not suitable for use via TASI or satellites because of its compelled nature.

9.10.3.1. Allocation of Signals

A two-out-of-six code is used since it can be checked to see whether a correct signal has been received (as is R1).

The receiving register can request the sender for information at any time during sending, independently of any chronological order. Thus, for example, a given selection code can be repeated many times on request.

Each frequency combination can have two and sometimes three distinct meanings. Thus backward signals always have two categories: A signals and B signals.

Meanings are changed by transmitting an A3 signal to the sender.

Forward signals are split into two groups called I and II, making it possible to attribute either of two meanings to each of the forward signals. A signal will have a category I and II meaning according to the instructions of a transit or incoming register through backward signals.

It is possible to add a third category of backward and forward signals for national use. This was done in some instances to reduce the system cost by reducing the number of A and B codes and hence frequency generators and receivers by reusing the codes as C codes.

9.10.3.2. Operation

A signal persists until an appropriate acknowledgment signal is received in almost all signaling situations. Figure 9.14 shows the basic principle of operation.

The originating exchange sends various types of signals forward; these include address information, country code and echo suppressor indications, category of calling party, and end of outpulsing. The distant exchange returns congestion signals, address information complete, condition of called line, and network signals. The next action of the system is determined by the signals in each direction which makes for flexible interactive signaling. The switching system transmission path can operate in the two-wire or four-wire mode. In the first case this is effected by the use of band separation filters in series with the sender and register to prevent false operation. This prevents forward signals reaching the originating register and backward signals reaching the originating register.

Typical signal sequence. The sender at the originating exchange, A, sends a signal indicative of the first digit to be sent. The register at the distant end decodes and acknowledges receipt of the signal by instructing the sender to return

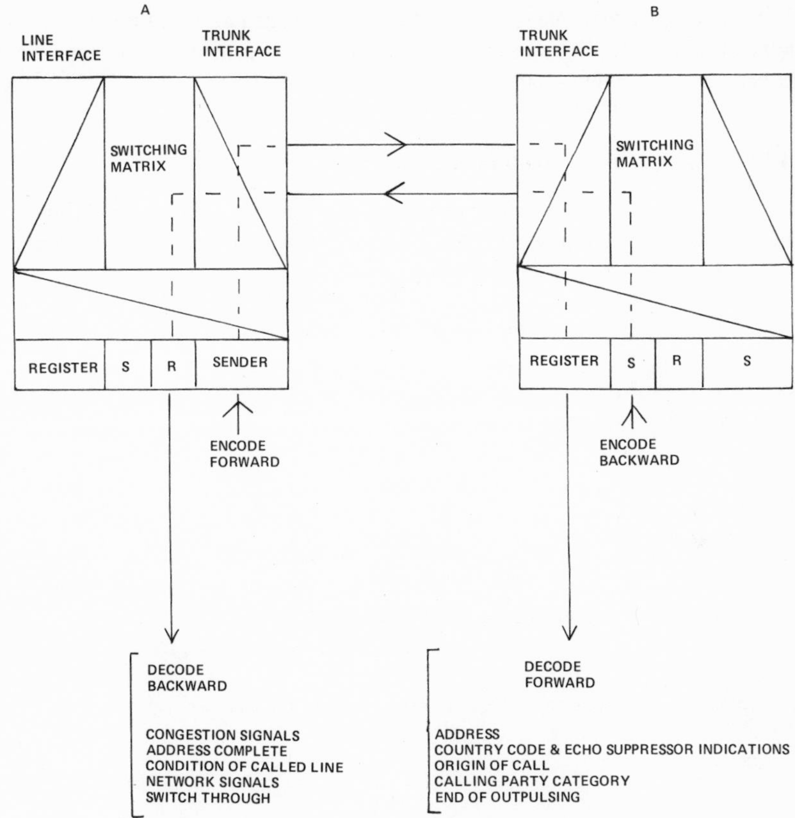

Figure 9.14. General basis of R2 signaling.

a "send next digit" signal. This interaction of signals continues until all the digits have been sent.

At this point exchange B, recognizing that all digits have been received, signals exchange A to change to a second set of signals (B signals). On receipt of this signal exchange A sends the category of the caller. If this is a regular customer, and hence no special treatment is necessary, exchange B indicates the state of the called line. If the line is free, the relevant signal is sent to exchange A, which switches the calling line to the outgoing trunk and releases the register. If the called line is engaged, the relevant signal is sent to exchange A and busy tone is returned from exchange A. Hence the signaling system interacts with the system control in response to special signals.

An example of this is the return of busy tone from the originating exchange, no matter where the busy condition is encountered in the network. Other signals may require routing the call to a different point from that originally dialed and others indicate special treatment that has to be afforded to the calling party. One such requirement for instance can be the restriction of calls in an exchange which has been put under line load control.

9.10.4. Position of Controlling Register

Figure 9.15 shows how the position of the controlling register can change in a telecommunications network. The R2 signaling system is designed to operate

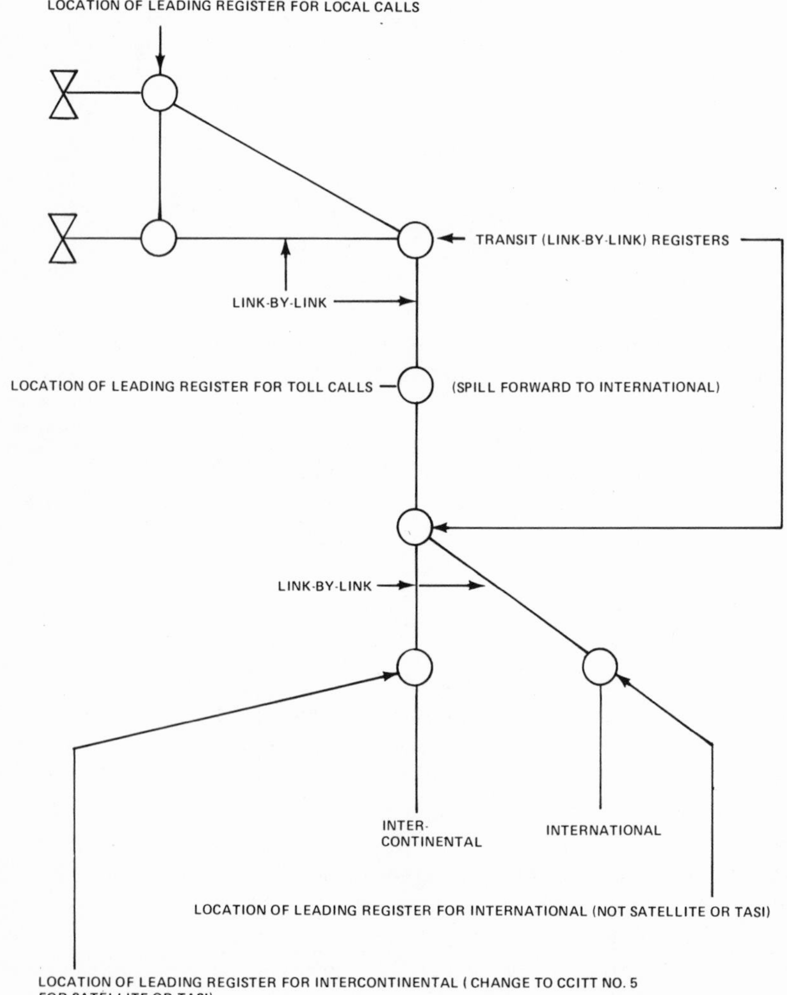

Figure 9.15. Position of controlling register R2 signaling system.

on an end-to-end signaling basis. This means that at least one register (that in the controlling exchange) must be retained in the connection until the setup of the call has been completed. Intermediate registers in the switching network, involved in the routing of a call, are normally only associated with the call long enough to connect the incoming trunk to the outgoing trunk via their transmission matrix. However, under certain routing conditions such as those shown in the figure the control of the setup of the call can be transferred to a higher-level exchange. In this case the originating register in the exchange to which the customer's line is connected is also retained in the connection until the called party answers. Transfer of information to the controlling register takes place on a link-by-link basis.

The figure shows the location of the controlling (leading) register for local calls, trunk (toll) calls, and international calls. The originating register is retained in the call so that busy tone may be returned to the caller from the nearest exchange. This condition is not possible if an automatic exchange of the direct control type serves the called party since such exchanges have no means of returning a called-line-busy signal except as a call progress tone.

9.10.5. Allocation of Signal Values

Figures 9.16 and 9.17 show the recommended values given to the signals in the Proposed CCITT National Signaling Standard code. It should be pointed out

Signal	I. First position Value	I. Other position Value	II. Calling parties' Category Value	
1	Digit 1	Digit 2	Subscriber	
2	Digit 2	Digit 2	Priority call	
3	Digit 3	Digit 3	Maintenance call	
4	Digit 4	Digit 4	Coin box	
5	Digit 5	Digit 5	Operator	
6	Digit 6	Digit 6	Data transmission	
7	Digit 7	Digit 7	International call I/C	
8	Digit 8	Digit 8	Data transmission international I/C	
9	Digit 9	Digit 9	Priority call international I/C	
10	Digit 0	Digit 0	Not required	
11	Not required	Not required	Spare	For national use
12	Not required	Request not accepted	Spare	
13	Test call	Test call	Spare	
14	Not required	I/C ½ echo suppressor	Spare	
15	Not used	End of pulsing	Spare	

Figure 9.16. Proposed R2 national forward signals.

Signal	A. Value Transfer Control Signals	B. Value Condition of Called Line
1	Send next digit (n+)	Connections under control of I/C side
2	Send last but one digit (n−1)	Subscriber transferred number change intercept
3	Address complete change to B signals (send calling-party category)	Subscriber line busy
4	Congestion national network	Congestion
5	Send calling-party category	Vacant national number or direction
6	Address complete through connect	Subscriber line free, charge
7	Send last but two digits (n−2)	Subscriber line free, no charge
8	Send last but three digits (n−3)	Subscriber line out of order
9	Send first digit where needed calling-number local	Spare } national use
10	Spare for national use calling-number national	Spare } national use
11	Not used	Not used
12	Not used	Not used
13	Not used	Not used
14	Not used	Not used
15	Not used	Not used

Figure 9.17. Proposed R2 national backward signals.

that the R2 system has been in use for some time; hence, apart from the values of the digit signals, there are many variations of the signal code allocation. A system has therefore to be designed so that the decoding of the signals can be flexible. What is involved is a flexible association of functions with codes. There are 15 different meanings possible in the two-out-of-six code used in each of the two forward and two backward signal groups.

It should also be possible to allocate a third group of six backward signals to accommodate some early system installations.

The control of the call requires functions for call routing, special handling, and dealing with switching data, etc. at the terminating exchange. It also requires functions such as type of call, automatic number identification, matrix control (switch through, return busy, intercept etc.), charging controls, and call disconnection.

9.10.6. Advantages of R2 Signaling System

The advantages claimed for this signaling system are as follows:

- A reduction of the total time for establishing a connection.
- A minimization of translation requirements at intermediate exchanges.
- The independence of the operating time of the receivers from the response times of the switching equipment and, especially, of the transmission circuits, since the durations of signals are automatically adapted to the propagation times of the circuits.
- The ease of interconnection with other signaling and switching systems and compatibility with any line-signaling system.
- Its high operational security and reliability.

9.10.7. Register Limits

Frequency tolerances. The frequency tolerances, in hertz, are as follows:

Forward signals:

1380 ±10	1740 ±10
1500 ±10	1860 ±10
1620 ±10	1980 ±10

Backward signals:

540 ±10
660 ±10
780 ±10
900 ±10
1020 ±10
1140 ±10

Signal level. The system must operate at −35 to −5 dBm and not operate at a level of ⩽ −42 dBm.

Signal duration. The system must respond within a period ⩽ 70 msec (receiver operate plus release time).

9.10.8. Sender Limits

Frequency tolerance. The system must respond to frequencies within ±4 Hz of their nominal frequency.

Level. This is −8 ±1 dBmO with a minimum difference between signals of 1.0 dB.

Transmit signal duration. This is compelled hence (except for overall time-out) the signal persists until it is acknowledged.

Tone coincidence must be less than 1 msec difference.

10

Separate-Channel Signaling

10.1. Introduction

The idea of separating signaling from the transmission path and allocating a separate path shared with other calls is not new. An early version of the system was investigated by engineers of the, then, Automatic Telephone and Electric Company and the British Post Office.[45] However, suitable components were not available and the work was discontinued. The present-day systems rely on the availability of stored-program techniques and of low-cost electronic components.

At the time of going to press two systems using the method existed. The first, using an analog transmission path with data modems, is called CCITT No. 6. This has country-specific versions for national applications which are basically compatible but have somewhat modified formats.

One of these variants is the AT&T common channel interoffice system (CCIS). This is being installed extensively in the Bell System. The second system uses a digital transmission path and is called CCITT No. 7. The No. 6 and No. 7 systems differ greatly in their organization and format and hence are not compatible.

10.2. CCITT No. 6 Signaling System

The CCITT system No. 6 was standardized in 1968. It is described in the *Green Book*.[45]

Signaling system No. 6 entirely removes the signaling from the speech path by using a separate common signaling link over which all signals for a number of trunks are transferred. For the purpose of explanation the control functions are considered to take place in a separate processor. In practice the functions will be performed in the system control.

The signaling link is capable of operating over standard voice bandwidths.

Signal messages (consisting of 28-bit signaling units) carry information to identify the trunks concerned. An outgoing trunk call being controlled by signaling system No. 6 will transfer the information relating to the call, and to the selected outgoing trunk, to the No. 6 processor. This proceeds to generate the necessary messages which are sent over the separate signaling link to the distant exchange. As a result of this information, the distant exchange identifies the trunk between the two exchanges and sends a return message to the originating exchange No. 6 processor. The necessary switching actions to route the call to this trunk are then made. All signals (including disconnection signals) are sent over the separate common signaling path. It will be seen that such a procedure involves interaction between the outgoing or incoming trunk and its processor. This occurs for all signals used to originate and terminate the connection as well as those used to transfer the data to and from the system controls and the common signaling path.

The transmission of information on a separate channel simplifies the trunk terminal equipment, thus reducing costs. This method also allows greater utilization of speech circuits and extends the opportunities for introducing information not economical in present systems.

Signaling system No. 6 is faster, more reliable, and handles more signals than No. 5 bits or any other existing signaling system.

System No. 6 can also operate in a nonassociated mode. In a nonassociated mode, signals between exchanges which are not connected by data link can be routed via another exchange which has data links between itself and the other exchanges. This mode of operation is used in the U.S.A. via a separate signaling network set up solely for the purpose. It minimizes the number of common channel systems required to serve a given telephone network and hence smaller trunk groups can be served economically.

10.2.1. Signaling over the Common Channel

The system is designed to operate with a standard terminal. The terminal should be capable of working with digital or analog data links at speeds of 2000, 2400, or 4000 bits/sec.

An analog channel requires a 2400 bits/sec modem to generate the data signals. It can also operate asynchronously to a 64-kbit/sec part of a PCM transmission system or synchronously to a 2-kbit/sec or 4-kbit/sec part of a PCM transmission system.

In the case of analog working, timing is derived from the modem. Timing is derived from the data link in the case of data-link working.

10.2.2. Line Signals

No line signals are sent over the selected trunk. Line signals are transmitted as signal units and the trunk circuit being used is identified by a code forming part of the signal unit.

10.2.3. Trunk Continuity Test

The trunk circuit has to be tested for continuity and a continuity signal generated before a path is established. A transceiver (check-tone transmitter and receiver) is connected to the "go" and "return" paths of the outgoing circuit at the first and at each succeeding exchange, excluding the last exchange, in that part of the international connection served by the signaling system No. 6. A check loop is connected to the "go" and "return" paths of the incoming circuit at each exchange.

The check tone frequency is 2000 ±20 Hz. The sending level is −12 ±1 dBmO. The check loop should have 0 dB loss.

Operating limits. The receiver must respond to a frequency of 2000 ±30 Hz.

Signal Level. The absolute power level N of the check tone must be within the limits $(-18 + n) \leqslant N \leqslant (-6 + n)$ dBm, where n is the relative power level at the receiver input.

Recognition time. This is 30 to 60 msec. Upon receipt of the continuity signal from the preceding exchange, the check loop will be removed.

Nonoperating limits for receiver. The receiver shall not operate under the following conditions:

- The signal frequency is outside the frequency band of 2000 ±200 Hz.
- The signal level is $\leqslant -22 + n$ dBm.
- The received level limit is 10 dB below the nominal absolute level of the check tone at the input of the receiver; if the level falls below this point, transmission is considered unacceptable.
- The signal duration is $<$ 30 msec.
- There is a failure to receive a signal after $\leqslant$ 2 sec.

Recognition of removal of check tone. The check tone is considered to have been disconnected if it is interrupted for 15 msec and the level is $< (-27 + n$ dBm).

Removal of check tone. The receipt of the signal unit identifying the trunk and containing the continuity signal will result in the removal of the check tone. The circuit can then be switched through.

Response time of receiver. This provides an indication of the signal removal in $<$ 40 msec.

A continuity check is considered successful when a tone is sent on the "go" path and is received on the "return" path.

10.2.4. Organization of Signaling Units

The data are organized into signal units consisting of 28 bits; the first 20 are information bits and the last 8 are check bits. These units in return are grouped in blocks of 12 signal units (Figure 10.1). The first 11 signal units of each block,

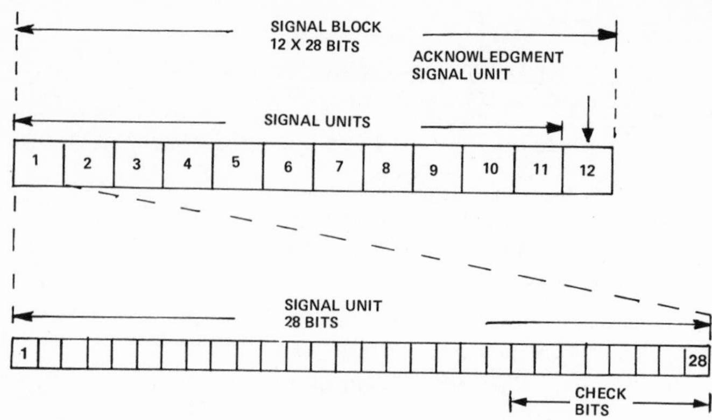

Figure 10.1. CCITT No. 6 signaling format signal block. Note: The 11 signal units can be made up of any combination of the following: one-unit message (signal or management), multiunit message, initial and subsequent signal units, synchronization signals.

used either individually or in combination, carry signal messages or idle signal units to maintain synchronization. The last signal unit of each block is used to acknowledge whether or not each of the 11 signal units in the last block was received with or without detected errors. If it turns out that an error is indicated for any signal unit in a message, the entire message will be transmitted.

The 11 signal units can be made up of any combination of the following:

- one-unit message (signal or management)
- multiunit message
- initial and subsequent signal units
- synchronization signals

Signaling messages are sent between offices in a specified format. An 11-bit trunk label, which associates the message with a particular trunk, will be sent as part of the first signal unit in the message. The initial address message will use from two to five signal units. Supervisory signals and single-digit address messages will be sent as one signal unit. Signals not directly associated with processing of individual calls (such as network management signals) will also be handled.

10.2.5. Abbreviations Specific to Signaling System No. 6

ACU	Acknowledgment signal unit
ADC	Address-complete signal, charge
ADI	Address-incomplete signal
ADN	Address-complete signal, no charge
ADX	Address-complete signal, coin box

AFC	Address-complete signal, subscriber-free, charge
AFN	Address-complete signal, subscriber-free, no charge
AFX	Address-complete signal, subscriber-free, coin box
ANC	Answer signal, charge
ANN	Answer signal, no charge
BLA	Blocking-acknowledgment signal
BLO	Blocking signal
CB1–3	Clear-back signal Nos. 1–3
CFL	Call-failure signal
CGC	Circuit-group-congestion signal
CLF	Clear-forward signal
COF	Confusion signal
COT	Continuity signal
CSSN	Circuit state sequence number
FOT	Forward-transfer signal
IAM	Initial address message
ISU	Initial signal unit
LOS	Line-out-of-service signal
LSU	Lone signal unit
MRF	Message-refusal signal
MUM	Multiunit message
NMM	Network-management and maintenance signal
NNC	National-network-congestion signal
RA1–3	Reanswer signal Nos. 1–3
RLG	Release-guard signal
SAM1–7	Subsequent address message Nos. 1–7
SCU	System control signal unit
SEC	Switching-equipment-congestion signal
SNM	Signaling-network-management signal
SSB	Subscriber-busy signal (electrical)
SST	Subscriber-transferred signal
SSU	Subsequent Signal Unit
SU	Signal Unit
SYU	Synchronization signal unit
UBA	Unblocking-acknowledgment signal
UBL	Unblocking signal
VNN	Vacant-national-number signal

10.2.6. Signal Unit Format (One-Unit Message)

The format of a one-unit message is shown in Figure 10.2. It is designed to transmit either an initial signal unit of a multiunit message, a single telephone signal, signaling system control signals, or management signals. Bits 1, 2, 3, 4, and 5 provide heading information. Bits 6, 7, 8, and 9 contain signal information, and bits 10–20 identify the trunk being used.

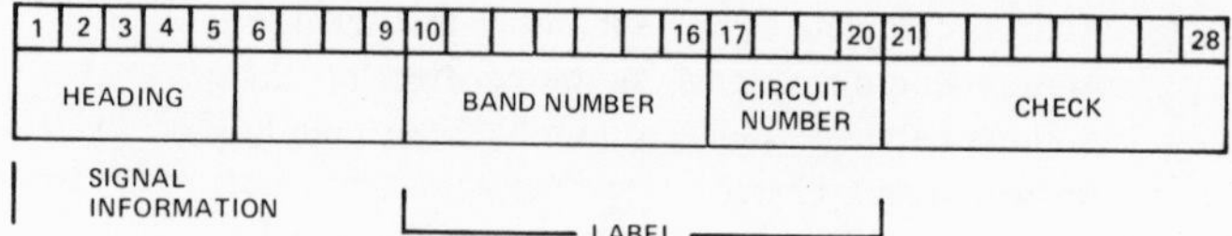

Figure 10.2. Basic format lone signal unit or initial signal unit.

10.2.6.1. Heading Code

Heading codes identify the types of signal unit being sent. The heading code (bits 1–5) is as shown in Table 19.

10.2.6.2. Signal Information

The signal information area, bits 6–9, is used either to define a particular signal in a group of signals, or to define a subgroup within a group, or to indicate that the subsequent signal unit (or units) contains a number of signals belonging to the group of signals defined by the heading code.

Combination of heading and signal information. The heading and signal bits are combined in order to increase the heading vocabulary. The combination with the signal information results in the code shown in Table 20. This uses the abbreviations listed in Section 10.25.

Table 19. CCITT Signaling System No. 6 Heading Codes

Code	Meaning
0 0	Subsequent signal unit
0 1 0 0 0	Spare (regional and/or national)
0 1 0 0 1	
0 1 0 1 0	
0 1 0 1 1	
0 1 1	Acknowledgment signal unit
1 0 0 0 0	Initial signal unit
1 0 0 0 1	Subsequent address message
1 0 0 1 1	
1 0 1 0 0	
1 0 1 0 1	
1 0 1 1 0	
1 0 1 1 1	
1 1 0 0 0	International telephone signals
1 1 0 0 1	
1 1 0 1 0	
1 1 0 1 1	
1 1 1 0 0	Spare (regional and/or national)
1 1 1 0 1	Signaling system control signals except acknowledgment
1 1 1 1 0	Spare (regional and/or national)
1 1 1 1 1	

Table 20. Allocation of Heading and Signal Information Codes[a]

Bits 6–9 \ Bits 1–5	0000X	0001X	0010X	0011X	01000	01001	01010	01011	011XX	10000	10001	10010	10011	10100	10101	10110	10111	11000	11001	11010	11011	11100	11101	11110	11111	Bits 1–5 / Bits 6–9
0000	SSU				ISU of MUM	ISU of MUM	ISU of MUM	ISU of MUM		ISU of IAM	ISU of SAM1	ISU of SAM2	ISU of SAM3	ISU of SAM4	ISU of SAM5	ISU of SAM6	ISU of SAM7	ISU of MUM	ISU of MUM	ISU of MUM	ISU of MUM	ISU of MUM	ISU of MUM	ISU of MUM	ISU of MUM	0000
Not 0000					LSU	LSU	LSU	LSU		ISU of MUM	Lone SAM1	Lone SAM2	Lone SAM3	Lone SAM4	Lone SAM5	Lone SAM6	Lone SAM7	LSU	LSU	LSU	LSU	LSU	LSU	LSU	LSU	Not 0000
0000	One SSU or five SSUs (IAM only)	Two SSUs	Three SSUs	Four SSUs	Reserved for regional and/or national use	Reserved for regional and/or national use	Reserved for regional and/or national use	Reserved for regional and/or national use	ACU													Reserved for regional and/or national use		Reserved for regional and/or national use	Reserved for regional and/or national use	0000
0001											Digit 1	Digit 1	Digit 1	Digit 1	Digit 1	Digit 1	Digit 1	RLG		COT	AFC		NMM			0001
0010										Reserved for regional and/or national use	2	2	2	2	2	2	2	ANC		CLF	AFN					0010
0011											3	3	3	3	3	3	3	ANN	SEC	FOT	AFX					0011
0100											4	4	4	4	4	4	4	CB 1	CGC		SSB					0100
0101											5	5	5	5	5	5	5	RA 1	NNO		VNN		SNM			0101
0110											6	6	6	6	6	6	6	CB 2			LOS		Regional and/or national			0110
0111											7	7	7	7	7	7	7	RA 2			SST					0111
1000											8	8	8	8	8	8	8	CB 3	CFL							1000
1001											9	9	9	9	9	9	9	RA 3								1001
1010											0	0	0	0	0	0	0				ADC					1010
1011																				BLO	ADN					1011
1100																				UBL	ADX		SCU			1100
1101																				BLA	ADI		SYU			1101
1110																			COF	UBA			Regional and/or national			1110
1111											ST	ST	ST	ST	ST	ST	ST			MRF						1111

[a] Note: All unassigned codes are reserved for international use. The interpretation of the abbreviations for signals is given in the list of abbreviations, Section 10.2.5.

10.2.6.3. Label

Only one label per message is used. This is sent in the first signal unit of the sequence. It defines a group of up to 16 speech circuits by use of the band number (bits 10–16). It identifies a circuit within a group of up to 16 speech circuits (bits 17–30). This identifies one of up to 2048 speech circuits (128 multiplied by 16).

10.2.6.4. Check Code

Bits 21–28 are used for checking the message information. The code received is compared with the code generated after all 28 bits have been decoded. The check bits are inverted for transmission and reinverted at the receiving end. This protects against a single bit slip.

The method is to send a 1 or 0 in each position 21 to 28 according to whether the least significant digit in the binary addition of selected bits is odd or even; for example,

- Bit 21 is 1 if the binary sum of bits 2, 3, 5, 7, 9, 13, 14 and 15 is odd and is 0 if the sum is even.
- Bit 22 refers to the sum of bits 1, 3, 4, 6, 8, 10, 14, 15, and 16.
- Bit 23 refers to the sum of bits 2, 4, 5, 7, 9, 11, 15, 16, and 17.
- Bit 24 refers to the sum of bits 1, 3, 5, 6, 8, 10, 12, 16, 17, and 18.
- Bit 25 refers to the sum of bits 2, 4, 6, 7, 9, 11, 13, 17, 18, and 19.
- Bit 26 refers to the sum of bits 3, 5, 7, 8, 10, 14, 18, 19, and 20.
- Bit 27 refers to the sum of bits 2, 3, 4, 5, 6, 7, 8, 11, 14, 19, and 20.
- Bit 28 refers to the sum of bits 1, 2, 4, 6, 8, 12, 13, 14, and 20.

A message consisting of the following information, in bits 1–20, could be

11011 0110 1101101 1011

which indicates an LOS (line-out-of-service signal) sent on circuit group 91 circuit number 13. Its check code would be 10101111.

Every signal unit uses bits 21–28 for a check area.

10.2.7. Signal Format Multiunit Message

A multiunit message consists of two to six signal units in tandem.

The first signal unit is called an *initial signal unit* and the second and any following signal units are called *subsequent signal units*. The format of an initial signal unit is the same as that of a one-unit message. The format of a subsequent signal unit differs for the initial and subsequent signal units.

10.2.7.1. First Subsequent Signal Unit

The format of the first subsequent signal unit of a multiunit message is shown in Figure 10.3.

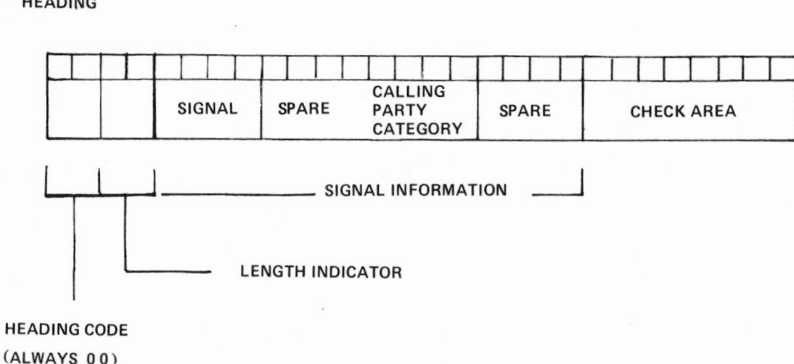

Figure 10.3. First subsequent signal unit.

Heading. This is always 00 and occupies bit positions 1 and 2 only.

Length indicator. Each subsequent signal unit of a multiunit message carries the same length indicator. The codes (which use bits 3 and 4) are as follows.

For one subsequent signal unit, if the signal unit is the first subsequent signal unit no indicator is given; if it is another signal unit then 00 is used. The respective codes for two subsequent signal units are 01 and 01; for three subsequent signal units they are 10 and 10; for four subsequent signal units they are 11 and 11; and for five subsequent signal units they are 00.

Call route information. This uses bits 5–8 in the following ways:

- Bit 5 is 0 if a country code is not included; 1 if it is.
- Bit 6 is 0 if a satellite is not included in the route; 1 if it is.
- Bit 7 is 0 if ½ an echo suppressor is not included in the outgoing route; 1 if ½ an echo suppressor is included.
- Bit 8 is spare, presently coded 0.

Spare bits. These are bits 9–12 which are coded 0000, and bits 17–20 which are not coded. Bits 9–12 are reserved for regional and/or national use.

Special call routing. This uses bits 13–16, which are coded as shown in Table 21.

10.2.7.2. Subsequent Signal Units 2–5

Subsequent signal units 2–5 use bits 1 and 2 for the heading code, which is always 00, and bits 3 and 4 for the length indicator, which is coded as described in Section 10.27.1 above. Bits 5–20 are used for four 4-bit words which provide signal information and are coded as shown in Table 22 for each word. The filler uses 0000 when no signal is being sent.

Table 21. Calling Party's Category Relative to Special Call Routing

Spare	0 0 0 0
Operator language French	0 0 0 1
Operator language English	0 0 1 0
Operator language German	0 0 1 1
Operator language Russian	0 1 0 0
Operator language Spanish	0 1 0 1
Language specified by administration	0 1 1 0 0 1 1 1 1 0 0 0
Reserved	1 0 0 1
Regular Subscriber	1 0 1 0
Subscriber with priority	1 0 1 1
Data call	1 1 0 0
Test call	1 1 0 1
Spare	1 1 1 0 1 1 1 1

10.2.8. Typical Call

The following illustrates the signals involved in two typical international calls. The check signals have been omitted to simplify the example, so messages include only bits 1–20.

10.2.8.1. From U.S.A.

The relevant dialing code to reach the gateway exchange in the U.S.A. are absorbed in routing the call, leaving the number 31-2150-43551 to be transmit-

Table 22. Address Information in Subsequent Signal Units 2–5

Filler	0 0 0 0
Digit 1	0 0 0 1
Digit 2	0 0 1 0
Digit 3	0 0 1 1
Digit 4	0 1 0 0
Digit 5	0 1 0 1
Digit 6	0 1 1 0
Digit 7	0 1 1 1
Digit 8	1 0 0 0
Digit 9	1 0 0 1
Digit 0	1 0 1 0
Call operator code 11	1 0 1 1
Call operator code 12	1 1 0 0
Spare	1 1 0 1
Spare	1 1 1 0
ST	1 1 1 1

ted. The country code is 31. This will be used at the London exchange (call A) to route the call to The Netherlands but will not be repeated to The Netherlands from London (call B).

10.2.8.2. To U.S.A.

The relevant dialing codes to reach the gateway exchange in The Netherlands are absorbed in routing the call, leaving the number 1-201-949-5813 stored in the gateway exchange. The combined world–country code which identifies the U.S.A. is 1. This will be used to route the call directly to the U.S.A. (call C).

10.2.8.3. Call to Amsterdam

A call to Amsterdam has two stages.

Signals involved in call from U.S.A. to London. The first stage is based on the following assumptions:

- ○ semiautomatic, English language
- ○ associated signaling links
- ○ New York–London satellite with echo suppressors

The signal unit codes and meanings are shown in Table 23. The check code has been omitted.

Call from London to Amsterdam. The second stage, from London to Amsterdam, assumes the same items as the first stage of the call except that this stage of the route was cable without echo suppressors. Table 24 gives the signal unit codes and meaning.

10.2.8.4. Call from Amsterdam to U.S.A.

A call from Amsterdam to the U.S.A. is based on the following assumptions:

- ○ automatic traffic, ordinary subscriber
- ○ speech path cable with echo suppressor
- ○ nonassociated signaling link
- ○ dialed information

Table 25 gives the first four messages with the signal codes and meanings.

Subsequent to this the main station code was received and a series of five subsequent signal units formed, one for each of the four digits and one for the end-of-dialing ST signal.

Subsequent messages. The subsequent messages are shown in Table 26.

At the end of this sequence all the information has been transmitted.

Table 23. Messages Involved in Call from New York to London

1st message	
Initial signal unit	10000
Initial address message	0000
Circuit identity	0000101 0011
2nd message	
Subsequent signal unit	00
4 SSUs	11
Country code included	1
Satellite in connection	1
Including echo suppressor	1
Spare (no meaning)	0
Spare (no meaning)	0000
Operator English language	0010
Spare (no meaning)	0000
3rd message	
SSU	00
4 SSUs	11
Digit 3	0011
Digit 1	0001
Digit 2	0010
Digit 1	0001
4th message	
SSU	00
4 SSUs	11
Digit 5	0101
Digit 0	1010
Digit 4	0100
Digit 3	0011
5th message	
SSU	00
4 SSUs	11
Digit 5	0101
Digit 5	0101
Digit 1	0001
ST end of sending	1111

10.2.9. Additional Signal Units

There are a number of other signal units which are needed to operate the system, some of which are described below. Reference should be made to the latest CCITT published recommendations for modifications or additions to codes, especially network management and network maintenance codes.

10.2.9.1. Test Call Signal Unit

The test call signal unit uses bits 1 and 2 for the heading code and 3 and 4 for the length indicator. Bits 5–8 are coded as follows:

Table 24. Messages Involved in Call from London to Amsterdam

1st message
ISU 10000
1st ISU of IAM 0000
Circuit identity 00000001010

2nd message
SSU 00
4 SSUs 11
Country code omitted 0
No satellite in connection 0
No echo suppressor 0
Spare (no meaning) 0
Spare (no meaning) 0000
Operator English language 0010
Spare (no meaning) 0000

3rd message

0 0	1 1	0 0 1 0	0 0 0 1	0 1 0 1	1 0 1 0
SSU	4 SSUs	Digit 2	Digit 1	Digit 5	Digit 0

4th message

0 0	1 1	0 1 0 0	0 0 1 1	0 1 0 1	0 1 0 1
SSU	4 SSUs	Digit 4	Digit 3	Digit 5	Digit 5

5th message

0 0	1 1	0 0 0 1	1 1 1 1	0 0 0 0	0 0 0 0
SSU	4 SSUs	Digit 1	ST	Filler	Filler

Continuity check system no. 6	0 0 0 0
ATME 2 SIG & XMSN	0 0 0 1
ATME 2 Signaling test only	0 0 1 0
The remainder are spare	

Bits 9–12 are coded 1111 and the remaining two 4-bit words are coded 0000 (filler).

10.2.9.2. Acknowledgment Signal Unit (Figure 10.4)

The first three bits are coded 011; the remaining bits are used as follows. This signal unit contains information on the 11 signal units in the block

Table 25. Initial Address Messages Involved in Call from Amsterdam to London

ISU	10000
1st message	
ISU initial AM	0000
Circuit identity	0010000
2nd message	
SSU	00
3 SUUs	10
No country code	0
No satellite	0
Echo suppressor	1
Spare (no meaning)	0
Spare (no meaning)	0000
Regular calling subscriber	1010
Spare (no meaning)	0000

3rd message

0 0	1 0	0 0 1 0	1 0 1 0	0 0 0 1	1 0 0 1
SSU	3 SSUs	Digit 2	Digit 0	Digit 1	Digit 9

4th message

0 0	1 0	0 1 0 0	1 0 0 1	0 0 0 0	0 0 0 0
SSU	3 SSUs	Digit 4	Digit 9	Filler	Filler

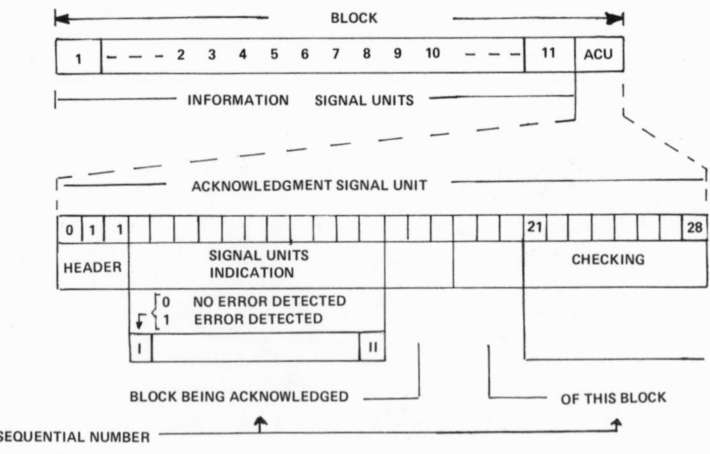

Figure 10.4. Block checking procedure.

Table 26. Subsequent Address Messages Involved in Call from Amsterdam to London

1st message	
1st SMU	10001
LSU digit 5	0101
Circuit number	001 0000 1001
2nd message	
2nd SMU	10010
LSU digit 8	1000
Circuit number	001 0000 1001
3rd message	
3rd SMU	10011
LSU digit 1	0001
Circuit number	001 0000 1001
4th message	
4th SMU	10100
LSU digit 3	0011
Circuit number	001 0000 1001
5th message	
5th SMU	10101
LSU "ST"	1111
Circuit number	001 0000 1001

which it services. Bits 4–14 each represent the received state of the signal unit; 0 indicates that no error was detected and 1 that an error was detected in the relevant signal unit. The sequence number of the block being acknowledged is indicated by bits 15–18, and bits 18–20 the sequence number of the block completed by this signal unit. The receipt of an error signal at the point of signal origination results in the retransmission of the block in which the error occurred.

10.2.9.3. System Control Signal Unit

Bits 1–5 are coded 11101. Bits 6–9 are coded 1100 and provide the signal information code. Bits 10–12 are coded 001; Bits 13–16 are coded 0001. Bits 17–20 are coded as shown in Table 27 and provide system control signals.

10.2.9.4. Synchronization Signal Unit

The synchronization signal unit is used to maintain synchronization between the two ends. Bits 1–5 are coded 11101. Bits 6–9 are coded 1101. Bits 10–16 are used in conjunction with bits 6–9, which proyide the signal information, and bits 1–5 to produce the synchronization pattern 1110111011100011. Bits 17–20 indicate the sequence number of the signal unit in relation to the position of that synchronization signal unit in the block of signal units.

Table 27. Control Information Codes

0 0 0 0	Spare
0 0 0 1	Changeover
0 0 1 0	Manual changeover
0 0 1 1	Spare
0 1 0 0	Standby ready
0 1 0 1	Spare
0 1 1 0	Load transfer
0 1 1 1	Emergency load transfer
1 0 0 0	Spare
1 0 0 1	Spare
1 0 1 0	Manual changeover acknowledgment
1 0 1 1	Spare
1 1 0 0	Standby-ready acknowledgment
1 1 0 1	Spare
1 1 1 0	Load transfer acknowledgment
1 1 1 1	Spare

10.2.9.5. Management Message Unit

Bits 1–5 are coded 11101. Bits 6–9 are codes as follows:

Network Management and Maintenance 0 0 0 1
Signaling Network Management 0 1 0 1
System Control Unit 1 1 0 0
Synchronizing Unit 1 1 0 1
All Others Spare

Bits 10–16 are used for the band number as described earlier. Bits 17–20 are coded as follows:

Transfer Prohibited 0 1 0 1
Transfer Allowed 0 1 1 0
Transfer Allowed Acknowledged 1 0 0 0
All Others Spare

10.2.10. Characteristics of Analog Link Receiver

10.2.10.1. Coding

The signals are sent with a modulation rate of 1200 bauds operating at 2400 bits/sec (one cycle per bit). Pairs of bits which are called dibits are represented by the following phase shifts:

Dibit:	00	01	11	10
Phase shift (deg):	+45	+135	+225	+315

10.2.10.2. Carrier Tolerance

The carrier frequency is 1800 Hz. The carrier envelope frequency is 600 Hz.

10.2.10.3. Receiver Sensitivity

The receiver sensitivity is -15 ± 8 dBmO.

10.2.10.4. Synchronization

Bit synchronization. Bit synchronization must take place within 150 msec of the receipt of the synchronization signal unit. Synchronization is maintained during a loss of the data carrier of 1 sec or less.

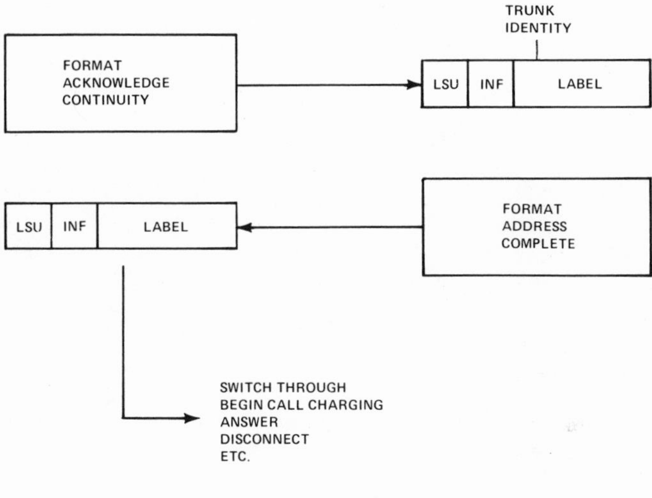

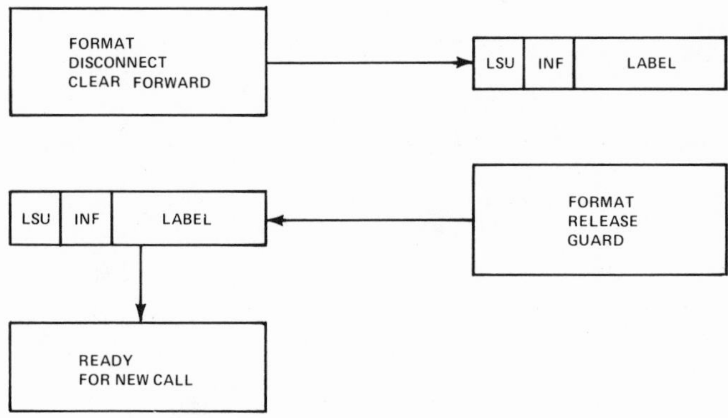

Figure 10.5. Sequence of address messages CCIS.

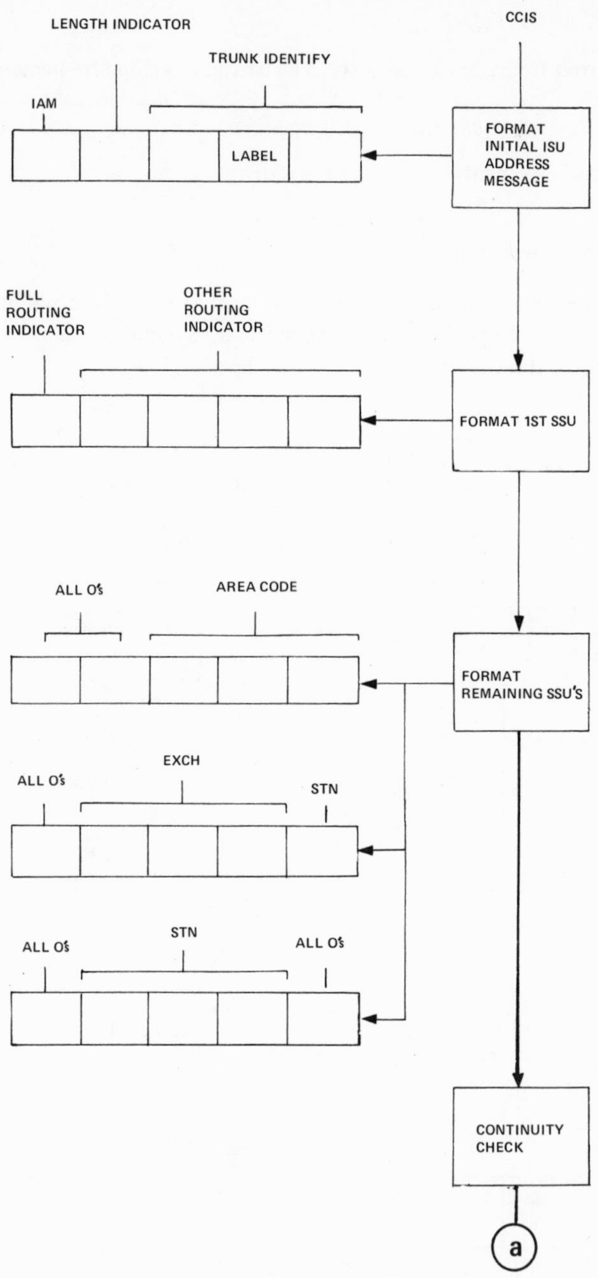

Figure 10.6. Part(i). Sequence of control messages CCIS.

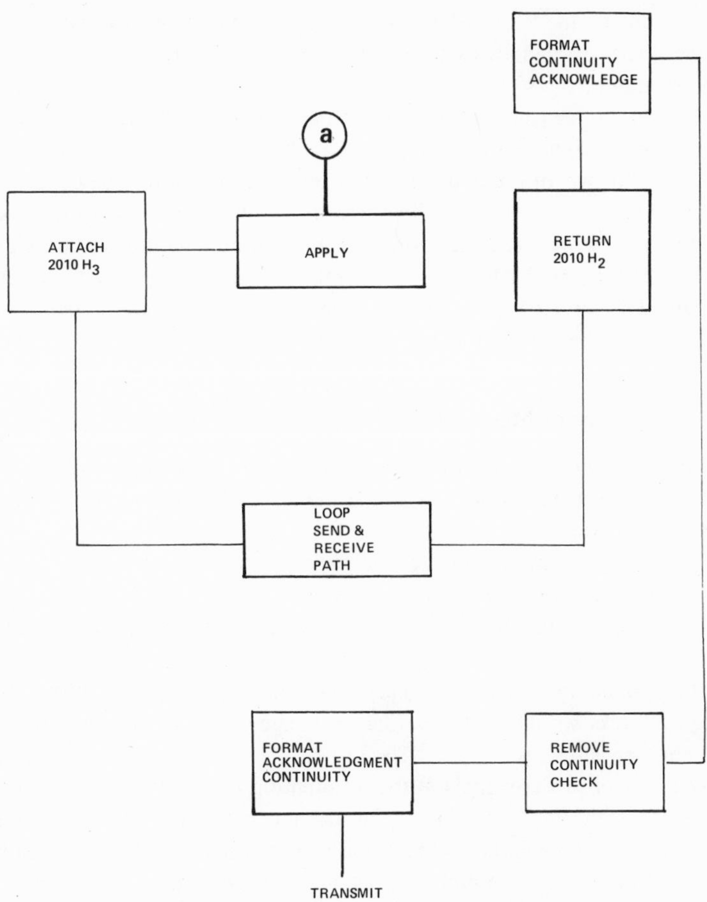

Figure 10.6. Part(ii). *See legend on previous page.*

10.2.11. Characteristics of Analog Link Sender

The analog link sender generates its frequencies plus or minus 0.005% at the level of -15 ± 1 dBmO. It generates its bit timing and carrier frequency from the same source.

10.3. Common-Channel Interoffice Signaling (CCIS)

The Bell System has introduced its own version of No. 6 for use in world zone 1. The principal difference is the increased label size necessary because of the large numbers of trunks encountered. This system is described in various Bell

System publications.[47] It utilizes nonassociated signaling links which connect from the CCIS exchanges to signal transfer points. It can also operate in the associated mode.

In general large trunk groups are served by associated data links and smaller trunk groups via signal transfer points. The number of signaling points in tandem is restricted to two in order to prevent excessive signaling delays. This could cause trouble if one of the signaling systems in the overall connection used R1 line signaling. However, under primary route failure conditions, a maximum of four signaling points in tandem is permitted. The signaling network is fully redundant. Figures 10.5 and 10.6 show the sequence of operation in setting up a call. This is similar to the sequence involved in system No. 6.

10.3.1. Sequence of Messages

The initial message identifies the selected trunk in the originating exchange and also indicates the number of signal units that are necessary to convey the address information.

The second message indicates the routing treatment necessary for the call, when full routing information is required.

The third contains the area code, assuming a call which terminates in world zone 1.

The remaining signal units carry the remaining address information. A ten-digit address would next send the exchange code and the first digit of the called number and so on.

The receipt of the signals at the terminating end of the trunk transmission path results in the looping of the send and receive paths. The originating end applies the check tone to its send path and it is returned over its receive path. This results in satisfactory continuity check and the generation of a signal which acknowledges continuity. The originating exchange removes the check tone.

The terminating exchange generates an acknowledgement signal with whatever control signals, such as ''switch through'' etc., are appropriate.

Subsequently the necessary signals to disconnect the call are generated.

10.3.2. Comparison of Signal Unit Format

This section provides comparisons between the CCITT No. 6 and Bell System formats for some of the principal signal units. It should be noted that the acknowledgment signal unit format is identical. CCIS uses an additional first subsequent signal unit which can provide expanded routing information.

Lone signal unit. The heading uses only bits 1–3. The signal information uses four bits but occupies bits 4–7. The label field is expanded to provide nine bits, 8–16. The remaining fields are the same as No. 6.

Subsequent signal unit. The heading uses only bits 1–3. Bit 4 is used as a route bit. Remaining fields are the same as No. 6.

System control signal units. Bits 1–3 are used as a heading code. Bits 4–7 are used for signal information, in lieu of 1–15 and 6–9, respectively, for CCITT No. 6. Bits 8–20 are used for control signals in lieu of bits 10–20.

One-unit management signal unit. This uses bits 1–7 in the same way as a system control signal unit. Bits 8–16 are used to provide band numbers, terminal numbers, etc., and bits 17–20 for management information.

Synchronization signal unit. This carries the same synchronization pattern as does system No. 6 except for bit 16, which can be 1 (odd) or 0 (even).

10.3.3. Comparison of Codes

The following is a comparison of some of the codes of CCITT signaling system No. 6 and CCIS.

Heading codes. System No. 6 uses bits 1–5, CCIS bits 1–3. The corresponding codes are as shown in Table 28a. It should be noted that the acknowledgement signal unit code is the same for both systems.

Table 28a. Comparison of CCITT No. 6 and CCIS Heading Codes

No. 6	CCIS	Type of Unit
0 0	1 1 0	Subsequent signal
0 1 0 0 0		
0 1 0 0 1		
0 1 0 1 0		
0 1 0 0 1		
0 1 1	0 1 1	Acknowledgment
1 0 0 0 0	1 0 1	Initial of IAM (1010) or MUM (1011)
1 0 0 0 1		
1 0 0 1 0		
1 0 0 1 1		
1 0 1 0 0	1 1 0	Subsequent address message
1 0 1 0 1		
1 0 1 1 0		
1 0 1 1 1		
1 1 0 0 0		
1 1 0 0 1		International telephone signals
1 1 0 1 0		
1 1 0 1 1		
1 1 1 0 0		
1 1 1 0 1	1 1 1	Signaling system control and lone (CCIS)
1 1 1 1 0		
1 1 1 1 1		
	0 0 0	
	0 0 1	
	0 1 0	Lone (telephone)
	1 0 0	

Length indicator codes. The following is a comparison of length indicator codes. CCITT No. 6 uses a SSU indication 00 as a heading code in bits 1 and 2 and bits 3 and 4 to indicate the number of SSUs except for the initial signal unit, which is coded 10000 (heading code) 0000 (signal information). CCIS uses a 3-bit heading code which is 101 in all cases, an initial signal unit code (1 if applicable, 0 if not), and 3-bit length indicator in bits 5, 6, and 7. The codes used are as shown in Table 28b.

Principal signal codes. Table 28c shows a comparison of the coding of some of the principal signals. The abbreviations are the same as those described in Section 10.2.5. CCITT No. 6 uses bits 1–5 for the heading code. The signal code includes the heading code and the signal information.

10.4. CCITT Signaling System No. 7

CCITT system No. 7 is a new signaling system intended for use in digital networks which is presently (August, 1980) being developed. The international specification is expected to be completed in 1980. A national application of this system will be used as standard in many future networks. The signaling system will be optimized for use with the signaling rate of 64 kbits/sec, but it is also suitable for lower rates such as 4.8 or 2.4 kbits/sec derived from analog circuits equipped with modems.

This system is likely to replace CCITT No. 6 in world use. This must inevitably lead to two noncompatible world standards: one CCIS in North America in the AT&T network, with a national variant of system No. 6 in Japan, and CCITT No. 7 elsewhere. CCIS is being applied extensively in the Bell System because of its predominantly analog network.

The introduction of CCITT No. 7 in other countries depends on the application of PCM carrier systems. CCITT No. 7 takes advantage of the state of the art in

Table 28b. Comparison of CCITT No. 6 and CCIS Length Indicator Codes

No. 6	CCIS	Signal
[a]	1 0 1 0/1 0 0 0	One SSU
0 0 0 1	1 0 1 0/1 0 0 1	Two SSU
0 0 1 0	1 0 1 0/1 0 1 0	Three SSU
0 0 1 1	1 0 1 0/1 0 1 1	Four SSU
0 0 0 0	1 0 1 0/1 1 0 0	Five SSU
—	1 0 1 0/1 1 0 1	Six SSU
—	1 0 1 0/1 1 1 0	Seven SSU
—	1 0 1 0/1 1 1 1	Eight SSU

[a] 1 0 0 0 0 0 0 0 0 CCITT ISU indication (length indicator in SSUs).

Table 28c. Comparison of Some of the Principal Signal Codes of CCITT No. 6 and CCIS

CCITT No. 6	CCIS	Signal
1 1 0 1 0 0 0 0 1	0 0 0 0 0 0 1	COT
1 1 0 1 0 0 0 1 0	0 0 0 1 0 0 0	CLF
1 1 0 1 0 1 0 1 1	0 0 0 1 0 0 1	BLO
1 1 0 1 0 1 1 0 0	0 0 0 1 0 1 0	UBL
1 1 0 0 1 1 0 0 0	0 0 1 0 0 0 0	CFL
1 1 0 1 0 1 1 1 1	0 0 1 0 0 0 1	MFR
1 1 0 0 1 1 1 1 0	0 0 1 0 0 1 0	COF
	0 0 1 0 1 0 0	NTC
	0 0 1 0 1 0 1	NSC
	0 0 1 1 0 0 0	RST
1 1 0 0 1 0 1 0 1		NNC
1 1 0 0 1 0 0 1 1		SEC
1 1 0 1 0 1 1 0 1	0 1 0 0 0 0 0	BLA
1 1 0 1 0 1 1 1 0	0 1 0 0 0 0 1	UBA
1 1 0 0 0 0 0 0 1	0 1 0 0 1 1 0	RLG
1 1 0 0 0 0 0 1 1	0 1 0 0 1 1 1	ANN
1 1 0 0 0 0 0 1 0	0 1 0 1 0 0 0	ANC
1 1 0 0 0 0 1 0 0	0 1 0 1 0 1 0	CB1
1 1 0 0 0 0 1 0 1	0 1 0 1 0 1 1	RA1
1 1 0 0 0 0 1 1 0	0 1 0 1 1 0 0	CB2
1 1 0 0 0 0 1 1 1	0 1 0 1 1 0 1	RA2
1 1 0 0 0 1 0 0 0	0 1 0 1 1 1 0	CB3
1 1 0 0 0 1 0 0 1	0 1 0 1 1 1 1	RA3
1 1 0 1 1 0 1 0 0	1 0 0 0 1 0 0	SSB
1 1 0 1 1 1 0 1 1	1 0 0 1 0 1 0	ADN
1 1 0 1 1 1 1 0 0	1 0 0 1 0 1 1	ADX
1 1 0 1 1 1 0 1 0	1 0 0 1 1 0 0	ADC
1 1 0 1 1 1 1 0 1	1 0 0 1 1 0 1	AD1
1 1 0 1 0 0 0 1 1		FOT
1 1 0 1 1 0 1 1 0		LOS
1 1 0 1 1 0 1 1 1		SST
1 1 0 1 0 1 1 1 0		UBA
1 1 0 1 0 1 1 0 0		UBL
1 1 0 1 1 0 1 0 1		VNN
	1 0 1 1 0 0 0	ISU 1 subsequently
	1 0 1 1 0 0 1	ISU 2 subsequently
	1 0 1 1 0 1 0	ISU 3 subsequently
	1 0 1 1 0 1 1	ISU 4 subsequently
	1 0 1 1 1 0 0	ISU 5 subsequently
	1 0 1 1 1 0 1	ISU 6 subsequently
	1 0 1 1 1 1 0	ISU 7 subsequently
	1 0 1 1 1 1 1	ISU 8 subsequently

integrated circuitry. CCITT No. 7 is designed to maximize the utilization of 64,000 bits/sec data links. It reduces the cost of terminal hardware through the use of microprocessors. It has the following characteristics.

10.4.1. Use of Packet Switching Techniques

CCITT No. 7 uses packet, labeled message switching with a format similar to the standard data format ×25 (HDLC) and uses the same error-detecting methods.

10.4.2. Messages Based on Bytes

The system uses 8-bit bytes with the least significant bit sent first.

10.4.3. Variable Length Messages

There is a "start flag" at the beginning of a message, which does not have a fixed length. The end of the message is indicated by the "start flag" of the next message. If there is no succeeding message then the "start flag" of a filler message serves the same purpose. However, a maximum length of 560 bits has been set for a message.

10.4.4. Organization of System

The system consists of a message transfer part (MTP) and two user parts. One of the latter is the telephone user part (TUP) and the other is the data user part (DUP). The user type is signaled by an appropriate code.

10.4.4.1. Message Transfer Point

The message transfer point operates in synchronism with the distant end. It generates messages of variable length; there are no multiunit messages. It provides overlap signaling. A message can contain a single digit. There are no blocks of units. It stores messages in a cyclic memory stack which erases messages after a predetermined delay which is proportional to the capacity of the memory stack. The memory stacks at each end operate with a dedicated serial number per memory area. There is no acknowledgment unit. It detects errors by analyzing a 10-bit cyclic code. An error is corrected by the retransmission of the data stored in the cyclic memory area. The receipt of the serial number of the memory area containing an error results in the retransmission of the message and all subsequent messages.

10.4.4.2. Application to Carrier Systems

System No. 7 can use time slot 16 of the CEPT carrier system. Its use in T1 carrier requires a reduction in the number of channels to 23. One channel is then dedicated to signaling. This could free time slot 8 for message transmission quantization use in frames 6 and 12 of the T1 carrier system. No path check is provided in digital networks.

10.4.5. Message Format

The final specification for the signaling system will be published by the CCITT. There are two basic message signaling units, and others will be used for such purposes as initialization.

10.4.5.1. Message Unit

The message unit consists of an 8-bit flag, followed by the following 8-bit words.

Backward indicator. This byte has a 1-bit backward indicator and a 7-bit sequence number.

Forward indicator. This byte has a 1-bit forward indicator and a 7-bit sequence number.

Length indicator. This byte has a 6-bit length indicator which indicates the number of signal information fields plus two spare bits. The use of this signal is not essential to the operation of the signaling system. It limits the maximum number of signaling information bytes to 63.

Service indicator. This byte has a 5-bit service indicator (only two service types existed at the time of going to press, "voice" and "data") and three spare bits.

Signaling information field. This consists of at least two bytes and consists of as many bytes as are necessary to transmit the relevant information.

Check bits. These consist of two bytes which checks the sequence of bits, including the check bits but excluding the flag bits. The check polynomial used is $x^{16} + x^{12} + x^5 + 1$.

10.4.5.2. Fill-in Signal Unit

This consists of the following bytes with the bit organization as described above: a flag byte, followed by a backward indicator, followed by a forward indicator, and possibly a length indicator (when a length indicator is used it indicates that there are zero signaling information bytes), and lastly the check bits.

10.4.5.3. Other Signal Units

A signal unit which can be used for such purposes as initialization has the same format as a message unit with the following exceptions.

Length indicator. This shows one or two status bytes.

Status field. This replaces the signaling information field and consists of one or two bytes.

10.4.6. Use of Signaling Information Field

The signaling information is organized so that information is sent in the following order for the telephone user part. Backward and forward signals (unacknowledged) are interchanged using the following format.

Label. The label consists of four bytes and contains the terminating exchange code and the originating exchange code; the label indicator may change in the final CCITT specification. The label indication identifies whether or not the call terminates in the same world zone.

Heading. This consists of four bits and identifies the specific function of the message, e.g., initial, answer, etc.

Indicator. This consists of four bits and indicates such things as the transmission means, e.g., satellite, whether echo suppressors are required, etc.

Calling party's category. This consists of four bits which identify any special treatment necessary for the call.

Number of address signals. This consists of four bits coded to indicate the number of address signals.

Address signals. These consist of as many bytes as is necessary to convey the address in binary-coded decimal-code, using four bits per decimal digit.

Table 29. Evolution of Signaling Systems

Purpose	R1	R2	Nos. 6 and 7
Connection control	Line	Line	
Address	Register	Register	
Calling identification	Applique	Register	Separate channel
Network control	Applique	Applique	
System control	Applique	Register	
Pulsed	Yes (register)	No	Yes
Acknowledged	Yes (line compelled)	Yes (compelled)	Yes (No. 6 only)
Checked	No	No	Yes

10.4.7. Other Formats

Subsequent address messages will not contain category and address signals but will exchange information through coding the heading and indicator bits. This permits signaling such things as address complete, called party free, charge call, etc.

10.5. Evolution of Signaling Systems

Table 29 shows a comparison of the signaling systems which provide interaction between the sending-receiving portions and system R1. The earliest system R1 requires a number of separate signaling systems for call charging identification, system control, and network control. R2 provided additional possibilities, but it was not until the emergence of signaling systems Nos. 6 and 7 that a universal signaling system was possible. These last two systems, which use stored logic controls, will make possible more efficient routing and promise many new services both to customers and administrations.

11

Maintenance

11.1. Introduction

Modern stored-program switching systems require more diagnostic and maintenance procedures than those required in earlier electromechanical systems, while still requiring line and trunk transmission equipment testing. This imposes new demands on the maintenance personnel since the number of faults to be expected will be small and hence there will be little opportunity for personnel to gain expertise in maintenance. For this reason, many administrations are using centralized maintenance as described in Chapter 2. It is essential, therefore, that, in addition to local maintenance facilities, the maintenance means provided allow for remote control. This should include means for the following:

- The remote readout of operational data and application of diagnostic programs.
- The monitoring of the system. The remote control of line and trunk testing and transmission testing.
- The removal of faulty equipment from service.

11.2. General Maintenance Requirements

General maintenance requirements are provided on a local and/or on a remote basis. They should include man–machine interfaces which provide displays indicating how the system is performing and means to store the relevant data. It must be possible to select specific functional units or traffic-carrying devices and subject them to simulated traffic and monitor the performance of their tasks. The maintenance test center must be simple to use and easily understood by maintenance personnel.

A printer should be a part of the maintenance test center. This will produce hard copy to document system data and to derive listings of administrative data.

It should be possible to make independent simultaneous tests of various switching system areas. This facilitates the initialization and emergency recovery procedures.

11.2.1. Local Maintenance

Local maintenance can be divided into the following two categories.

Upkeep. This includes both scheduled time needed for the testing of lines, trunks, etc., and unscheduled time needed for fault location and repair.

Evolution of system. This includes cross-connection work on frames and the data base modification resulting from new and terminated line service and modifications in trunk routes.

11.2.2. Personnel Requirements

Personnel requirements are usually expressed in employees per 10,000 working lines. Modern systems should require less than two persons for upkeep and one for changes per 10,000 lines. The requirement for changes is based on 2% of the working lines requiring some action each month.

11.2.3. Removal of Working Equipment

The packaging design of the system must be such that the removal or loosening of any plug-in component is signaled to the level-three or equivalent control automatically. This control must remove the unit, or units, from service, substitute redundant equipment if relevant, and generate a local or remote fault record.

11.2.4. Fault Correction

Corrective maintenance action will normally be effected by the replacement of plug-in components. However, some faults will not be cleared by replacement of equipment and will require skilled personnel using complex diagnostic programs and special equipment.

11.2.5. Maintenance Facilities

The system maintenance facilities must include diagnostic programs, recovery programs and programs to produce messages to indicate the system status and initiate alarms. In addition, it must include programs to initiate tests and maintenance functions. The actions may be initiated either by means of a manually controlled keyboard or by recovery programs.

They will require the use of portable test equipment. Adequate documentation of fault repair procedures must be available including cooperation with higher levels of maintenance centers. The resultant maintenance plan must be carefully structured to indicate the steps necessary for fault analysis and rectification. This is essential in order to achieve maintenance at the specified MTTR (mean time to repair) objectives.

11.2.6. Provision of Redundant Equipment

Additional equipment (which preferably should be substituted automatically for equipment removed from service for maintenance) shall be provided if the equipment serves more than 100 lines.

11.3. Line Testing

A local exchange must be capable of working with an existing line testing equipment already in use at remote locations via trunk interfaces (switching of control). Such interfaces may serve trunks which are dedicated for testing or which are also used for regular traffic. It must be possible to switch the control function from the local exchange test equipment (when it is not in use locally) to the remote facility. The arrangement should not insert any transmission-impairing components into the path to the customer's line so that a metallic path of resistance due only to the wiring and crosspoints is extended.

The local or remote test equipment may be of various types and capacity but all employ the same basic principles of testing.

11.3.1. Test Panel

A test panel is usually provided for exchanges having less than 500 lines. This provides facilities for connecting to lines by manually plugging extension cords from the test panel. It usually only provides basic testing of line resistance and insulation, ringing, and dialing. It also provides for testing new customer lines. It provides the above tests by the use of the following:

- A means of performing voltage, current, and resistance tests, including resistance tests using a low-voltage power supply (Wheatstone bridge), from local test equipment.
- Performing voltage, current, and resistance tests from a remote test facility, either utilizing a dedicated physical pair for measuring or using commercially available remote test equipment which permits use of nonmetallic interoffice facilities.
- A means of performing tests which stimulate the generation of calls.

- A means of determining the approximate number of ringers on the line.
- A means of applying a variable current to the line to test transmitters and DTMF dials.
- A means of testing rotary dials and DTMF calling devices.

11.3.2. Local Test Desk

A local test desk is provided for most exchanges with as many test positions as the administration requires.

In addition to the facilities described above a test desk can reach lines automatically by the use of a line test access network. It has the following general requirements.

- Test access circuits (which are provided in the quantities required by the user).
- An incoming test trunk or equivalent from a remote office which is able to access the test circuits on a ''switching-of-control'' basis.
- A means to control the line test network access.
- A means to originate calls from a line interface to which the line access network has been connected.

11.3.2.1. Centralized Test Consoles

The advent of low-cost microprocessors has made possible program-controlled test equipment. This allows the use of control consoles which can initiate and control system testing over data links to the controlled office. The local office level-three control directly interfaces with such consoles. Data sent to the consoles is analyzed and displayed at the console which is capable of performing many logical functions. The data should be coded using standard data formats such as the American Standard Code for Information Interchange (ASCII).

11.3.2.2. Interpretation of Fault Displays

Information can be generated so as to relate a fault readout to the probable equipment failure. This is usually developed to the point where faults can be identified as being within a small number of modules. The fault is then cleared by the replacement of the faulty modules. This is the objective, but fault location is based on the failure of components in a wired circuit, hence changes in circuitry or components may require a modification in the analysis program. If the readout data are not interpretable at the first maintenance level it may be necessary to transfer the data, corresponding to the stage in call processing at which the fault occured, to a higher-level maintenance center. This center will be capable of a detailed analysis of the program and will be able to generate lists of probable causes of the fault.

The availability of satellite signaling channels offers the possibility of referring such information to the equipment manufacturer when the number of stored-program systems in a particular country does not justify such a maintenance center locally.

11.3.2.3. Stored-Program Aids to Maintenance

Special diagnostic programs control the functions of the test call generation and equipment monitoring. These programs should be separated from the normal call-processing functions of the machine as far as possible. This avoids the duplication of the diagnostic programs in level-two controls and hence reduces their size. A test program will be initiated as a result of indication from the level-two control comparator function or from traffic-carrying or level-one functional units. The functions involve monitoring frequency generating equipment such as clocks, frames, etc., as well as test points which monitor normal responses that are being made by level-one controls (''all seems well'').

A recovery program is initiated when a fault is detected and in addition to identifying the faulty module the program should remove it from service and switch in a redundant unit.

Routine test programs should be applied during low-traffic intervals. The hierarchical structure of test programs is designed to give preference to recovery programs. Other programs have lower priorities.

The maintenance center enables the maintenance personnel to select any test program or sequence of test programs and to obtain a visual indication of the progress and the results of such tests. It will not permit access to or any means of modifying the basic call-processing or diagnostic-processing programs. It will allow updating of dimensioning and customer data with an authorization code.

11.3.3. Test Access for Lines

A test access network is provided which can connect to all line interfaces in the local exchange. Access can be made to all lines, whether they are free or busy, unless they are marked with a ''data line security'' or other restricted class of service and the line is busy.

The test access should connect to, or provide, the following features:

- A means of testing ringing and ringtrip functions.
- A manually controlled means of disassociating the station loop from the associated line interfaces and connecting it to the test equipment.
- The application of high-level progress tones such as receiver-off-hook tone if this feature cannot be provided by the transmission matrix.
- An indication of the line status to the local or remote test equipment.

- ○ Automatic testing for
  - •Detection of short circuits, crosses, and extraneous potentials within the network.
  - •Check of the continuity of the test path when it is first established.
- ○ Permit the connection of the test access either to the line side of a line circuit or to the line side of a line adapter such as a coin control circuit, a loop extender, etc.

11.3.4. Test Access for Trunks

A means of test access to any trunk interface and its associated signaling and transmission equipment is necessary.

This arrangement replaces the test jack field appearances of all trunks provided by earlier trunk-testing equipment but permits the performing of all tests, monitoring, talking, busying, etc., which are normally made from such test jack access. Modern equipment should include a display to indicate such things as the following:

- ○ The trunk interface address
- ○ The type of trunk, i.e., register signaling using R2, No. 5, or loop disconnect, or line signaling using E&M etc.
- ○ The trunk impedance
- ○ The trunk transmission levels and the state of the trunk

11.3.4.1. Types of Line Access

Earlier types of switching systems utilized the concentration network for test access. Modern systems can employ the concentration stage if they use metallic or other low-impedance high-power cross-points. Alternatively they can use a separate access network of the metallic crosspoint type.

If such a network is based on reed contacts, or other contacts not designed to make or break direct current circuits, the access network must be connected before any connections are made to test means. This can usually be accomplished by providing separate contacts of the mercury wetted type which complete the loop after the test access contacts are actuated.

11.3.4.2. Line Access Control

The control of the access to the line for testing is usually provided from colocated or remote test desks or consoles. It should also be possible to control special remote test equipment using in-band interoffice test control and measurement transmission techniques and automatic line insulation test equipment.

11.4. Network Integrity Tests

A test to check the transmission path on all calls should be part of the call processing routine. It should test for continuity and the absence of signal coupling to a test path.

A diagnostic routing should be initiated whenever trouble is encountered and a second attempt to complete the call made, using other paths if possible. The diagnostic routing should identify the fault preferable to an individual module. The data regarding the fault, as well as the identity of the traffic-carrying paths and functional units involved in the faulty connection should be recorded for analysis. An indication should be given whether this is a first or second attempt to process the call.

Line integrity tests. When a call is extended to a line, the called line should be tested before ringing is applied to the line and before ring tone is returned to the calling party. Two tests should be made: one to check the continuity from the line interface of the switching matrix to the station, and the second to check that the line will not trip ringing falsely due to the presence of a low-resistance loop.

The continuity test includes the line and terminal equipment. This test checks only that there is an alternating current path available, not its parameters.

11.5. Test Lines and Maintenance Communications Circuits

Test lines or trunks are provided as an aid to maintenance. They allow access to special test equipment or provide loop-around and other testing aids. A numbering allocation responsive to the country numbering plan is used.

The following are typical test lines. The first three test lines described are for use in trunk exchanges. However, these test lines may also be arranged to be seized by ringing and to trip ringing so as to be usable in local exchanges.

Communications test line. This is a standard communications circuit providing access from any trunk or line to a test position.

Transmission measuring and noise checking. This provides a connection to a transmission-measuring and noise-checking circuit at the local exchange. The measuring circuit is designed to transmit the following to the toll switching office originating the test: test power, information on transmission measurement, and a rough check of noise measured at the local exchange location.

Automatic transmission-measuring system (ATMS). This provides a connection to an automatic transmission-measuring system responder. It permits two-way transmission loss and noise measurements to be made on trunks through the use of an ATMS director and suitable test equipment at the trunk exchange.

Data transmission test line. This provides a connection to a signal source which allows for one-way testing of data and voice transmission parameters.

11.6. Intraoffice Maintenance Communications Circuits

The number and type of intraoffice maintenance communications circuits depend on the size and equipment layout of an exchange. The following two-way maintenance communications test lines are typical (they terminate on keys at the test desk or where appropriate on maintenance consoles):

- Test desk to test desk.
- Test desk to maintenance consoles or maintenance panels.
- Test desk to line and trunk interface frames.
- Test desk to transmission terminal equipment racks.
- Test desk to MDF.
- Maintenance console to equipment bays.
- Intra- and interequipment frames.
- Maintenance consoles to line circuit.

The circuits terminate on headset jacks at frames and equipment racks and are arranged for two-way signaling.

11.7. General Maintenance Features

The test access also provides a means of accessing ports which do not serve station lines. This permits the monitoring of the signals entering or leaving the system by way of these ports. This includes both station signaling and transmission monitoring. Test using portable test equipment is also possible from these access points. Transmission monitoring must be done by high-impedance circuits which do not impair transmission.

It is necessary to provide a means for tracing calls through the switching network to assist in fault location. The control center provides a means to identify lines, trunks, traffic-carrying units, and functional units that are not available for use. It indicates the nature of the removal from service such as equipment taken out of service automatically by the action of diagnostic programs, traffic-carrying devices which have been allocated to, or special terminations such as permanent-signal, line lockout of other services.

The control center also indicates the status of control and functional units as well as providing service-observing equipment to check the quality of service.

All data and signal distribution buses have a means to isolate the bus at each frame with test access points to permit fault isolation and tracing. This allows the connection of portable test equipment. They also provide a means for the connection of dc voltages to portable test equipment.

They also provide an on-demand printout of any and all existing alarm conditions. This includes delayed alarms which have not attained the preset delay limit.

11.8. Reliability

Switching systems must meet the required reliability levels at the time of cutover to commercial service following the success of the specified precutover testing procedures. The overall system reliability can be expressed in the following parameters.

11.8.1. Down-Time

The total system down-time during which all customers are without service should not exceed 10 hr during the entire 40-yr life of the machine.

11.8.2. Lost Calls

Incorrectly handled calls which result from troubles, faults, or errors of the switching equipment should average not more than 0.01% of all calls processed, i.e., one lost call in 10,000. This excludes calls lost or delayed due to call blockage within the network and controls caused by traffic congestion.

11.8.3. Customer Trouble Reports

Customer service reports attributable to troubles in the exchange should be less than 0.1 per 100 connected telephones per month.

11.8.4. Typical Requirements

The system should be capable of meeting the following requirements or those specified by the country of use.

Redundancy of equipment. The system should provide for duplication of any functional units or components, the failure of which would reduce the service of 100 or more lines by more than 25% of the traffic-carrying capacity (where such circuits would otherwise be provided one per group).

Method of path selection. The system should also provide selection algorithms which ensure that faults that result in the failure to reach an early choice trunk shall not prevent subsequent calls reaching lower choice trunks.

Second attempt. When a failure is encountered on any control operation, a second attempt must be made to complete the operation, using alternative facilities insofar as this is possible.

11.9. Human Factors

The two principal man–machine interfaces in telephone systems work at the MDF and the identification and clearing of faults.

Maintenance personnel work at the MDF virtually every day; therefore, it must be designed to meet human engineering requirements in respect to clearances, reach, freedom from danger, etc. Human engineering is well documented in the literature and it will not be discussed here.

Documentation of equipment. Identification and dealing with faults in the least possible time requires that clear, unambiguous, and visible equipment location information be employed (using the characters and language specified by the customer). This information should be easily obtained from the fault printouts and the relevant position on the floor plan identified. Rows and frames should carry information identifying the location of the equipment.

11.10. Materials

In general, structural members should be of steel. All materials used in the system packaging should be suited to their proposed use and life expectance.

Unmodified bright tin plating should not be used in the vicinity of printed wiring board assemblies or similar equipment.

The materials used and the types of finishes, coatings, etc. employed—especially plastic materials—should not constitute a fire hazard. Certain plastic materials which emit toxic gases when flaming should not be used.

The system should be designed so that it cannot fail in such a manner as to create a fire hazard and should not use any operating mode which could create a fire hazard.

11.11. Safety

The system should be so designed that it will meet or exceed the safety standards of the countries to which it will be supplied.

11.12. Alarm and Supervisory Requirements

The alarm and supervisory requirements for the maintenance of earlier systems were provided by lamps and alarm panels. In addition audible alarms were provided.

These systems were equipped with visual alarm indicators located so as to guide maintenance personnel to the location of the fault. However, the increased diagnostic capabilities of the stored-program control system can provide a fault printout, and hence only limited fault lamp displays are necessary. This provides a printed fault record for evaluation. It also identifies faults that would not be practical with light indicators (unless the light indicators were provided extensively).

Audible alarm signals in earlier exchanges provided only two alarm classifications, defined as either major or minor; stored-program switching systems provide many more alarm levels. This is due to the need to distinguish between call processor faults, which affect the entire exchange, faults in line or trunk interfaces, which affect only a part of the exchange, and faults which affect a major part, but not all of the exchange.

Modern systems use audible alarms only to draw attention to the existence of the condition. These produce alarm conditions which are processed by the level-three control which produces the relevant fault printout. The control also produces the necessary visual and audible alarm signals to alert personnel.

11.13. Timing-out Calls

Time-out calls are called ''permanents.'' These typically result from a request for service which has not been followed by complete address information within a predetermined time interval. Time-out can occur for the following reasons:

- Delayed dialing or no dialing within a specified time. This is normally an interval of digit 25 to 30 sec during normal traffic, or 12 to 15 sec, with no dialing for 6 to 8 sec, following a digit during heavy traffic; however, other timing may apply because of local requirements.
- Failure of the caller to disconnect 30, 60, or 90 sec after the called line disconnects, according to local requirements.
- Failure of the called party to disconnect 30, 60, or 90 sec after calling disconnect, or according to local requirements. If the call is to an operator or an emergency service or is a malicious call, then there will be no time-out period, and the release of the connection can be under the control of the called party, if this is indicated by the class of service.

11.13.1. Recording and Readout of Line Permanents

When a line is identified as permanent it is placed in a locked-out state. A record of the condition is made which includes the identity of the interface by equipment number and by directory number with the date and time of occurrence. When the cause of the permanent is cleared and the line restored to service, the record is erased.

11.13.2. Cumulative Permanent Alarms

A permanent alarm signal is provided. This indicates when a predetermined number of lines within a group are in this condition. The number of permanents required before an alarm is given depends on the type of transmission means serving the group. An alarm condition occurs if the condition persists (typically 5 min for a local alarm and 30 min for an alarm extended to a remote center).

11.13.3. Incoming Trunk Permanent

Incoming trunk permanents also result from trunk seizure with line or register signaling delayed for a predetermined time interval. This occurs for the following reasons:

- A trunk is seized and ready to receive register signals but none are received or there is a delay between digits of 16 sec, during normal traffic. During heavy traffic the timing remains the same except that the delay following a dialed digit reduced to 10 sec after the digit and 5 sec after subsequent digits. The delay following an R1 digit is reduced to 10 sec for all digits.
- The failure to receive a calling disconnect signal 25–30 sec after a called disconnect signal has been received. In this case, the called transmission path is disconnected.
- A signal from carrier equipment which indicates that the trunk must be locked out immediately.

Recording to trunk permanents. A permanent incoming trunk condition must be recorded. The record will identify the time and date and the trunk group and trunk number. At the end of the time-out period the trunk circuit is locked out of service.

When the cause of the permanent is cleared, the trunk is automatically restored to normal service and the record erased.

12

Administration

12.1. Introduction

The administration of telecommunications systems relates to three main areas. These are traffic engineering, exchange and networks, and operations.

12.1.1. Traffic Engineering

Traffic engineering provides for meeting the demands of growth of service and evaluation of service. It issues estimates of trunk requirements, monitors the trunk traffic, and prepares plans for future system expansion.

12.1.2. Exchange and Networks

The area of exchange and networks provides for data collection, balancing traffic, analysis of service, assignment of equipment, the instruction of customers for new services, line load control, and network management.

12.1.3. Operations

To a large extent operator services are being centralized with service provided by highly automated stored-program control consoles. It involves the staffing of consoles and the provision of such items as centralized toll investigation which checks disputed bills. It provides budgets, controls operator training and operator supervision, provides business services to multiline customers, and also provides for assistance services, interception, etc.

12.1.4. Applicable Areas

Telecommunications systems administration applies to traffic measurement and analysis, load balancing, line and terminal assignments, and quality of service. It is based on the collection and manipulation of data. Stored-program systems should provide as much of this information as possible.

The specific requirements of a particular country of use must be met but the following provides an indication of the type of service that will be necessary.

12.2. Exchange Data

Stored-program systems provide a great deal of data which must be maintained and retrieved when required for reference.

12.2.1. Requests for Service

Requests for service are normally originated at the business office. Some typical requests are as follows: requests for new and additional service, service location changes, terminating service requests, requests for unlisted directory number, changes or additions to station equipment, and requests for auxiliary station apparatus.

The collection, maintenance, and distribution of data to meet customer requests for service affects functions concerned with directory listing, information, maintenance, repair service and service history, assignment of cable and terminal facilities, installation department, and accounting and billing.

Older systems used separate equipment for these purposes, either manual card files or data bases resident in commercial computers. Modern switching systems can store much of the required data in their memory banks.

12.2.1.1. Manual Processing

Early types of systems which lacked a data base could require that 50% of all operating personnel work in customer service. This required reference to card files to provide the data. A system with 50,000 stations can receive an average of about 2,000 service requests per month.

Manual processing requires a multicopy form for each service request from the business offiice. Copies are sent to the areas concerned with installation, accounting, maintenance, traffic, and directory listing. This is dealt with in the following way.

Installation information. This contains details of the customer's name and address, line equipment number, cable pair, and residence terminal pair, together with, if applicable, the party number and ringing frequency and the class of service.

In the case of new service, line transmission records must be referred to to identify vacant facilities. In the case of transferred or termination service, the relevant records must be modified.

Accounting information. This is used for billing purposes. It identifies the type and class of service and any applicable installation or removal charges and determines whether there are extra charges for special station equipment, such as a DTMF calling device or special telephone. It should also indicate if the station equipment is owned by the customer.

Directory listing information. This is used for adding to or changing the master directory.

12.2.1.2. Stored-Program Processing

A support program is used to assemble the relevant exchange records for transfer to the relevant memory areas for subsequent call processing purposes. This is used in conjunction wuith manual processing where necessary.

12.2.2. New Office Initialization

It is necessary that the administration provide the information relating to directory numbers, traffic classes of service, and routing and charge listings. This is used to produce the information to dimension the exchange and provide the required services.

12.2.3. Security of Data

It is necessary to protect the exchange records from loss or tampering.

Hence local and remote access to the equipment is restricted to authorized personnel. It should also be only accessible by the use of computer passwords. Records can be changed by machine or teletypewriter. Teletypewriter-instigated changes should be stored in an intermediate memory for verification and checking before the data base is changed.

The support software system should provide listings of translation and call processing information. Typical are as follows: directory numbers to equipment numbers, equipment numbers to directory numbers, categories of classes of service to directory and equipment numbers and a numerical list of unassigned directory and equipment numbers. Other stored information may be necessary for administrative purposes.

12.2.4. Traffic

The administration of an exchange requires that it provides the required grade of service. This requires that the system accommodate the anticipated line and trunk traffic. Traffic-measuring functions which are a part of call processing provide the necessary data to verify satisfactory operation.

The operation of the control functions must also be monitored to ensure that they are not overloaded. This is especially important in the early periods of service when the behavior pattern of the system is being determined. This information will enable realistic values to be obtained for false and peak traffic and may lead to a larger allowable capacity for normal traffic.

12.3. Quality of Service

Determination of the quality of service is based on random selection of calls originating from customer lines and random selection of incoming trunk calls. It is not satisfactory to monitor traffic from specific line or trunk interfaces as a measure of overall service. Hence such line access is only necessary to observe lines which have been the subject of customer complaints.

Dial service measurement is based on such data as calling line identity, data and time of call, address information, answer time, and involved traffic-carrying devices. The call-generating process is initiated from the level-three control which makes the approproate records.

12.4. Service Analyzation

The need for traffic data has increased because of the need to meet specified standards of performance. It requires up-to-date data bases. It is necessary to analyze the data to determine the results of various calls and their results.

This requires a record of ineffective call attempts due to all causes such as blocked attempts, percent equipment faults, "no answer" calls, etc. Data are also needed on repeated attempts and lack of traffic-carrying or functional units and on incorrect customer calling procedures due to dialing a partial or incorrect code. A total of calls to wrong numbers due either to the customer or the equipment must also be obtained. The total number of matured calls is necessary as are called line busy and no answer calls and the number of calls routed to intercept. Analyzing this information and comparing it on a periodic basis will indicate how well the system is performing.

12.5. MDF (Main Distribution Frame)

Main distribution frames are labor intensive items. They have been virtually unchanged for a century. It is still necessary to cater to wiring changes at the MDF and to provide for practices relating to the assignment of short MDF jumper wire connections and practices which make the reuse of disconnected jumper wires possible. The operation of a MDF requires the provision of a data base which is accessible by all or any of the personnel.

Attempts to automate the MDF function, and hence reduce the amount of manual effort, have been made from time to time. There were no such systems in operation at the time of going to press. When switching systems are introduced which have low blocking probabilities in the line concentrator stage then some of the effort at the MDF will be reduced. This follows since the need for load balancing which requires relocation of lines into different line groups will then largely be obviated. However, some relocation of subscribers' lines will be necessary under fault conditions.

12.6. Dial Tone Delay

The dial tone delay interval is a common way of checking the quality of service. Hence the design of the level-one control must provide service. These data are normally accumulated during the busy hours on a monthly basis and are used as part of the analysis of traffic. All administrations specify this parameter, which is often stated as "No more than 1½% of calls shall have a dial tone delay greater than 3 sec." However, more and more administrations are requiring that the means to obtain the value are included in the basis system design.

12.7. Network Management

Network management has been defined by AT&T as "The supervision of the telecommunications network, to assure the maximum flow of traffic under all conditions."

This requires monitoring, measuring, and, when necessary, controlling the flow of traffic at the various local and trunk exchanges and transmission equipment.

The objective of network management is to ensure optimum utilization of facilities at all times. The capacity of the network is limited by the traffic-carrying capacity of the switching systems and the trunk transmission equipment. When channels are occupied with messages, the network is performing its function. Under normal day-to-day operation, the network provides a high quality of service. However, under certain overload conditions, such as increased calling, switching system congestion, or faulty equipment not yet repaired, the quality of service can deteriorate. Network management procedures and systems compensate for this state of affairs thus assuring the maximum utilization of all available switching and trunk transmission facilities.

Modern switching systems and signaling systems are making possible the inclusion of network control features as part of the switching system call-routing process. Hence any new switching system design must either provide, or be capable of providing, network control according to the requirements of its point of installation.

12.7.1. Cause of Overloads

An overload exists in the network whenever the demand for service exceeds the ability of the equipment, switching or transmission, to provide satisfactory service. This may be due either to heavy traffic or to equipment failure. The diagnostic routines of the exchange will identify the cause of the condition.

12.7.2. Control Measures

During an overload, network management must control the flow of traffic so that traffic-carrying units are fully utilized and network congestion is minimized. This is accomplished by the following:

- Rerouting traffic from heavily loaded trunk groups to lightly loaded trunk groups.
- Restricting new call attempts.
- Controlling the direction of traffic flow.
- Changing the route selection procedures in the relevant exchanges.

The earlier switching systems accomplished these measures by auxiliary equipment. The new switching systems should be designed so that they provide two basic types of network control. The first is concerned with the performance of the telephone exchange, the second with the performance of the network.

12.7.2.1. Measures at Exchanges

Switching systems experiencing unusual traffic conditions will signal to higher-level exchanges. These exchanges will instigate the necessary routines to correct the situation by instructing the subtending exchanges to modify their routing. This change of routing follows a predetermined program. This implements control measures such as the following.

Reduction of operator traffic. This is accomplished by various signals to the operator console systems which notify the administration of the nature and, if possible, the duration of the delays. This is important since, when callers experience delays, more calls are routed to operators. Hence such operators can control the flow of traffic and the spacing of subsequent attempts to reach a requested delay. They may also be able to inform the customer of the reason for the delay.

Recorded announcements. These are used on final trunk groups. The announcement replaces the busy tone with an announcement which hopefully deters repeated customer and operator attempts to reach the same line. This is used when a group is encountering a traffic overflow higher than a preset level.

Alternate route cancellation. This keeps alternate routed traffic from congested trunk groups and overloaded exchanges. In the case of the Bell System the actuation of a procedure for regular automatic alternate route cancellation gives the greatest relief to the home regional trunk group. This may be used when the regional office is in serious overload or, as might occur when there is a facility failure, the group to that office is severely congested. When the cancellation is in effect it prevents traffic "from" high usage trunk groups overflowing to the final trunk group. Alternatively it may only prevent a group or groups of trunks from overflowing traffic to the final trunk group.

Directional reservation. This is used to control the direction of the flow of traffic on two-way circuits. It is used at a low-level exchange to give preference to calls which have penetrated further into the network. This it does by reserving some of the trunks for incoming calls from a higher-level exchange under heavy traffic conditions. Both-way access to the trunk group is resumed when the traffic drops to a prearranged level.

This arrangement is effective for both the higher- and lower-level exchanges. It helps the higher-level exchange by reducing traffic from its subtending exchange. It prevents the lower-level exchange from routing calls that will probably encounter congestion. This increases the availability of the control for those calls that are not affected by the congested routes.

Modifications to call processing. The switching system control should be arranged so that, under heavy traffic conditions, the amount of routine diagnostic processing is reduced to the minimum needed to check the level-two controls. In addition, timing intervals, which control register and sender holding times, should be reduced to a minimum. Also the selection routines may be modified and the routing of calls modified. All of this depends upon the environment for the exchange and the nature of the network that it serves.

12.7.2.2. Performance of the Network

The performance of the network is affected by failure of transmission equipment as well as failure of telephone exchanges. The latter case has unfortunately become more frequent as more and more of the call processing has been allocated to a limited number of controls. Failure of both regular and standby controls is a mathematical probability. Such failures should not be of a prolonged nature unless they are caused by physical damage due to a disaster.

Measures to correct such situations are relatively easy to accomplish in stored-program systems, since, assuming that alternate routing possibilities to bypass the affected exchange exist, only program changes are necessary to reroute calls. If the affected exchange is out of service for a prolonged period, alternate physical rerouting may be necessary. If this is not the case then inevitably there will be a degradation of service due to the effective removal of service of some groups of trunks. This results in a reduction of the traffic-carrying capacity of the network.

12.7.2.3. Handling Disaster Areas

When a disaster occurs, the exchanges serving the affected area become overloaded. Hence, means have to be provided to restrict incoming calls (which have little or no possibility of success) and to give preference to outgoing calls from the area which have a good possibility of success.

This is effected by the use of such arrangements as directional reservations or recorded announcements. It also requires the cancellation of the relevant alternate routing routine and the invoking of emergency routing instructions or programs.

When the extent of the disaster and its effect on telephone traffic are better known, additional measures may be necessary. In any case, provision must be made for the storage of all relevant data. This is used to assess the action to be taken and to produce short-term and long-term means of alleviating the situation.

12.8. Services

The administration provides two types of services, one for customers and one for operating the telephone exchange. In addition to the basic telephone services new customer services are being offered. These include speed calling, add-on conference, call forwarding, call waiting, and wake-up service. Administration services include ring back, operator recall, directory inquiry, line load control, observation, intercept, malicious call hold, and operator-handled calls.

These services, which are described below, are allocated to lines by a relevant class of service. A system must provide a minimum of 15 different originating and terminating classes of service as standard with provisions for expansion of at least 225. These are used to provide special services to some lines while denying them to others. Where a combination of several services are to be provided to one or more lines, that combination is assigned a single class of service. These special services may be service enhancement such as speed calling, call forwarding, call transfer, etc., or they may be service restricting by limiting dialing to certain codes. They may also be user descriptive such as coin boxes and private branch exchanges.

12.8.1. Custom Calling Features

The new services that are being offered to telephone customers are called custom calling features. These custom calling services are generally intended for application to individual lines in local exchanges.

Application to PBX service may not always be possible. For example, the features of call waiting and three-way calling are incompatible with call transfer in a PABX if hookswitch flash is used.

12.8.1.1. Call Waiting

When a line with call-waiting service is called while it is busy, a momentary pulse of a distinctive tone is applied to that line to indicate that another call is waiting to be answered. The calling party will receive ring tone instead of busy tone.

In the following description it is assumed that line A has call-waiting service and is busy on a connection with line B when a subscriber on line C calls line A.

- If there is another call already waiting for line A, line C receives busy tone and the call cannot progress further, otherwise line C receives ringback tone.
- An alerting tone (typically consisting of two 300-msec spurts of 440 Hz precise tone 10 sec apart) is applied to line A but not to line B.
- The customer on line A may choose to answer a waiting call with or without releasing line B.
- If the connection between lines A and B is released with the call from line C waiting and unanswered, line A is rung and the call dealt with in the usual manner.
- If line A wishes to take the waiting call from line C without releasing line B, then line A depresses the hookswitch for a period of 1–2 sec or more. This causes line B to be placed on hold, ring tone to be removed from line C, and a transmission path established between lines A and C.
- Subsequent momentary hookswitch depressions on line A change the connection, substituting line B for line C or vice versa. The call is disconnected when line A replaces the handset for a period greater than 2 sec.

The call-waiting feature is provided on a modular basis usually to handle an ultimate capacity of 15% of the system.

12.8.1.2. Three-Way Calling

Three-way calling permits a customer to add a third party to an existing connection on which he may be either the calling or called party.

In the following description, it is assumed that line A having three-way calling is connected to line B and wishes to add line C to the connection.

- The subscriber on line A depresses the hookswitch for 1–2 sec. This causes line B to be placed on hold.
- Line A now receives a special dial tone, such as three spurts of dial tone 100 msec on and 100 msec off followed by steady dial tone.
- Line A then signals the directory number or the abbreviated speed calling code, if applicable, of the desired third party C. If line C is busy, line busy tone is returned to line A. Two hookswitch depressions restore the connection between lines A and B.
- If line C is free, it is rung and ring tone returned to line A.

- If line C answers the call, ringing is tripped and a connection established between lines A and C with line B still being held.
- A hookswitch depression on line A causes lines A, B, and C to be connected to a three-way conference.
- A hookswitch depression on line A releases line C from the conference.
- If line B or C disconnects from the conference the remaining line remains connected to line A. When line A disconnects, the connection is released.

Three-way calling is provided on a modular basis with a capacity usually up to 5% of the ultimate capacity of the system.

12.8.1.3. Call Forwarding, Station-User Activated

Call forwarding provides the means for a user to have incoming calls automatically transferred to another destination. This often permits a customer with call forwarding to transfer his or her calls to any place in the dialable area. Any extra trunk call charges for forwarding such calls are charged to the customer's directory number.

While call forwarding is activated, the home station receives a spurt of ringing each time it is called but cannot answer the call. However, calls can be originated from the home station. Call forwarding can be activated or deactivated only from the home station.

Activation of call forwarding. Call forwarding is activated by dialing a special activating code and then, following the receipt of dial tone, dialing the number to which calls are to be transferred.

This may be a seven or ten digit number preceded by an access code. If the called line is free, ring tone is returned and call forwarding will be activated when the call is answered.

If there is no answer, or the called line is busy, call forwarding can be activated by repeating the above sequence within 2 min. In this case, two spurts of dial tone, 100 msec on, 100 msec off, and 300 msec on, indicates that call forwarding has been activated.

Forwarding calls. When call forwarding is activated, any call to the home station is processed as follows:

- A short ringing signal is applied to the home station whether the line to which calls are to be forwarded is busy or free. This call cannot be answered by the home station.
- The system automatically generates a call to the stored replacement station number and the call proceeds in a normal way.
- If the call is answered, a transmission path is completed from the calling station to the new station. Any incurred charges are accrued and applied to the home station directory number.

Deactivation of call forwarding. Call forwarding remains in effect until it is deactivated from the home station by dialing the deactivating code. When this has been accomplished a special tone will be returned to the caller.

12.8.1.4. Speed Calling (Abbreviated Dialing)

A subscriber with speed calling service may place calls from a list of 8 or 30 frequently called numbers by dialing one or two digits, respectively, instead of say seven or ten digits and any access codes.

Entering numbers into speed calling memory. A subscriber with speed calling may enter or change numbers in his or her speed calling list by dialing the activating code. The customer then dials the selected one-digit code (typically 2 through 9 for an 8-number list or two-digit code 20 through 49 for a 30-number list) followed by the access or prefix code if required, and the required number.

Using speed calling codes. Speed calling codes are used in originating calls by dialing the required code. The scanner obtains the calling line identity. The register function obtains the relevant called number from the calling line identity list and processes the call.

The memory provided for call forwarding and speed calling must be expandable. Each address needs a storage capacity of at least 12 digits. The memory addresses may be assigned as required to serve the two services.

12.8.1.5. Wake-up Service

Wake-up service provides the automatic means for ringing a subscriber station at a prearranged time if that line is free. Ringing is applied until the call is answered or for a preset number of ringing cycles if the call is not answered. The number of ringing cycles should be adjustable on an individual application basis from three to six. If the wake-up call is answered, the called party is connected to a recorded announcement for one cycle of the announcement and then immediately released to lockout. If the wake-up call is not answered, the called line is automatically released to normal after the preset number of rings. If a called line is found busy when a wake-up call is attempted, the wake-up equipment will usually cancel the wake-up call. The service normally provides a printed record of the call including the directory number of the called line, the time called, and the result of the call (line busy, sub answered, etc.).

12.8.1.6. Third-Party Charges

The use of credit cards is increasing the need for operator assistance, and systems are being studied to allow the direct dialing of calls charged to a third party. One application being considered requires the use of a personal identification number (PIN), perhaps consisting of four digits, on such calls. The number dialed would then consist of the access code (in world zone 1 this would be 0) followed by the called number, then the number to which the call is to be charged, then the PIN code. This requires the availability of common channel signaling and the availability of stored-program processors at the called end of the call to control the data base and provide validity checks of PINs, etc.

12.8.1.7. Special Services for Customers

The following are typical special requirements.

Broadband switching. Some terminals, such as those serving computers, require greater than normal bandwidth. This bandwidth may be up as high as 1 MHz, which is required for videophones. This frequency range is beyond the transmission range of the switching matrix and hence it may be necessary to provide a separate switching matrix for broadband service if analog transmission is used. In the case of a digital switching system access could be provided at the superhighway stage.

Dial teletypewriter service. Dial teletypewriter communication service provides for the connection of TWX stations to telephone exchanges by means of voice-band telephone subscriber loops. Each station is given a distinctive telephone directory number. This provides customer-dialed communication with all other TMX stations in the country over the telephone switching network.

Off-hook service to operator (manual lines). Off-hook service provides telephone service for disabled subscribers. The switching equipment is so arranged that calls originated by such lines will be automatically routed to an operator.

Wide-area telephone service. Inward wide-area telephone service provides for calling designated lines directly from any station in specified areas or zones at no charge to the calling station. Charges for this service are made to the called station at a fixed or timed rate.

Restricted direct dialing. This restricts the dialing area of a calling station such as a coin box or private branch exchange extension. It can be accomplished by assigning a class of service or by a special indication from the station. It is often called *toll barring* or *toll restriction.*

Toll barring by class of service prevents marked lines from originating any one or a combination of types of trunk calls. These include station to station and person to person direct dialed or operator completed. If a restricted type of call is attempted from such a line, the call is either diverted to a recorded announcement, routed to an operator, or has a progress tone applied.

Lines marked at the station for toll barring will be marked by a diode in series with the calling telephone. Each line is tested for the presence of this diode before dial tone is returned. This is usually done by momentary reversal of the battery feed. An unauthorized call is treated as above.

12.8.2. Special Services for the Administration

The first emphasis on the provision of new services was aimed at the telephone customer, but recently a large priority is being placed on services to assist in operating the system. Stored-program control and interactive signaling systems such as R2, No. 6, and No. 7 have increased such services by offering traveling class of service indications and providing a larger signaling repertoire. These have extended and enlarged the operating services; however, many of the following typical operating services have been in use in switching systems for some time.

Manual interception service. Manual interception service intercepts calls to lines which have changed, or unused station codes, or calls on which unallocated area, exchanges, or other codes are dialed. Calls to local changed numbers may be routed to a local or centralized manual switchboard, or these calls may be routed to a recorded announcement, or to a special tone source. Calls to unassigned station or trunk codes are usually routed to a recorded announcement or to a special tone source.

When intercepted calls are routed with other traffic over a common trunk group, using associated signaling, to an operator, a distinctive tone is applied to the trunk (usually for 1.5 sec) when the operator answers. This identifies the type of call.

Register signaling systems, such as R2, provide backward signaling from an intermediate or terminating point when a vacant code is received by the processing register. This causes the diversion of the call to the applicable intercept method in the originating exchange. The outgoing trunk connection in the originating exchange is disconnected. Earlier systems often used the trunk network for the return of the intercept tone or announcement. This tied up expensive trunk transmission equipment on non-revenue-producing calls (it not being a normal administrative policy to charge the caller for such calls).

Automatic intercept service. Automatic intercept provide a matching generated automatic voice response to the receipt of a called number. This requires the repetition of the called number to the intercept system automatically. Automatic intercept systems normally serve a number of exchanges.

Absent subscriber intercept service. Absent subscriber intercept service provides for the temporary diversion of calls for identified lines to a special operator or recorded announcement to inform a calling party that the called subscriber is temporarily absent from his or her premises.

Line-load control. Line-load control provides the means for reducing the locally originated traffic in a telephone exchange during emergency conditions. This can be done by dividing each line group into three classes. The first of these will be able to originate calls in the event of an emergency. The second group will only be able to originate calls as the emergency subsides. The third class will not be able to originate calls until the emergency has ended. However, it should be possible to reach any of the lines with a terminating call, irrespective of the class of service for originating calls.

Malicious call tracing. The local methods of applying this feature may vary according to the administration policy. The following is a typical way of providing the service which is identified by a called class-of-service.

Malicious call tracing provides a means for the automatic number identification of all local calls to lines having the class of service. It records all calls to the line and provides a printout indicating the calling number identity, the called number identity, time of day, and date. The release of the call is under the control of the called party. Some administrations use special line interfaces to enable the called party to hold the incoming call and also to signal to an operator.

Some line signaling arrangements provide a backward signal to hold trunks. Some register signaling systems such as R2 indicate that a called number is marked for malicious call tracing. These signals are interpreted in the originating exchange which provides a printout indicating the calling number, outgoing trunk, identity, time of day and date. The terminating office with the facility provides a printout indicating the called number, incoming trunk identity, time of day, and date for all such calls.

Protection against defective trunks. This requires that a nonhoming selection method be applied to a trunk group or that the first choice rotate over the first three choices. This prevents successive call attempts originating during low-traffic periods from reaching the same defective trunk. The rotation of choice applies especially to two-way trunks in a group which is graded at the remote end. This latter condition applies if the distant exchange selectors have limited availability.

Automatic trunk busying. Automatic trunk busying is provided with one-way loop signaling trunks. While the trunk is free, the trunk circuit at the outgoing end monitors the loop with a high-resistance (approximately 20,000 Ω) sensor in series with a diode for current interruption or reversal. If either event occurs the trunk is marked busy at the line interface.

Pad control. When a switching system handles incoming trunk traffic which is routed to other exchanges, as opposed to terminating in this exchange, transmission pads may be required to meet the transmission objectives. Exchanges having this requirement are arranged so that calls originating or terminating locally via long-distance trunk are routed through pads which insert a fixed loss into the connection. Calls through the office between trunks usually do not include pads.

Blocking calls. The routing and selection methods must block the completion of unauthorized dialing of calls in the base rate local area by using trunk routes and also prevent calls incoming over a trunk group being routed back over the same group.

All calls switched to a termination. When the control completes its functions in setting up a call, the calling line must have been either connected to the desired destination or routed to a service or be placed in a locked-out condition. It must always be connected to a termination during and after call processing.

Manual holding. Manual holding is used on connections to operators to provide operator control of disconnection. Thus switch-hook signals are transmitted through the system to the operator but the call is not released until both the calling party and operator disconnect.

Alternate routing. This is the arrangement by which calls to another office which cannot be completed over the direct route because of traffic congestion or other reasons are automatically routed over alternate trunk group. The number of alternate groups depends on the point of application in the network. Usually four or five such groups are the maximum.

Busy hour service priority. This is an arrangement which limits the processing of local-originating traffic after it has reached a predetermined busy-hour level for local traffic, while retaining the required grade of service for trunk traffic. This can be done by giving priority to incoming trunk calls and by scanning interfaces more rapidly than line interfaces.

Busy out of vacant equipment positions. This arrangement marks as unusable vacant equipment position. It is also applied in the case of blown fuses or other causes of power disconnection.

Operator busy verification. This permits an operator to override the busy condition and connect to an established connection. A special tone warns the parties to an established connection of the intervention by an operator.

In earlier systems a separate access means was provided and the access equipment was only available to the operators over special trunks. Modern systems use the normal concentration network with a special class-of-service identifying operator-originated calls. The service is used, in response to a customer's request, to check the condition of a line that the customer has experienced difficulty in reaching. It is not used to extend calls.

Trunk offering. Trunk offering provides the means for an operator to override a line busy on a call and follow a procedure to extend the new call to the called line. This is used when an operator who is attempting to complete a trunk call receives a busy signal. The operator manually applies a signal, which connects the operator to the established call. The operator then offers the trunk call to the required called party. If the called party desires to accept the trunk call, he or she disconnects from the preexisting call, releasing that connection.

When the called line becomes free, ringing is applied. The call then proceeds normally. The sequence of operations in a digital system is shown in Figure 12.1.

This assumes that a call exists between lines A and B. The operator dials the A line number to the register (1). Busy tone is returned to the operator (2); the register path is disconnected but reserved. Busy tone is removed and the register reconnected to the operator, who keys the trunk offering activization code (3).

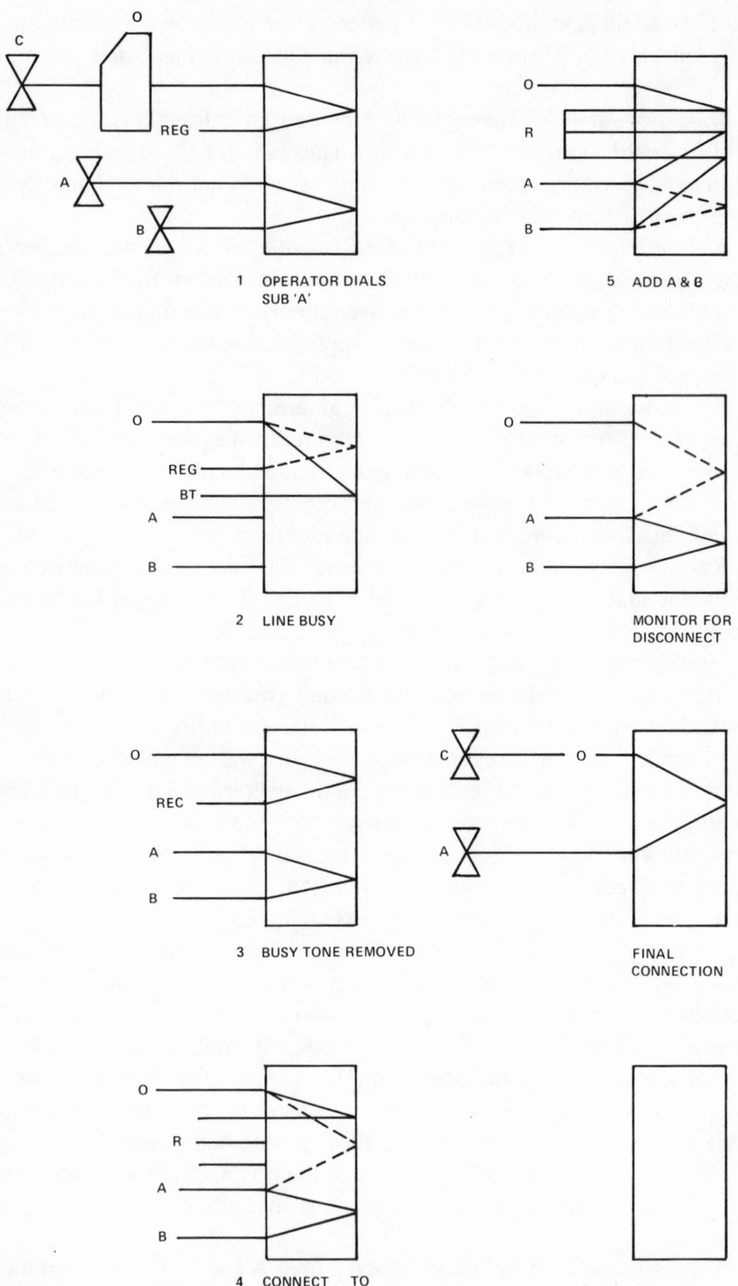

Figure 12.1. Sequence of connections for trunk offering in a digital switching system.

The existing path from A to B is temporarily disconnected and reserved and A and B and the operator are connected in a conference.

A new path is reserved from the operator to line A (4). Line A is now monitored for release (5). When line A releases, a ringing condition is applied to it. When line A answers, the reserved connection to the operator is used to complete the connection. In this method the service is used to extend a semiautomatic or manual trunk call.

Time, weather, and other recorded announcements. Recorded announcement machines are used to provide this type of information which is provided from special trunk interfaces. Alternatively the service can be provided by special programs in the level-one controls. The information is obtained by dialing the relevant special service code or station code.

CO centrex. This service provides business customers with all private branch exchange services from the local telephone exchange. No switching equipment is located at the customers' premises, only attendant consoles. It provides service for a number of separate businesses, each of which has its completely separate classes of service etc. The customers' premises are then served by a separate line for each of the extensions and all call processing is accomplished at the serving local exchange.

Such an arrangement requires a very large number of classes of service and a large call processing capability. The arrangement is well adapted to providing service for large office buildings, which often have a heavy turnover of tenants. This is because all services are provided centrally and the installation and removal of private branch exchanges at the premises are obviated. The association of the lines and the provision of services can be accomplished largely by software changes to the local exchange data base.

13

System Evaluation

13.1. Introduction

An administration requiring a new switching system produces a specification which describes the requirements as fully as possible. This provides, with varying degrees of completeness, the initial and ultimate size of the exchange, its anticipated traffic, its signaling methods, numbering schemes, services, environment, and so on.

Any system which is proposed to the administration must meet, or give a satisfactory explanation of exceptions to, this specification and all applicable national and international interworking requirements and standards. It must meet the safety and other requirements of its point of application and any regulations regarding its staffing.

Manufacturers must provide complete documentation and training, perhaps in a language foreign to them. They must also provide the necessary maintenance services to deal with faulty modules and provide additional equipment as the system expands. Some of the specified services will be mandatory, others will be negotiable. The administration will require complete details, schedules, and costs of the various stages in the provision of the exchange. There are a number of equipment manufacturers who are competent to meet any specification.

What we are concerned about is how a technical evaluation of a proposed system can be made. This assumes that the system is cost effective since pricing of systems is related to many market factors as well as to current and projected manufacturing levels at the potential suppliers' plants. The objective of the system architect is to produce a system which is cost effective over its projected range of sizes and services and for its life cycle. This presents the marketing department with a solid basis for meeting its goals.

Basis for Costing System

The methods of making cost comparisons between competing alternative solutions have been well documented. A telephone administration has to build some kind of a rate base to develop its call-charging policy. Typically the cost of the local exchange and its line and trunk transmission equipment plus the operating cost is used. In making comparisons the first cost of the equipment and the annual costs are used. This provides a true indication of the cost of providing service since it includes costs reflecting return on investment, depreciation, maintenance (both hardware and software), power and utilities, taxes, etc.

Unfortunately claims of lower annual costs can only be sustained by the history of systems in service. This poses problems of credibility for vendors of new systems with little or no service data. Sometimes this seems like a vicious circle to the vendors of such new systems. Often they have to enlist the help of some sympathetic authority, which may be motivated by national pride, to make their initial field test.

The vendors of the system have to be able to provide verifiable data to support their claims, hence the conduct of field trials must include means of obtaining and recording the system's performance. This should include the verification of action taken to correct malfunctions. Subsequent installations must also provide these data. Without it marketing is compromised into submitting objectives rather than results.

13.2. System Evaluation

In addition to the ability of the system to meet the specification, system evaluation is concerned with the following:

- The line and trunk capacity and the busy hour call-handling capacity.
- The number and classes of service available both for the customer and the administrations.
- The minimum and maximum number of groups of trunks or lines.
- The minimum and maximum number of trunks in a group.
- The maximum traffic handling without deloading, both the maximum traffic per line and the maximum traffic per trunk.
- The system growth pattern, both the size of units and the packaging of hardware equipment and software equipment.
- Its service performance such as the grade of service provided during normal and overload traffic and corresponding internal blocking for local–local calls, local–trunk calls, and trunk–local calls.
- Its inherent reliability such as the degree of redundance, the size of modules, the effects of single faults, and the effects of multiple faults.
- The software programs including the programming language, the construction

of the program, and the control of modifications to the program whether generic or modular programs are used. The flexibility of the data base, ability to add features, ability to add capacity. Its ability to meet signaling systems.

- Its first costs including installation and its needs including space requirements, floor loading, power, environmental control and special protection.
- Its annual costs including buildings, outside plant, utilities, personnel, and integrating switching and transmission.

Increasingly, administrations are submitting detailed questionnaires in an attempt to probe the strengths and weaknesses of competing systems. They produce detailed answers in the following areas of application to their needs.

13.2.1. System Architecture

The assumption is made that any modern system must operate in the stored-program mode. If it does not it cannot meet the long-term requirements for signaling, network management, data services, and system operation.

13.2.1.1. Distribution of Functions

Does it have enough control levels to separate real time usage such as customer signaling from the level-two control? Does it separate diagnostic programs from operating programs? If the level-two control includes the diagnostic program, as opposed to being able to accept it as an overlay program, then the majority of the program it carries will be diagnostic. This means that if the diagnostic program is faulty the operating program on the level-two control carrying it may be inoperative. It also greatly increases the memory requirements from the single unduplicated memory necessary with a level-three control.

13.2.1.2. Type of Switching Network

It will be necessary to determine to what extent the concentration and selection networks are compatible with the existing and projected line and trunk transmission equipment. If one assumes the system is cost effective, on a stand-alone basis, then the projected plans may be more important. These can include a degree of combination of digital switching and digital transmission equipment.

13.2.1.3. Type of Multiplexing

Modern local and trunk exchanges are more likely to use pulse code modulation than pulse amplitude modulation for the reasons described earlier. They can use CEPT or T1 parameters. Ideally they should be readily adaptable to either.

Highway bit rate. Is the first-level highway 2.048 or 1.544 Mbitsec? What is the maximum bit rate in the system and in the superhighways? What cable lengths correspond to the maximum permissible propagation time? What and how are the clock rates obtained?

Coding law. Is this the CEPT A law or T1 μ law? Is this compatible with the digital carrier systems in use or projected for use?

Format. Is the bit stream in the highway organized in word or bit order? This has an impact on the length of the buffer store when the highway directly interfaces with the digital carrier.

13.2.1.4. Type of Memory

What types of memory are used—volatile or nonvolatile? How are they organized? What module sizes are in use? What back-up systems are provided? How are they initalized and updated? What security measures are used to prevent unauthorized access?

Addressing structure. Is the addressing system open ended or limited? If it is limited, how does this affect the required numbering scheme, ultimate capacity, and so on?

13.2.1.5. Control Operation

What functions are controlled by software programs and by what wired logic circuitry? This indicates the degree of flexibility in modifying the system operation. However, it should be remembered that software is abstract and that the performance of functions requires the provision of the relevant wired logic circuit elements and interfaces with other functional units. This includes the switching network, the line and trunk interfaces, and other traffic-carrying devices. In addition to this the new program, possibly with its support memory, must be capable of being added to the existing program.

Modular controls. Is the addition of features, functions, or additions to a system provided by modular hardware and software or does it need modifications to a number of separate areas?

Redundancy of controls. How much processing capability, if any, is lost by the failure of functional units? How many lines are denied service or how many trunks are not usable due to the failure of functional units or traffic-carrying devices?

Mean time to repair. What is the mean time to repair a fault and how many and what types of faults are probable? What assistance is available and what skills are required for the maintenance personnel? Must spares be shipped for great distances or are they available locally? How far away is the nearest support facility or manufacturing plant of the vendor? Is there access, by satellite if necessary, to a permanently manned support center which can analyze software problems and accept data from call processing memories in bulk?

Availability of components. What kind of components are used, who makes them, and how long will they be available? Is local manufacture possible and permissible in order to ensure their long-term availability?

Upward compatibility. If the vendor uses a family of processors are they upward compatible? Do simple procedures exist for the replacement of processors when the system expansion dictates? Preferably the initial processor should be expandable to the ultimate size without replacement.

13.2.1.6. Programs

What is the word size—the number of instructions for each major call category? What programing language is used? What host computer does it need? Is it a generally used language or a special language which is limited to the vendor's products? Can it be applied by local personnel? Does the warranty depend on its modification by the manufacturer only? Is adequate documentation available? Is the program structured, with modules, or is it a generic type? It should be remembered that a modular program has a precise interface but easy management, and a generic program may not. To what extent has the vendor planned the expansion of the program? Will it run on all the vendor's processors?

13.2.1.7. Auxiliary Equipment

What auxiliary equipment is necessary to provide services or operate the exchange?

13.2.1.8. Available Services

How many classes of service are provided? What are they? What percentage of customers can they serve? Can all present administrative and maintenance services be performed by the system with a minimum of auxiliary equipment? Does it accommodate common channel signaling—both No. 6 and No. 7? Do proven operating programs exist for all or only some of the services? Can it provide for automated billing by raw data or collated data and can it work with automatic multifee coin boxes?

13.2.1.9. Site Requirements

Answers to the following questions will usually be required:

- What ambient temperature limits apply for short-term (how long) and normal operation?
- What are the tolerable humidity levels?
- Does the system require forced-air circulation to ensure its operation? Is this provided centrally or on critical equipment frames?

- What maintenance procedures are necessary for the environment control equipment?
- What is the floor loading?
- What is the aisle width?
- What are the frame sizes?
- What arrangements exist for cabling?
- Do all interframe cables plug in?
- Do equipment modules plug in?
- What are the power requirements for operating the system and any auxiliary equipment?
- What is the floor layout for the subject system?
- What battery voltage variation will the system tolerate?
- What provisions are necessary for maintenance centers?

13.2.1.10. Line and Trunk Transmission Limits

The ability of a new system to provide longer station loops, better transmission, etc. may be reflected in a reduction of the auxiliary equipment previously needed for long lines. This has to be studied and stated in the response to a bid.

13.3. Traffic-Carrying Ability

Any blocking-type switching network has a limit based on its maximum size and its switching matrix architecture. This means that the traffic sources to the switching system are limited. It is usual to divide the sources into originating lines and incoming trunks; a corresponding arrangment is made for terminating lines and outgoing trunks. The traffic that the switching network carries is due to the total of local–local, local–trunk, trunk–local, and trunk–trunk calls. This is the switched traffic. Sometimes system manufacturers specify the traffic by taking the traffic that an input or output can carry and multiplying it by the number of inputs and outputs. Switched traffic is revenue-earning traffic; maximum traffic capacity is not. Care should be taken to determine which basis was used for defining capacity.

13.3.1. Traffic per Port

The amount of traffic that a port can carry is a function of the system design. The cost of providing the port includes in addition to the cost of the interface a prorated cost of the complete exchange. This is the constant cost of providing the equipment. To this must be added the costs due to actually handling traffic.

Hence the cost per port has a fixed plus a variable cost. The revenue to support the system is based on a prorated cost per switched Erlang or part thereof.

Hence the traffic that a port can be offered should approach the traffic that it can carry. For example a station will originate and receive about 0.1 to 0.2 Erlangs. If it is assigned to a port which can carry 0.7 Erlangs, which represents the average trunk traffic, there is unusable traffic capacity.

System evaluation should consider how the traffic per port limits suit their application. Will it result in free but unusable ports because of group traffic limitations? Will it result in unassignable traffic when all the ports in a group are assigned? It is important that if free ports exist, but cannot be used because of traffic loading, their use is inhibited by the level-one control. The number of possible busy hour calls per line may also be limited for the reasons described below.

13.3.2. Traffic per Control

It is important to study the impact of the local traffic on a proposed switching system. Areas to consider are the following.

13.3.2.1. Use of Customer Calling Devices

There may be differences in the customer dialing habits from those on which the provision of directly controlled functional units, such as originating registers, are based. This is especially true when automatic service is extended to areas unfamiliar with it. If initial reaction to the new service results in slow dialing, the switching system may time-out calls before dialing is completed. Hence the system supplied must be capable, in the early period of service, of providing an adequate number of register functions and allowing for delays in dialing to suit the local requirements.

13.3.2.2. Duration of Calls

Now traffic is calculated based on the use of the number of calls and their duration. Traffic-carrying devices are not affected by the number of calls, except as it affects their mean time to fail. On the other hand, controls can be affected by the number of calls. This is because each step in controlling a call has a finite duration which is not related to the holding time of the call. Hence a control has a traffic-carrying limit based on the number of calls.

For example if we assume a system handles 1000 Erlangs of switched traffic from 10,000 lines each originating 0.1 Erlangs of traffic then we know what the total traffic is. However, the 0.1 Erlangs of traffic could typically be made up of two calls of 180 sec average duration or it can be made up of six calls of 60 sec duration. The first case will produce 20,000 busy hour calls, the second case 60,000 busy hour calls. Because of this it may not be possible to serve 10,000 lines depending on the maximum busy hour processing ability of the system.

13.3.3. Distribution of Traffic

A switching system carries traffic to and from lines and trunks. The system has a capacity for switching Erlangs which depends on its architecture. However, local exchanges have to cater to both lines and trunks. This means that their mechanical packaging sets limits on the number of lines and trunks that it can accommodate. This is based on anticipated percentages of local and trunk interfaces and traffic distribution.

For example a local exchange could have an equal originating and terminating traffic per line. Assume that the system can have a maximum of 10,000 lines each with 0.1 Erlang originated and 0.1 Erlang terminated traffic. If we assume 30% of the local terminated traffic terminates locally then each line generates 0.07 Erlangs of outgoing trunk traffic and the total incoming trunk traffic which terminates locally will generate 0.07 Erlangs per line. This corresponds to 10,000 multiplied by 0.07 Erlangs of trunk traffic and 700 Erlangs of outgoing trunk traffic. Taking a broad brush approach, i.e., assuming an average trunk traffic of 0.7 erlangs per trunk, this requires 1000 incoming and 1000 outgoing trunks. This assumes one-way trunks.

A system designed to cater to this kind of traffic distribution would require at least 2000 trunk interface ports as well as 10,000 line interfaces. It can be seen that if the percentage of local traffic is greater, more trunks will be required. There is of course also the need to provide all the necessary service trunks and to cater to small trunk groups which can carry less traffic per trunk than can larger groups. Hence in evaluating competing systems the constraints due to packaging must be identified and related to the traffic distribution and to the local trunk transmission equipment and its routing pattern.

13.3.4. Ultimate System Size

The availability of very powerful stored-program processors has increased the capacity of individual telephone exchanges by 50,000 to 100,000 lines. Usually such exchanges have to be equipped with a large number of lines for them to be cost effective. It is basically a matter of modularity, and the increments by which groups of lines and trunks and processing power can be increased.

There are of course finite limits for any system technique. They may be due to physical limits such as cable lengths due to propagation delays. They may be due to capacitance buildup leading to cross talk or to data bank access or processing capacity. Because of this multiple unit exchanges may be used to meet a particular size requirement.

13.3.4.1. Multiunit System

The principle of operation, when applied to a digital system, is that two units A and B which are served by CEPT primary multiplex bit streams are interconnected by such bit streams. Calls are distributed between the two units and the address of the call included in the initial setup of the connection. This address information is added and extracted by the level-one controls. In this arrangement each unit has its own controls and call-processing structure but each loses part of its traffic capacity for lines and trunks since part of the traffic is used for interunit calls.

Distribution of traffic in multiunit system. If we assume that a system can switch 1400 Erlangs of traffic then we can divide the units into incoming trunk units and outgoing trunk units with lines distributed over the units. The line traffic has been omitted to simplify the example which therefore refers to a trunk exchange. Expansion of the system to four units as shown in Figure 13.1 results

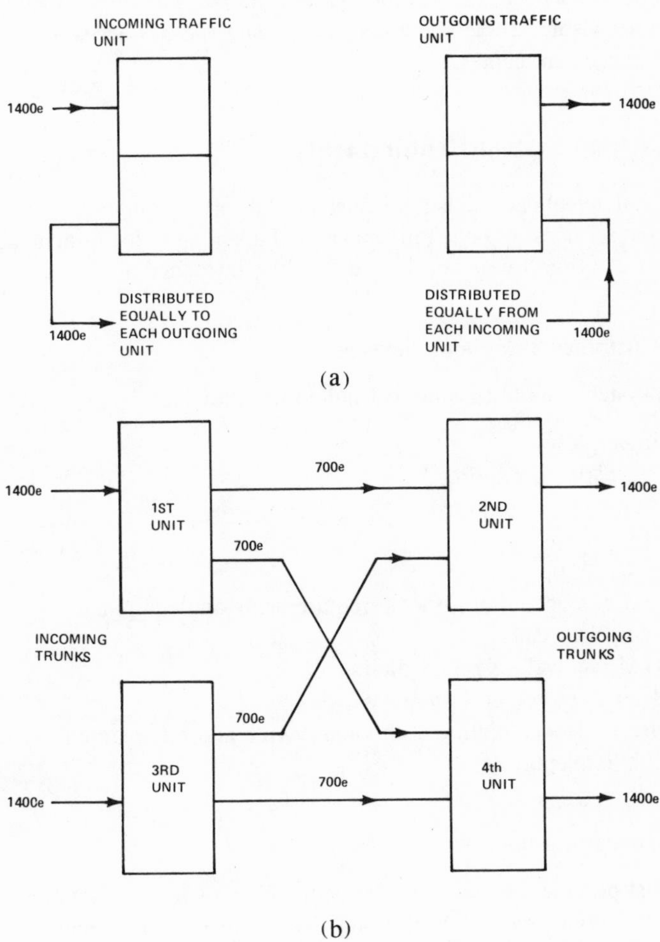

Figure 13.1. Principle of multiunit system. (a) Distribution of traffic. (b) Interconnection of units for 2800eI/C and 2800 O/G.

only in a doubling of the switched traffic. The remainder of the traffic is interunit traffic. It can be seen therefore that the traffic gained is offset by the interunit traffic needed for routing between units. However, the number of trunk interfaces is equal to the number of external trunks only, since all interunit traffic is switched in undemodulated bit streams. Hence the cost per trunk, while it is higher than for a single unit, is not four times as high. This is because items such as battery power costs, floor space diagnostics, and maintenance do not increase linearly. This follows because the greatest users of space and power are the trunk interfaces. Hence a decision to use multiunits deserves detailed study if the initial and ultimate size of an installation go beyond the capacity of a desired switching system.

Centralized maintenance. Each of the units connects to a buffer–access circuit. This connects to the common level-three control which in addition to its access to a console, a teletypewriter, and a data file has access to a local and remote management center.

13.4. General System Requirements

A local telephone exchange must provide as a minimum the following features and services, or be readily capable of adapting to them, at an acceptable cost within the time frame considered for their introduction.

13.3.1. Customer Telephone Service

The system should be able to handle the following:

- Single party
- Two party, if required
- Multiparty, as required
- PBX
- PBX (DID)
- Coin box according to local requirements and equipment
- Emergency calls
- DTMF as well as rotary dialing
- Meter at customer's premises
- Direct distance dialing both countrywide and international
- Mobile telephones

13.4.2. Charging Functions

It must provide the ability to derive the charges for international calls, via operators or automatically, for toll (long distance) calls automatically using pulse metering for bulk billing and producing detailed charging information when required. It must be able to derive charges for local calls using the applicable method such as: pulse metering, flat rate, bulk billing, or detailed charging.

The information should be produced in a form suitable for use with data processing machines and commercial computers. Older methods employing electromechanical meters at the exchange should not be considered as viable.

13.4.3. Operator Functions

The system must be capable of working with the following services provided by operators as required: call setup assistance, directory information, call charge information, call interception, absentee subscriber service, trunk offering, and call holding.

13.4.4. System Operation Administration and Maintenance Functions

The system must provide the necessary functions to interwork with the administration's maintenance centers both local and centralized. It must provide the means for obtaining traffic data either for local recording or centralized recording for both traffic-carrying devices and controls.

It must include the means for local recording or centralized recording of all charging data. It should provide the means for automatic subscriber line testing and for automatic interexchange circuit testing. It should provide an automatic means of service interception of routine calls such as no-such-number. It must be able to interwork with the existing or projected network management systems. It must be responsive to and produce the required data for automatic overload control, manual overload control, automatic trouble location, stand-alone operation, and automatic call generation.

13.4.5. Added Features

It should be capable of including custom services such as call waiting, recorded message, call transfer, call forwarding, abbreviated dialing, call restricting, automatic routing (hot line), alarm wake-up, and add-on conference.

It should provide administration features such as CO centrex, malicious call tracing, calling line identification by directory number and, when the data base exists, by location and service observation.

13.5. Integration of Switching and Transmission

A digital carrier system has the potential for reusing existing analog cable pairs more effectively, thus reducing new construction costs. A digital switching system can integrate into a digital carrier system. Hence it is important to investigate whether there are savings in cost to be gained by adding the two to an existing network. The following shows how this may be an advantage.

13.5.1. Separation of Switching and Transmission

For the greater part of the telephone era the transmission channels and switching systems have been separated by the main distribution frame into independent entities. Switching systems which are largely of electromechanical types as far as switching is concerned interconnect cable pairs which carry analog signals. All switching techniques that have been introduced, whether they use crosspoints spatially divided or time-divided multiplexing of the voice path, and whether or not they have stored-program control have to switch analog inputs to analog outputs. This has required that, until the advent of digital systems, digital carrier systems have had to be separated into individual channels and demodulated into an analog form at switching centers. These switching centers use metallic crosspoints and were designed to be compatible with wire lines.

The digital switching technique encodes the voice signals in a binary way and hence it allows direct connection into PCM carrier systems obviating the modulating and multiplexing carrier terminal equipment.

Installed carrier systems return each channel to analog at the connection to switching systems. However, digital bit stream switching operating at the CEPT bit rate (2.048 megabits) and having 30 speech and two signaling time slots can directly interface at the carrier group level by simply omitting the demultiplexing and demodulating stages and taking the bit stream directly to the cable pairs. A control on a group basis is necessary for the insertion of line signals synchronizing, etc.

This results in reduced equipment costs and space since trunk interfaces are obviated together with their associated cells and frame space per carrier group. The carrier equipment is also simplified since it no longer needs multiplexing and modulating equipment.

13.5.2. Base for Evaluation

A comparison of the costs and savings in the following areas is necessary and detailed information must be obtained at the intended location. The location may already have digital carrier which is compatible with the switching system installed. In this case the demodulating equipment at the proposed digital exchange can be salvaged. The trunk interfaces with the existing system may also be usable elsewhere.

13.5.2.1. Analog Trunks

Costs must be obtained for each route for the switching system interfaces the ducts and cables, frames, both on a first-cost and annual-cost basis.

13.5.2.2. Digital Trunks

For new PCM carrier system application, costs must be obtained for each route identifying the cost of the analog paths used and also those made available for other purposes. The savings resulting from the obviation of individual trunk interfaces by the use of direct digital interfaces must be identified with their support costs. The cost of providing demodulated channels must be compared with the cost of direct interfacing. These may well show that the present cable routes may be converted from analog to digital with advantage. This could be especially true if the present cable ducts were full since such a conversion, if practical, could avoid costly construction. Thus there is no one solution but only a decision based on the alternative possibilities of the exchange due to its location.

13.5.2.3. Elements of Channel Costs

A number of first costs apply in an analog-to-analog carrier system using PCM carrier. These include on a per channel basis, the costs of the trunk interface, intraexchange cabling to the carrier equipment, prorated frame costs, and the cost per channel unit. In addition to this there are common charges per system. This includes the common multiplexing equipment, span equipment (for redundant path provision), repeaters (provided every 2 km or so), and repeater housings.

The system will use at least two cable pairs per system, hence the cost of these must be included. It should also include a prorated value for the cable including the ducts installation and other materials per kilometer. When the system connects to a digital exchange without demodulation the direct digital interface replaces the multiplexing and individual channel units and trunk interfaces, etc.

13.5.2.4. Example of Application

The following costs are typical. They will vary from location to location and be subject to economic factors, etc. They serve only as a general example of the first stage of a comparison. These assume a CEPT carrier system with 30 revenue-earning channels.

It is assumed that a trunk circuit costs \$150, a channel unit \$2000, span equipment \$450, repeaters with prorated housing \$192, wire pair value \$59, and a direct digital interface costs \$3500. (This is very conservative.)

Then if we assume a direct analog connection of 100%, a PCM carrier system, analog to analog connection, is 66%. A PCM carrier system with a direct trunk interface at one end is 56% and with direct trunk interfaces at both ends is 47% of the cost. This assumes a route length of 20 km. Hence applications of competing systems should be considered in relation to their possible savings in trunk transmission equipment and trunk interfaces and not on an isolated stand-alone exchange basis.

13.6. Assessing a System Application

The engineer who is involved in assessing the degree to which the system that his company would offer in response to a specific application has to determine what has to be added or modified in a system to meet the requirements. Usually this is carried through a number of successive stages of increased detail.

The initial assessment can be made using a check list based on his system's characteristics and features. This will ensure that the items that require design or adaptation or cannot be met in the financial or time frame constraints are identified. Such a check list should form the input to a computer program with the traffic and dimensional data. The program should be capable of producing a list of equipment for a proposal to meet the application.

References

1. IEEE, *Standard Definitions of Tones for Communication Switching,* IEEE Standard 312—1977, The Institute of Electrical and Electronics Engineers, Inc., New York (1977).
2. S. Lederman and N. H. Petschenik, "Automated Handling of International Calls through TSPS No. 1," *Conference Record IEEE International Conference on Communications,* 11 (1974).
3. G. D. Crider and W. K. Foster, "Automatic Telephone Dial Directory," *IBM Technical Discipline Bulletin* (December, 1976).
4. J. S. Ryan, "The Role of the ITU and CCITT in Telecommunications," *International Switching Symposium* 121/1–121/7, (1974).
5. R. Syski, *Introduction to Congestion Theory in Telephone Systems,* Oliver and Boyd, London (1960).
6. Functional Specification and Description Language (SDL). CCITT *Orange Book* Vol. V. 1.4, International Telecommunications Union, Geneva (1977).
7. *ISO Standards for Flow Charts,* ISO/R, 1028 (1969).
8. *Notes on Distance Dialing,* AT&T New York (1975).
9. H. E. Vaughan, "Introduction to No. 4 ESS," *International Switching Symposium* 19–25 (1972).
10. K. E. Fultz and D. B. Penick, "The T1 Carrier System," *Bell Systems Technical Journal* **44,** (September, 1965).
11. CCITT *Green Book,* Vol. III, Part 1, Sect. 1, International Telecommunications Union, Geneva (1973).
12. H. Nyguist, "Certain Factors Affecting Telegraph Speed and Certain Topics in Telegraph Transmission Theory," *Transactions of American Institute of Electrical Enginners* (1924 and 1928).
13. B. Smith, "Instantaneous Companding of Quantised Signals," *Bell System Technical Journal,* **36,** 653–709 (May, 1957).
14. CCITT *Green Book,* Vol. III, (Q 2 D), International Telecommunications Union, Geneva (1973).
15. CCITT *Green Book,* Vol. VI-1, 25–40, International Telecommunications Union, Geneva (1973).
16. W. F. B. Wood, "Keg Telephone Systems: The Latest Chapter," *Bell Laboratories Record,* **44,** 85–88 (March, 1966).
17. G. A. Backman and R. F. Penn, "Automatic Identified Outward Dialing for PBX's: Central Office Facilities," *Bell Laboratories Record,* **46,** 112–116 (April, 1968).
18. J. G. Pearce, "The New Possibilities of Telephone Switching," *Proceedings of the IEEE* 65 (September, 1977).
19. J. G. Pearce, "Logic Circuits," *Telephony,* 49–55 (October 17, 1964).
20. J. G. Pearce, "Logic Circuits (Continued)," *Telephony,* 56–61 (November 14, 1964).

21. E. Brockmeyer, H. L. Halstom, and A. Jensen, "The Life and Works of A. K. Erlang," *Transactions Danish Academic Technical Science* No. 2, Copenhagen (1948).
22. J. Atkinson, *Telephony,* Vol. 2, Sir Isaac Pitman and Sons Ltd., London (1949).
23. CCITT *Green Book,* Vol. VI-1 Rec. Q45, 45, International Telecommunications Union, Geneva (1973).
24. *Notes on Transmission Engineering,* United States Independent Telephone Association, 1st Edition (1963).
25. A. E. Spenser and F. S. Vigilante, "No. 2 ESS System Organization and Objectives," *The Bell System Technical Journal,* **48,** 2615 (October, 1969).
26. A. Feiner and W. S. Hayward, "No. 1 EES Switching Network Plan," *The Bell System Technical Journal,* **43,** 2215 (September, 1964).
27. Y. Rapp, "The Economic Optimum in Urban Telephone Network Problems," *Ericsson Techniques,* **49,** 1-132 (1950).
28. V. Batra, "Grounding Electronic Switching Systems," *Telephony* (May 6, 1974).
29. E. C. Molina, *Poisson's Exponential Binomial Limit,* Van Nostrand, New York (1947).
30. R. Biddalph, A. H. Budlong, R. C. Casterline, D. I. Funk, and L. F. Goeller Jr.: "Lines, Trunk Junctor and Service Circuits for No. 1 EES," *The Bell System Technical Journal,* **43,** 2323 (September, 1964).
31. H. Kegl and G. Neovius, "Electronic Private Branch Exchange ASD 551," *Ericsson Review,* **4,** 109 (1974).
32. A. E. Joel Jr., "An Experimental Switching System Using New Electronic Techniques," *The Bell System Technical Journal,* **37,** 1091 (September, 1958).
33. CCITT *Green Book,* Vol. VIII, Rec. V. 3, International Telecommunications Union, Geneva (1973).
34. J. R. Donaldson, "Structured Programming," *Datamation,* 52 (December, 1973).
35. L. J. Murray, "One-at-a-time Operation in Telephone Exchanges," *ATE Technical Journal,* **14,** 256 (1958).
36. A. Feiner, "The Ferreed," *The Bell System Technical Journal* **43,** 1 (January, 1964).
37. CCITT *White Book,* Vol. 1, International Telecommunications Union, Geneva (1969).
38. CCITT *Red Book,* Vol. VI (5), International Telecommunications Union, Geneva (1960).
39. CCITT *Green Book,* Vol. VI (IX), International Telecommunications Union, Geneva (1973).
40. CCITT *Green Book,* Vol. VI (X), International Telecommunications Union, Geneva (1973).
41. CCITT *Green Book,* Vol. VI (XII), International Telecommunications Union, Geneva (1973).
42. CCITT *Green Book,* Vol. VI (3), Chap. 3, International Telecommunications Union, Geneva (1973).
43. ATT *Blue Book,* Sec. 5, Par. 5, International Telecommunications Union, Geneva (1975); also CCITT *Green Book,* Vol. VI–XV, International Telecommunications Union, Geneva (1973).
44. CCITT *Green Book,* Vol. VI–XVI, International Telecommunications Union, Geneva (1973).
45. N. H. Robinson, "Trunk Line Switching with Special Reference to the Separate Signaling System," ATM Technical Society Paper, No. 206 (1953).
46. CCITT *Green Book,* Vol. VI/XIV (6), International Telecommunications Union, Geneva (1973); also Special Issue of *Telecommunication Journal* (February, 1974).
47. C. A. Dahlbom, "Common Channel Signaling—A New Flexible Interoffice Signaling Technique," *International Switching Symposium,* 421–427 (1972).

Index